Lakes of the Eastern Mediterranean and Balkans

Konstantinos Stefanidis · Anthi Oikonomou ·
Elias Dimitriou
Editors

Lakes of the Eastern Mediterranean and Balkans

Biodiversity, Ecosystem Services, and Conservation Strategies

Springer

Editors
Konstantinos Stefanidis
Department of Biology
University of Patras
Patras, Greece

Elias Dimitriou
Institute of Marine Biological Resources
and Inland Waters
Hellenic Centre for Marine Research
Anavyssos, Greece

Anthi Oikonomou
Institute of Marine Biological Resources
and Inland Waters
Hellenic Centre for Marine Research
Anavyssos, Greece

ISBN 978-3-032-02167-0 ISBN 978-3-032-02168-7 (eBook)
https://doi.org/10.1007/978-3-032-02168-7

This Springer imprint is published by the registered company Springer Nature Switzerland AG
The registered company address is: Gewerbestrasse 11, 6330 Cham, Switzerland

Preface

Lakes are among the most valuable ecosystems on Earth, offering essential ecological, cultural, and economic benefits. In the Mediterranean region, lakes are particularly notable for their ecological importance and their contribution to sustainable livelihoods. However, they are increasingly threatened by climate change, pollution, water extraction, and other human activities, underscoring the need for a deeper understanding of their ecological processes and the services they provide. This book, a compilation of research chapters by leading experts, presents a wide range of studies addressing key aspects of Eastern Mediterranean lacustrine ecosystems. Covering key aspects such as biodiversity, hydrology, and socio-cultural significance, the book emphasizes the interconnected nature of these elements and underscores the importance of holistic interdisciplinary research in understanding and preserving these unique ecosystems.

Chapter "Key Biodiversity Areas in Mediterranean Lakes" introduces the application of Key Biodiversity Area (KBA) criteria for the first time in KBA assessments in the Mediterranean lakes aiming to identify priority areas for the conservation of freshwater species in the Mediterranean Basin. At the national level, the case of the KBAs in the Mediterranean region of Türkiye, with a particular focus on ten ecologically significant lakes is assessed in Chapter "Freshwater KBAs in Türkiye's Mediterranean Region: Challenges and Conservation Strategies for Fish Fauna", emphasizing the need for integrated management and adaptive strategies. Chapters "Lakes of Albania—Ecology, Biodiversity and Conservation Approaches", and "Lakes of Albania from Origin to Current Threats to Their Functional Ecosystems" turn to the biodiversity hotspot of the Mediterranean region, the Balkan Peninsula and the transboundary lakes of Albania, providing insights into biodiversity trends, habitat fragmentation, and the impacts of climate change. These chapters underscore the interconnectedness of these aquatic systems and their regional importance for biodiversity and ecosystem functioning. Chapter "Current Status of Greek Lakes and Future Perspectives: Results from WFD Monitoring" focuses on Greece, showcasing the advancements in lake monitoring and assessment through the EU Water Framework Directive, which have enhanced our understanding and informs conservation measures for Greek lakes.

In Chapters "Satellite Remote Sensing (RS): A Valuable and Efficient Implement for Optically Diverse Greek Lakes' Water Quality (WQ) and Trophic Status Monitoring", "Utilization of the Self-organizing Map for Limnological Modelling" and "Hydromorphological Analysis for the Assessment and Evaluation of the Environmentally Minimum Lake Level", we delve into innovative approaches and methodologies for monitoring and managing lake ecosystems. Satellite remote sensing is highlighted as a powerful tool for assessing water quality and trophic status, while hydrological modelling provides valuable insights into sustainable lake-level management. Neural networks are a subset of machine learning algorithms that are increasingly being used in various applications, including water quality prediction in lake management. Together, these chapters underscore the potential of science-driven strategies for informed decision-making and conservation.

Chapter "Ecosystem Services of Mediterranean Lakes: Management Implications" shifts the focus to the ecosystem services provided by the Mediterranean lakes and their critical role in achieving Sustainable Development Goals (SDGs). It emphasizes the need for adaptive management strategies that integrate the Ecosystem Services Approach, ensuring the sustainability of these vital ecosystems in the face of growing environmental and human pressures. In Chapters "The Use of Methodological Participation Tools in the Investigation and Evaluation of the Development Perspectives of the Transboundary Prespa Park, Based on the Perceptions of the Area's Primary Sector Professionals" and "Cultural-Historic Heritage of the Lake Ohrid (Albanian Part)", the spotlight turns to the transboundary and cultural dimensions of lake management. The cooperative efforts within the Transboundary Prespa Park showcase participatory processes as a means to foster sustainable practices and stakeholder engagement. Meanwhile, the rich cultural heritage of the Lake Ohrid basin highlights the interplay, between the environmental and the cultural component, illustrating the profound connections between ecological and cultural landscapes.

As we face the multiple challenges caused by environmental degradation and climate change, the need to protect and sustain freshwater ecosystems has never been more urgent. This book provides a comprehensive narrative on the challenges and opportunities for freshwater ecosystem conservation in the Eastern Mediterranean. By integrating cutting-edge science, policy insights, and cultural perspectives, it aims to inspire collaborative efforts toward safeguarding these invaluable ecosystems for future generations. Yet, while the focus of this book is on Eastern Mediterranean lakes, the insights and methodologies presented herein have broader applicability. Freshwater ecosystems worldwide face similar challenges, and the approaches discussed here can serve as models for other regions. By showcasing successful case studies and identifying areas for improvement, this volume advances the global dialogue on freshwater conservation and management.

Patras, Greece — Konstantinos Stefanidis
Anavyssos, Greece — Anthi Oikonomou
Anavyssos, Greece — Elias Dimitriou

Contents

Key Biodiversity Areas in Mediterranean Lakes

Michalis Pournaras, Konstantina Spiliopoulou, Anthi Oikonomou, Kostas A. Triantis, and Maria Th. Stoumboudi

Abstract Freshwater ecosystems are among the most biodiverse yet severely threatened ecosystems in the world. The Mediterranean Basin is a biodiversity hotspot for freshwater biodiversity and lacustrine species inhabiting ancient lakes. Key Biodiversity Areas (KBAs) represent by far the largest network of important sites for biodiversity that have been systematically identified using standardized quantitative criteria, contributing to the global persistence of biodiversity. In this chapter, we compiled a spatial dataset containing the global range maps of 793 freshwater fish mollusks, plants, insects, crustaceans, fungi, and worm species from the IUCN Red List of Threatened Species. Data were collected for the 50 largest natural lakes in the Mediterranean region. We tested for the first time, KBA Criteria to 169 sub-catchments, in order to identify potential KBAs (pKBAs). We determined the threats posed to the species that trigger KBA criteria, alongside the coverage of pKBAs from protected areas. One hundred and forty-four sites (85%) hold 227 species that trigger KBA Criteria (28%) in total. In addition, 78% of the species triggering KBA criteria are found in the Balkans. Only 36% of pKBAs is protected, with the lowest protection found in Turkey and Levante (6%). Our results highlight the importance of undertaking comprehensive KBA assessments for freshwater species and addressing the threats posed to freshwater biodiversity in policies and frameworks, while providing the areas that Mediterranean countries should prioritize when expanding protected area networks to achieve the 30% protected area target by 2030 of the Kunming—Montreal Biodiversity Framework.

Supplementary Information The online version contains supplementary material available at https://doi.org/10.1007/978-3-032-02168-7_1.

M. Pournaras (✉) · K. Spiliopoulou · K. A. Triantis
Department of Ecology and Taxonomy, Faculty of Biology, National and Kapodistrian University of Athens, Athens, Greece
e-mail: michpour@biol.uoa.gr

M. Pournaras · A. Oikonomou · M. Th. Stoumboudi
Institute of Marine Biological Resources and Inland Waters, Hellenic Centre for Marine Research, Anavyssos, Greece

K. Stefanidis et al. (eds.), *Lakes of the Eastern Mediterranean and Balkans*,
https://doi.org/10.1007/978-3-032-02168-7_1

Keywords Key biodiversity areas · Mediterranean biodiversity hotspot · Freshwater biodiversity · Biodiversity loss · Lacustrine ecosystems

1 Introduction

Freshwater ecosystems occupy 0.8% of the Earth's surface (Dudgeon et al. 2006). However, the diversity of species in freshwater systems is exceptionally high. More than 38,000 freshwater species are globally assessed in the IUCN Red List of Threatened Species—mainly Actinopterygii (13,173 species), Arthoropoda (9287 species), and Mollusks (3690 species; IUCN 2023). While the current number of all known fish species is 18,075 (WWF 2021), other estimates suggest that there are more than 18,000 fish (Fricke et al. 2023), 5000 mollusk (Pyron and Brown 2015), and 250 turtles (Bour 2008).

Freshwater species do not cross land or saltwater. This isolates their populations and therefore intensifies endemism rates (Oikonomou et al. 2014). Conservation practices for freshwater biodiversity are oftentimes poor and biased towards terrestrial species (Darwall et al. 2011; Di Marco et al. 2017a; Spiliopoulou et al. 2023). Only 37% of the world's rivers are free flowing, due to dam construction even within Protected Areas (hereafter PAs; Grill et al. 2019; Thieme et al. 2020), 70% of global rivers lack PAs in their upstream catchments, and only 11.1% achieve full integrated protection (Abell et al. 2017). In many cases, rivers delineate the PA boundaries instead of being conservation targets themselves (Thieme et al. 2007; Abell et al. 2007). Human activities impose threats on freshwater species inhabiting lakes, streams, and reservoirs on a global level (Reid et al. 2019). These threats exacerbate the global freshwater biodiversity loss, which seems to be more severe than in terrestrial systems (Dudgeon et al. 2006; Harrison et al. 2018; IPBES 2019; WWF 2022), with one quarter of all freshwater species being threatened with extinction (Sayer et al. 2025). In addition, the number of freshwater species has decreased by 83% since 1970, a number far greater than in terrestrial or marine ecosystems. Riverine fishes in Europe and the US suffer extinction rates 112 times higher than expected (Dias et al. 2017), while wetland loss occurs 3 times higher than forest loss (Ramsar Convention on Wetlands 2018).

The Mediterranean basin is one of the first 25 Global Biodiversity Hotspots (Myers et al. 2000). In terms of freshwater biodiversity, the rates of freshwater fish and mollusk species richness and endemism are exceptionally high (Tierno de Figueroa et al. 2013; Darwall et al. 2014; CEPF 2017; Máiz-Tomé et al. 2017). There are 914 freshwater species threatened with extinction in the Mediterranean Basin Hotspot (IUCN 2023). In particular, many Mediterranean lakes—especially in the Balkans—are ancient. Ancient lakes act as ecological refugia for species, thus intensifying the endemism and diversification rates (Frogley et al. 2001; Frogley and Preece 2004; Albrecht et al. 2007, 2009, 2012).

All countries through the Kummning-Montreal Global Biodiversity Framework are obliged to achieve global biodiversity targets to combat the global freshwater biodiversity loss by 2030 (CBD 2022). In addition, goal 6 of the Sustainable Development Goals focuses on the "protection and restoration of water-related ecosystems" and 15 on the "conservation, restoration and sustainable use of terrestrial and inland freshwater ecosystems and their services" (United Nations 2016).

Global biodiversity targets mandate the expansion of the protected area network towards areas of particular importance for biodiversity. Key Biodiversity Areas (KBAs) represent by far the largest network of important sites for biodiversity that have been systematically identified using standardized quantitative criteria (IUCN 2016a). They are defined as "sites that contribute significantly to the global persistence of biodiversity". In the *Global Standard for the Identification of Key Biodiversity Areas* (hereafter Global Standard), 11 quantitative criteria are introduced (Table 1), that can be applied at all levels of biodiversity, i.e., genetic diversity, species, and ecosystems and across all systems. When a single species or ecosystem triggers at least one KBA criterion, an area can be identified as a KBA site.

So far, the majority of identification and delineation efforts for KBAs in freshwater ecosystems come from IUCN reports in various regions: Africa (Holland et al. 2012; Spiliopoulou and Starnes 2021), the Mediterranean basin (Darwall et al. 2014; Máiz-Tomé et al. 2017), the Eastern Himalayas (Allen et al. 2010), the Indo-Burma biodiversity hotspot (Smith et al. 2012), Canada (Tognelli et al. 2017), the Lower Mekong River Basin (Máiz-Tomé 2019), the lake Malawi/Nyasa/Niassa catchment (Sayer et al. 2019), Madagascar (Máiz-Tomé et al. 2018), and Uganda (Plumptre et al. 2019). Darwall et al. (2014) and Máiz-Tomé et al. (2017) were the first to identify freshwater KBAs in the Mediterranean Basin Hotspot. Apart from these reports, KBAs for freshwater species have been identified in other means: based on the existing IBA network of Iraq (Rubec et al. 2009), or the application of the KBA criteria by including freshwater species along with terrestrial in an effort to define

Table 1 Key biodiversity areas (KBA) criteria (IUCN 2016a)

Category	Criteria
A. Threatened biodiversity	A1. Threatened species
	A2. Threatened ecosystem types
B. Geographically restricted biodiversity	B1. Individual geographically restricted species
	B2. Co-occurring geographically restricted species
	B3. Geographically restricted assemblages
	B4. Geographically restricted ecosystem types
C. Ecological integrity	C. Intact ecological communities
D. Biological processes	D1. Demographic aggregations
	D2. Ecological refugia
	D3. Recruitment sources
E. Irreplaceability through quantitative analysis	E. Complementarity-based quantitative analysis of irreplaceability

trigger species (Ambal et al. 2012; Tordoff et al. 2012; Eken et al. 2016). Many of the aforementioned works were based on non-standardized and in many case qualitative criteria (Rubec et al. 2009; Allen et al. 2010; Ambal et al. 2012; Holland et al. 2012; Tordoff et al. 2012; Darwall et al. 2014).

Lakes constitute biogeographical islands within terrestrial ecosystems (Hortal et al. 2014) and thus are of particular importance. At the same time, the human footprint and the human modification indices show that roughly one-third of the world's lakes undergo significant human pressures, which increase within PAs (Mammides 2020). In this work we perform for the first time a KBA scoping analysis using the Global Standard (IUCN 2016a) to identify pKBAs on lakes in the Mediterranean Basin hotspot. By performing a KBA scoping analysis, we prioritize the natural lakes that are the most important for the global persistence of biodiversity in a standardized methodology that utilizes objective, transparent, and repeatable criteria. This approach is important, because it provides the scientific basis to allocate conservation resources and expand PA networks, while addressing global biodiversity. We use HydroBASIN level 12 sub-catchments to capture the smaller freshwater ecosystems found in many Mediterranean countries (Pešić et al. 2022; Maasri and Bonada 2024), and we determine the degree of protection and the prevailing threats associated with the pKBA sites we identify.

2 Materials and Methods

2.1 Study Area

We used sub-catchments in the Mediterranean hotspot to scope KBAs for freshwater species, while including information on the hydrological connectivity (Máiz-Tomé et al. 2017). We used 169 sub-catchment polygons that include the 50 largest natural freshwater lakes (Table 2; Fig. 1). The sub-catchment polygons were obtained from the HydroBASINS database (Lehner and Grill 2013) delineated at level 12, which is the recommended resolution for the KBA identification process in freshwater systems (KBA Standards and Appeals Committee of IUCN SSC/WCPA 2022). The total area of all sub-catchments was calculated at 20,684.2 km^2. We separated the countries in the Mediterranean Basin Hotspot into four sub-regions: Balkans, Turkey and Levant, North-western Mediterranean, and North Africa (sensu Darwall et al. 2014).

2.2 Species Distribution Data

We downloaded species range maps from the IUCN Red List website for all assessed freshwater species present in the 50 largest natural lakes in the Mediterranean (IUCN 2023) (Table 2). We selected only the polygons that are coded with Presence = 1

Table 2 50 largest natural freshwater lakes of the Mediterranean hotspot

Country	Lake	Area (km^2)
Albania, Montenegro	Skadar	381
Algeria	Oubeira	20
Albania, North Macedonia	Ohrid	354
Albania, North Macedonia, Greece	Prespa	263
Bosnia and Herzegovina	Deransko	1.5
Croatia	Vrana	5
	Prološko	2
France	Scamandre	6
	Charnier	5
Greece, North Macedonia	Doiran	35
Greece	Trichonida	92
	Volvi	65
	Vistonida	43
	Vegoritida	39
	Small Prespa	38
	Koroneia	36
	Orestiada	28
	Yliki	20
	Amvrakia	12
	Lysimachia	10
	Petron	9
	Ozeros	9
	Paralimni	7.5
Israel	Tiberias	162
Italy	Trasimeno	119.5
	Bolsena	113.5
	Bracciano	57
	Vico	12
	Massaciuccoli	7
	Albano	6
	Chiusi	3
	Martignano	2
Montenegro	Malo Blato	6
	Šas	3
Portugal	Fermentelos	2
Spain	Sanabria	3
Tunisia	Ichkeul	87

(continued)

Table 2 (continued)

Country	Lake	Area (km²)
Turkey	Beyşehir	627
	Eğirdir	456
	Kuş	150
	Ulubat	118
	Marmara	54
	Köyceğiz	51
	Işıklı	49
	Kovada	7.5
	Gala	6.5
	Akgöl	5
	Sığırcı	5
	Karataş	5
	Gölhisar	3

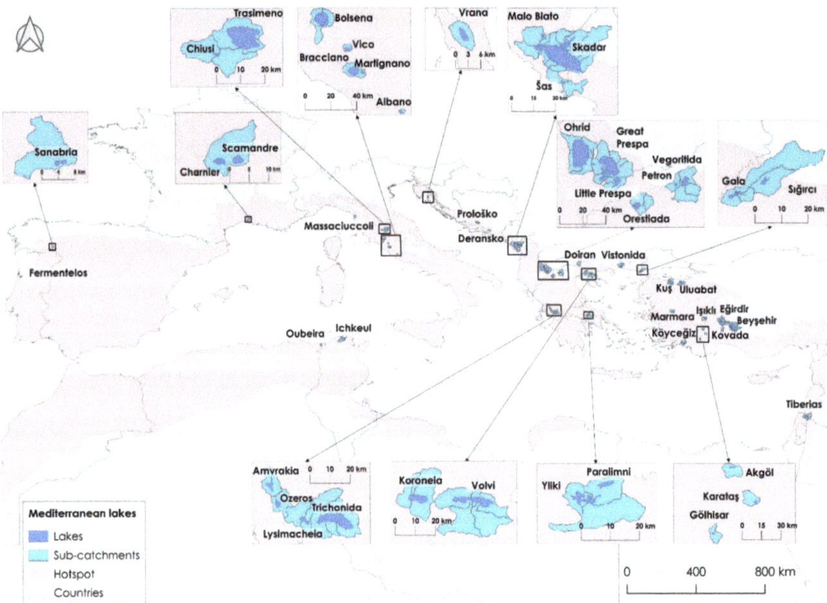

Fig. 1 The 50 largest natural Mediterranean lakes and their HydroBASINS level 12 sub-catchments in the Mediterranean Biodiversity Hotspot. Grey represents the national boundaries of all Mediterranean countries, pink the Mediterranean Biodiversity Hotspot boundaries according to CEPF (2017), light blue the HydroBASINS level 12 sub-catchments, and blue the lakes

Table 3 Counts of freshwater species per taxonomic class, assessed in the IUCN Red List of threatened species in the Mediterranean and the studied lakes

Class	No. of species	No. of species in lakes	No. of endemics in lakes
Actinopterygii	635	229	154
Gastropoda	530	201	141
Magnoliopsida	225	131	36
Liliopsida	223	136	30
Insecta	81	28	5
Malacostraca	52	22	10
Bivalvia	44	25	4
Polypodiopsida	15	10	2
Lycopodiopsida	14	3	1
Cephalaspidomorphi	9	4	1
Bryopsida	9	3	1
Branchiopoda	5	–	–
Charophylaceae	3	–	–
Chondricthythes	2	–	–
Sarcopterygii	1	–	–
Lecanoromycetes	1	1	–
Jungermanniopsida	1	–	–
Clitellata	1	–	–
Agariomycetes	1	–	–
Total	1852	793	385

The numbers of endemics are listed in last column

(Extant) and 2 (Probably Extant) and Origin = 1 (Native). We finally used 793 species that represent freshwater fishes, worms and leeches, snails and mussels, crustacea, fungi and plants, while freshwater fishes and snails represent more than half of the species (430) and have the highest percentage of lake endemics (Table 3). Almost half of the species used were endemics of the Mediterranean Hotspot (49%). More specifically, 67% of all Actinopterygii species and 70% of all Gastropoda species were endemics in the Mediterranean.

2.3 Analysis

We performed a KBA scoping analysis in the Mediterranean using the R package KBAscope (Spiliopoulou et al. 2024). We applied only species-specific criteria, namely A1 (a–e), B1, B2 and B3 (a–b), which refer to threatened and geographically restricted biodiversity (IUCN 2016a, b). Every HydroBASIN level 12 that includes at least one species that triggers KBA criteria is considered a potential KBA (pKBA). We identified the number of species that trigger KBA criteria in all pKBAs; the taxonomic groups (appropriate taxonomic levels to apply KBA criteria B2 and B3 are available in the KBA website Tools page) with the highest number of species triggering KBA criteria; the most common threat types—according to the IUCN Red List of Threatened Species (IUCN 2016a); and the level of protection for the identified pKBAs, by overlapping protected area boundaries (UNEP-WCMC and IUCN 2023) and pKBAs. We used threats on the second level of classification according to the IUCN Threat Classification Scheme (IUCN 2016b).

3 Results

3.1 Potential Key Biodiversity Areas

We identified 144 pKBAs, which correspond to 85% of the 169 HydroBASIN level 12 sub-catchments used in our analysis. The resulting pKBAs cover an area of 18,866.94 km^2, or 91% of the 20,684.2 km^2 area analyzed. These pKBA sub-catchments represented 43 of the 50 lakes in our analysis (86%; Fig. 2; Table 4).

The majority of pKBA are present in the Balkans and Turkey. The resulting 144 pKBAs are distributed as follows: 65 in Turkey and Levant (45%), 64 in Balkan countries (45%), 10 in northwestern Mediterranean (7%), and 5 in northern African (3%). Within each region we examined, more than half of the sub-catchments analyzed trigger KBA Criteria. Specifically, 93% of the Balkan, 84% of the Turkish and Levantine, 67% of the northwestern Mediterranean, and 63% of the northern African sub-catchments qualify as pKBAs. The countries with the highest number of pKBAs are Turkey and Greece, hosting 54 and 37 pKBAs respectively. The number of pKBAs in the rest of the Mediterranean countries ranged from 18 to 1 (Table 4). The lakes with the largest pKBAs are: Beysehir and Egridir in Turkey, Skadar in Albania and Montenegro, Tiberias in Israel, and Ohrid in Albania and North Macedonia (Table 5).

PA networks cover one third of the total pKBA area (6374 km^2 out of 18,866.94 km^2) (Table 5). The overlap analysis showed that the degree of protection varies for each geographic region: 58% of the Balkan, 6% of the Turkish and Levantine, 58% of the northwestern Mediterranean, and 28% of the northern African pKBA area is protected. Moreover, pKBAs in Turkey and Levante, along with north Africa exhibit the lowest levels of protection. Especially in Turkey, only the lakes Kus and Uluabat have almost 30% of their respective area protected. All other Turkish lakes lack protected area overlap (Table 5).

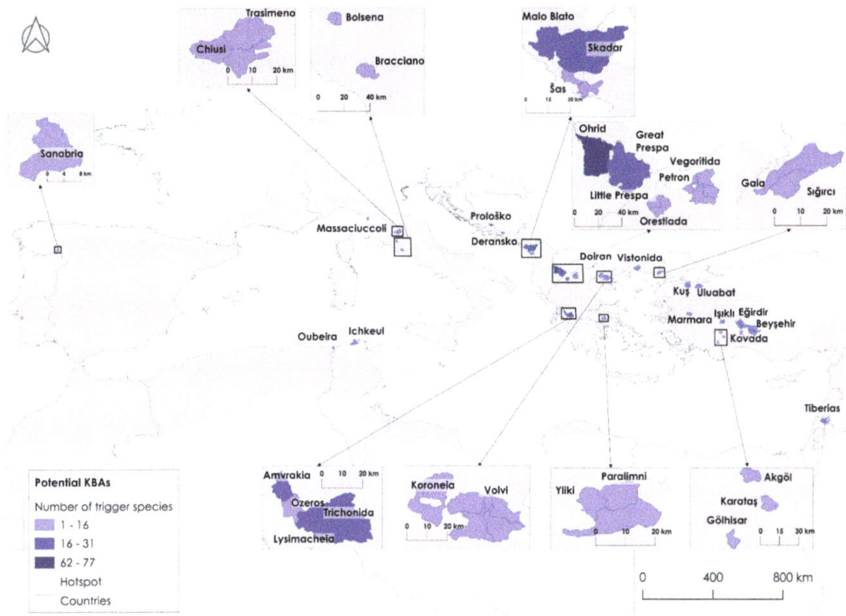

Fig. 2 The 144 potential Key Biodiversity Areas (pKBA) in HydroBASINS level 12 sub-catchments in the Mediterranean basin hotspot. Grey represents the national boundaries of all Mediterranean countries, pink the Mediterranean Biodiversity Hotspot boundaries according to CEPF (2017), and purple the pKBA sub-catchments. The darker the purple hue the higher the number of species triggering Key Biodiversity Area criteria in pKBAs

3.2 *Species that Trigger Key Biodiversity Area Criteria*

Out of the 793 species used in the KBA scoping analysis, 227 (29%) trigger KBA criteria. Most species triggering KBA criteria belong to the taxonomic groups of Gastropoda (49%) and Actinopterygii (44%). The remaining 7% is shared among plants (3%), decapods (2%), bivalves (1%) and insects (1%; Table 6). Criterion A1 (related to threatened biodiversity) was the only criterion met in every pKBA (Supp. Table 1) and was triggered by 175 species (77%; Supp. Table 2). Sub criterion A1e (related to the last living population of a critically endangered or endangered species) was met only in lakes Bracciano and Martignano (Italy), for the plant species *Isoetes sabatina*.

Criteria B1 (relating to individually geographically restricted species) and B3 (relating to geographically restricted assemblages) were met by 130 and 195 species respectively. The majority of the species in Criteria B1 and B3 belong to Gastropoda (89 species for B1; 101 species for B3) and Actinopterygii (36 species for B1; 83 for B3) taxonomic groups (Supp. Table 2).

The majority of the species triggering KBA criteria are found in the Balkan sub-region,176 out of 227 species (78%); 39 species (17%) are present in Turkey and

Table 4 Number of sub-catchments analyzed, pKBAs identified, and trigger species per lake

Sub-region	Country	Lake	No of Hydrobasins analyzed	No of pKBAs	No of trigger species
Balkans	Albania, Montenegro	Skadar	15	14	33
	Albania, North Macedonia	Ohrid	5	5	77
	Albania, North Macedonia, Greece	Great Prespa	9	7	28
	Bosnia and Herzegovina	Deransko	1	1	14
	Croatia	Vrana	1	0	–
		Prološko	1	1	20
	Greece, North Macedonia	Doiran	3	3	1
	Greece	Trichonida	3	3	21
		Volvi	5	5	2
		Vistonida	5	5	2
		Vegoritida	4	3	2
		Little Prespa	2	2	25
		Koroneia	3	3	3
		Yliki	3	3	5
		Amvrakia	1	1	18
		Lysimachia	2	2	18
		Petron	1	1	2
		Orestiada	3	3	3
		Ozeros	1	1	16
		Paralimni	1	1	5
	Montenegro	Malo Blato	1	1	27
		Šas	2	2	11
Turkey and Levant	Israel	Tiberias	11	11	7
	Turkey	Beysehir	21	17	12
		Egridir	16	15	11
		Kus	5	5	2
		Ulubat	4	4	1
		Marmara	3	3	6
		Köyceğiz	5	0	–

(continued)

Table 4 (continued)

Sub-region	Country	Lake	No of Hydrobasins analyzed	No of pKBAs	No of trigger species
		Işıklı	3	3	5
		Kovada	2	2	5
		Gala	3	1	1
		Akgöl	1	1	3
		Sığırcı	1	1	1
		Karataş	1	1	4
		Golhisar	1	1	3
Northwestern Mediterranean	France	Scamandre	1	0	–
		Charnier	1	0	–
	Italy	Bolsena	3	2	1
		Bracciano	1	1	2
		Vico	1	0	–
		Massaciuccoli	1	1	2
		Albano	1	0	–
		Chiusi	2	2	3
		Martignano	1	1	2
		Trasimeno	3	3	5
	Portugal	Fermentelos	1	0	–
	Spain	Sanabria	1	1	1
North Africa	Algeria	Oubeira	1	1	4
	Tunisia	Ichkeul	7	4	5
Total			169	144	227

The country and geographic region of the lake is provided. The total numbers do not represent sums of their respective columns, since some lakes share the same sub-catchment and some species trigger more than one pKBA

Levant; seven species (3%) in Northwest Mediterranean; and five species (2%) North Africa. Mollusk species trigger KBA criteria mainly in Balkan lakes (110 species), while fish species in lakes of the Balkan and Turkish and Levantine sub-regions (61 and 37 species respectively).

Individual pKBA in the Balkans hold a high number of species that trigger KBA criteria in comparison to the rest of pKBAs in the Mediterranean. There are 77 trigger species in Lake Ohrid, which constitute approximately one-third of the 227 trigger species across all pKBAs (Fig. 2). The lakes in the northern Balkan countries and southwestern Greece host from 18 to 33 trigger species (Fig. 2).

According to the IUCN Red List of Threatened Species, dams and water management/use, agricultural and forestry effluents, and invasive non-native/alien species/

Table 5 Area that pKBAs cover in km^2 and extent of this area that overlaps with Protected Area networks; shown in km^2 and overlap percentage per lake

Sub-region	Country	Lake name	pKBA Area (km^2)	Protected Area overlap (km^2)	Overlap percentage (%)
Balkans	Albania, Montenegro	Skadar	1486.24	594.27	39.98
	Albania, North Macedonia	Ohrid	857.45	760.63	88.71
	Albania, North Macedonia, Greece	Prespa	754.06	541.88	71.86
	Bosnia and Herzegovina	Deransko	130.12	91.15	70.05
	Croatia	Prološko	130.76	16.06	12.28
	Greece, North Macedonia	Doiran	42.07	40.98	97.41
	Greece	Trichonida	531.47	208.71	39.27
		Volvi	786.68	780.17	99.17
		Vistonida	645.53	320.91	49.71
		Vegoritida	593.33	225.11	37.94
		Small Prespa	157.67	156.42	99.21
		Small Prespa, Prespa	294.31	244.87	83.20
		Koroneia	389.44	296.91	76.24
		Yliki	312.21	48.53	15.54
		Amvrakia	115.20	70.16	60.90
		Lysimacheia	200.53	135.11	67.38
		Petron	113.24	65.84	58.14
		Orestiada	317.55	65.03	20.48
		Ozeros	107.70	55.04	51.10
		Paralimni, Yliki	120.23	50.21	41.76
	Montenegro	Malo Blato, Scutari	213.86	87.29	40.82
		Šas	247.11	105.01	42.50 lePara>
Turkey and Levante	Israel	Tiberias	1017.37	119.96	11.79
	Turkey	Beysehir	2369.74	0.00	0.00
		Egridir	1683.51	0.00	0.00
		Kus	736.07	209.72	28.49
		Ulubat	670.14	191.35	28.55
		Marmara	436.73	0.00	0.00

(continued)

Table 5 (continued)

Sub-region	Country	Lake name	pKBA Area (km^2)	Protected Area overlap (km^2)	Overlap percentage (%)
		Işıklı	429.88	0.00	0.00
		Kovada	265.86	0.00	0.00
		Gala	240.67	0.00	0.00
		Akgöl	158.28	0.00	0.00
		Sığırcı	113.87	0.00	0.00
		Karataş	141.33	0.00	0.00
		Golhisar	103.24	0.00	0.00
Northwestern Mediterranean	Italy	Trasimeno	268.15	169.02	63.03
		Bolsena	114.23	107.85	94.41
		Bracciano	181.92	149.86	82.38
		Massaciuccoli	145.75	67.00	45.97
		Chiusi	159.94	9.92	6.20
		Chiusi, Trasimeno	146.63	17.94	12.23
		Martignano	0.97	0.97	100.00
	Spain	Sanabria	152.35	151.49	99.44
North Africa	Algeria	Oubeira	122.62	117.40	95.74
	Tunisia	Ichkeul	660.99	101.21	15.31
Total			18,866.94	6373.98	33.78

The country and geographic region of the lake is provided. Values with two lake/country names refer to a shared pKBA between two lakes/countries.

Table 6 The number of species that trigger Key Biodiversity Area criteria in the 50 largest natural lakes in the Mediterranean per taxonomic group

Taxonomic group	Number of trigger species
Gastropoda	111
Actinopterygii	101
Decapoda	4
Bivalvia	3
Caryophyllales	2
Syrphoidea	2
Poales	1
Myrtales	1
Lycopodiopsida	1
Asparagales	1
Total	227

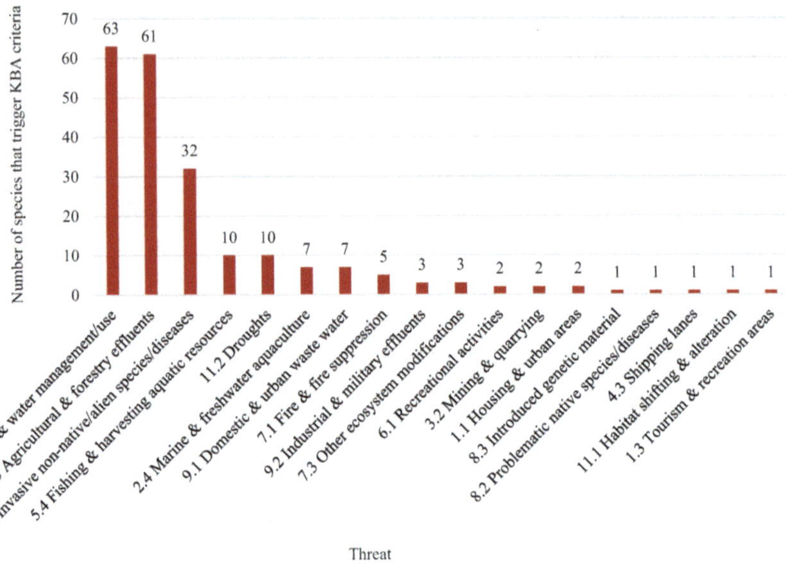

Fig. 3 Threats identified in the species assessments from the IUCN Red List of Threatened Species per species that trigger Key Biodiversity Area (KBAs) criteria. Threats are not identified only for 15 species out of the total 227 species that trigger KBA criteria

diseases are the most prominent threats among trigger species, impacting 28%, 27%, and 14% of the species respectively. All other threats affect a range of 4% to 0.4% of the trigger species. We did not include the threats of 15 species that trigger KBA criteria, because their threats were not stated in the assessments (Fig. 3).

4 Discussion

We found that 86% of the 20,684.2 km^2 area and 86% of the 50 lakes studied are identified as pKBAs, and 227 of the 793 species trigger KBA criteria. This result highlights the importance of the Mediterranean Basin as a Biodiversity Hotspot, especially for the freshwater species it hosts (CEPF 2017). The highest number of species that trigger KBA Criteria within one sub-catchment is 77. In other biodiversity hotspots or regions of interest where the Global Standard was applied for freshwater species, fewer species trigger KBAs in a single site: 20 in Western Africa (Starnes and Spiliopoulou 2021), 23 in Madagascar (Máiz-Tomé et al. 2018), 46 in the lake Malawi/Nyasa/Niassa catchment (Sayer et al. 2019), 55 in Canada (Tognelli et al. 2017), and 60 in the Lower Mekong River Basin (Máiz-Tomé 2019). As a result, the Mediterranean Biodiversity Hotspot stands out as a significant region for freshwater biodiversity.

The Balkans appear to be the most important region in the Mediterranean basin for lacustrine species, being a hotspot within the hotspot. This sub-region holds the highest percentages of species triggering KBA criteria (78%; mainly fish and gastropods) and sites qualifying as B3 (41; areas that hold important species assemblages of a taxonomic group). We believe this to be a result of the region's rich biogeographic history, especially for freshwater fish and gastropod species. The Balkans hold complex species compositions of fish because significant geological events resulted in ecological barriers among basins that hinder fish dispersal, thus leading to restricted long-distance dispersal and the occurrence of biotic isolation (Skoulikidis et al. 2009; Oikonomou et al. 2014). It has been well documented that isolation phenomena have further increased species diversification and promoted local endemism events (Heino 2011; Oikonomou et al. 2014). The Balkan sub-region holds the highest number of ancient lakes in Europe (Albrecht and Wilke 2008). Indeed, ancient lakes are hotspots for mollusk biodiversity because they exhibit high horizontal and vertical lake zonation, habitat heterogeneity, and have acted as ecological refugia (Frogley and Preece 2004; Albrecht and Wilke 2008; Albrecht et al. 2009, 2012). The ancient lake Ohrid is in the Balkan sub-region and holds the highest number of species that trigger KBA Criteria among the lakes we studied (77 species). Many studies have documented the exceptionally high mollusk species' diversification and endemism rates within the lake (Albrecht and Wilke 2008; Föller et al. 2015; Stelbrink et al. 2020).

We identified significant gaps in the protection of pKBAs. Only one third of all pKBAs in Mediterranean lakes are protected. The area of Turkey and Levante stands out as the primary sub-region with inadequate protection, as only 5% of Turkey's and 12% of Israel's pKBA area is covered by protected areas. This result aligns with the scoping analysis performed in Turkey, where Eken et al. (2016) found that less than 14% of Turkey's KBAs (both terrestrial and freshwater) are legally protected, while freshwater ecosystems are the most vulnerable. It should be a priority in Turkey and Israel to expand the protected area network to cover freshwater species that trigger KBA criteria. We also found that pKBAs in lake Yliki in Greece, lakes Tiberias in Israel and Chiusi in Italy, and lake Prološko in Croatia have less than 20% of their area legally protected. Protected Area networks should expand by prioritizing these lakes in their respective countries. However, lake Bracciano, which triggers sub-criterion A1e holding more than 95% of the global population of the critically endangered plant *Isoetes sabatina*, has more than 80% of its respective areas legally protected.

Freshwater biodiversity has faced the biggest declines since 1970 and freshwater ecosystems are shrinking with the highest rates (Harrison et al. 2018; IPBES 2019). Many new threats and challenges are emerging (Reid et al. 2019), and this emphasizes the urgency to protect freshwater biodiversity. The freshwater lake species triggering KBA criteria are mainly affected by dams and water management or use (28%), agricultural and forestry effluents (27%), and the presence/introduction of invasive alien species (14%). Other works have documented these threats, especially for freshwater fish in the Mediterranean. Dam construction has reportedly deteriorated 36% of all Mediterranean freshwater fish extinction rates, while promoting biological invasions (Freyhof et al. 2020), and threatens 39% of freshwater species globally (Sayer et al. 2025). Invasive species cause an array of impacts to native freshwater fish biodiversity of the Mediterranean (Ribeiro and Leunda 2012) with an ongoing taxonomic homogenization being the most prominent (Marr et al. 2013). Invasive species also cause ecosystem alterations in Balkan lentic ecosystems (Piria et al. 2018), which we found to be the sub-region with the highest numbers of species that trigger KBA criteria. Agricultural and forestry effluents often lead to eutrophication phenomena, which many other works having highlighted it as a persistent threat that affects the native molluscan fauna of ancient Balkan lakes (Frogley et al. 2001; Albrecht et al. 2007, 2009, 2012).

This is the first work to perform a scoping analysis for Mediterranean freshwater species by applying the KBA Global Standard. Previous work on KBAs using freshwater species in the Mediterranean basin include Darwall et al. (2014) and Máiz-Tomé et al. (2017). Our results are not comparable, because both studies used HydroBASINS level 8 as spatial units to scope for KBAs for rivers and lakes. Darwall et al. (2014) also used criteria defined in Holland et al. (2012), which do not allow for an objective, transparent and repeatable identification process. The thresholds for the KBA criteria are calculated as the proportion of the site to the global range size of a species. Therefore, there are two parameters that affect whether the KBA criteria are met, (i) the size of the spatial unit and (ii) the global range size. Decreasing the size of the spatial unit used to scope pKBA sites would result in fewer species triggering KBA criteria. In this study, we used the smaller HydroBASINS level 12 spatial units, and this is a first step towards targeting more efficiently the most important places for freshwater biodiversity, instead of prioritizing the potential KBA sites according to the number of species that trigger KBA criteria. In any case, KBAs should not be prioritized and regardless of the number of species that trigger KBA criteria, they all are equally important. In addition, Mediterranean countries contain smaller lakes and catchments compared to central Europe (Solheim et al. 2019; Stefanidis et al. 2021). Higher resolution HydroBASINS 12 can capture the attributes of the Mediterranean small, fragmented hydrological network, more adequately. Determining the appropriate spatial units for applying the Global Standard on freshwater species calls for further exploration. Sub-catchments may account for the connectivity among freshwater systems—which is key to the management of freshwater systems (Dudgeon et al. 2006)—but they do not represent a species' real distribution—which is key to the KBA identification process (IUCN 2016a, b). When using sub-catchments as planning units, the resulting KBAs may encompass more terrestrial than freshwater area, creating misleading results and requiring the repetition of the scoping process a few times.

Certain aspects of our study have limitations. In our species' distribution dataset, the number of trigger species in Turkey does not represent the country's freshwater fish biodiversity. Recent reviews estimate 409 freshwater fish species within the boundaries of Turkey (Çiçek et al. 2018). However, only 297 freshwater fish species have been assessed in the Turkish land region according to IUCN Red List (IUCN 2023). We believe that this is a source of bias within our dataset and influences the resulting pKBAs. Assessing more freshwater species is necessary to encompass the country's freshwater biodiversity and obtain a more representative KBA network.

Lake Ohrid (holding the highest number of species triggering KBA criteria) has been extensively studied in comparison to other Mediterranean lakes during the last decade (Albrecht and Wilke 2008). This can partially explain why we found the highest number of species that trigger KBA criteria in Ohrid. Further field data and assessments in other Mediterranean lakes will probably increase the number of species that trigger KBAs. Freshwater species' range is mapped in lower resolution (HydroBASINS level 8; IUCN Red List Technical Working Group 2021) from what is suggested to be used when scoping KBAs (HydroBASINS level 12; KBA Standards and Appeals Committee of IUCN SSC/WCPA 2022), covering a broad, coarse geographic area. Therefore, IUCN range maps minimize omission errors (not covering areas where the species is present) but tend to increase commission errors (covering areas where the species is absent; Di Marco et al. 2017b), which can lead to identified KBAs in areas where the species may be absent.

KBAs are a valuable prioritization tool that can help identify and safeguard the most important places for biodiversity globally. However, the mere coverage by protected areas will not be enough to safeguard biodiversity (Coad et al. 2019). Using the KBA approach can strategically guide the designation of new or expansion of already existing networks of protected areas or other effective area-based conservation measures (OECMs). This is a necessity for conserving freshwater biodiversity (Darwall et al. 2011) and reaching the 2030 target set by the Kunming—Montreal Biodiversity Framework, where at least 30% of all terrestrial, inland, and marine areas need to be legally protected (CBD 2022). KBAs can also act as a consultation tool that informs early conservation planning strategies and aids in priority setting by informing stakeholders and policymakers on which sites are crucial for the persistence of biodiversity on a global scale (Dudley et al. 2014).

The pKBAs for freshwater lake species we identified are only the first step towards the KBA identification. The delineation of sites and the official proposal of KBAs in the World Database of KBAs requires consultation with relevant stakeholders and knowledge holders, in a process usually led by the KBA National Coordination Group (NCG) of a country. The primary responsibilities of the NCG involve coordinating the identification, mapping, and monitoring of KBAs, while engaging the relevant national stakeholders and the influence of national policies in order to safeguard KBAs (KBA Secretariat 2019; KBA Standards and Appeals Committee of IUCN SSC/WCPA 2022). So far, Greece is the only country with an established KBA NCG in the Mediterranean and the only European country that has acknowledged KBAs in its legislation (Law 5073/2023). Italy and Spain are in the process of establishing their own NCGs, and Turkey has expressed interest in forming its own NCG. It is crucial that more Mediterranean countries take the requisite measures to establish NCGs and undergo rapid KBA analyses to identify sites that represent multiple taxonomic groups.

5 Conclusions

Most of the largest natural lakes of the Mediterranean Basin have at least one pKBA, with the Balkan sub-region having the highest number of species triggering KBA criteria. Our analysis showcases the importance and the level of threat that freshwater biodiversity faces. Finding the appropriate spatial units to apply the Global Standard and represent freshwater biodiversity should be further explored methodologically. It is important for Mediterranean countries to proceed to the formation of NCGs for the coordination of KBA identification and official proposals to the World Database of KBAs. There are gaps in protection of lake sub-catchments, as well as emerging threats to freshwater biodiversity, and conservation efforts to minimize the losses can be guided by the identification of KBAs. Current protected area networks and management plans fail to represent freshwater biodiversity. At the same time, threats are expected to rise and affect more species in the following years. Therefore, future policies and frameworks should be revised in a way that encompass freshwater biodiversity and conservation, alongside stakeholder deliberations; and especially in the Balkan region to foster transboundary conservation efforts and public engagement actions (Albrecht et al. 2012; Oikonomou et al. 2014). Protected area networks should be expanded by prioritizing areas that qualify as KBAs to enable all Mediterranean countries to reach the 30% protected area cover by 2030 as mandated by the Kunming—Montreal Biodiversity Framework.

References

Abell R, Allan JD, Lehner B (2007) Unlocking the potential of protected areas for freshwaters. Biol Cons 134(1):48–63

Abell R, Lehner B, Thieme M, Linke S (2017) Looking beyond the fenceline: assessing protection gaps for the world's rivers. Conserv Lett 10(4):384–394

Albrecht C, Wilke T (2008) Ancient Lake Ohrid: biodiversity and evolution. Hydrobiologia 615:103–140

Albrecht C, Schultheiß R, Kevrekidis T, Streit B, Wilke T (2007) Invaders or endemics? Molecular phylogenetics, biogeography and systematics of Dreissena in the Balkans. Freshw Biol 52(8):1525–1536

Albrecht C, Hauffe T, Schreiber K, Trajanovski S, Wilke T (2009) Mollusc biodiversity and endemism in the potential ancient lake Trichonis, Greece. Malacologia 51(2):357–375

Albrecht C, Hauffe T, Schreiber K, Wilke T (2012) Mollusc biodiversity in a European ancient lake system: lakes Prespa and Mikri Prespa in the Balkans. Hydrobiologia 682:47–59

Allen D, Shi Y, Holland R, Smith K (2010) Regional synthesis for all taxa. In: Allen DJ, Molur S, Daniel BA (compilers) The status and distribution of freshwater biodiversity in the Eastern Himalaya. IUCN, and Coimbatore, India: Zoo Outreach Organisation, Cambridge, UK, Gland, Switzerland, pp 68–78 (Chapter 6)

Ambal RGR, Duya MV, Cruz MA, Coroza OG, Vergara SG, de Silva N, Molinyawe N, Tabaranza B (2012) Key biodiversity areas in the Philippines: priorities for conservation. J Threat Taxa 2788–2796

Bour R (2008) Global diversity of turtles (Chelonii; Reptilia) in freshwater. In: Balian VE, Lévèque C, Segers H, Martens K (eds) Freshwater animal diversity assessment. Developments in hydrobiology, pp 593–598
CBD (2022) Kunming-Montreal global biodiversity framework. https://www.cbd.int/doc/c/e6d3/cd1d/daf663719a03902a9b116c34/cop-15-l-25-en.pdf
CEPF (2017) Ecosystem profile, mediterranean basin biodiversity hotspot. BirdLife International. https://www.cepf.net/sites/default/files/mediterranean-basin-2017-ecosystem-profile-english_0.pdf
Çiçek E, Fricke R, Sungur S, Eagderi S (2018) Endemic freshwater fishes of Turkey. FishTaxa 3(4):1–39
Coad L, Watson JE, Geldmann J, Burgess ND, Leverington F, Hockings M, Knights K, Di Marco M (2019) Widespread shortfalls in protected area resourcing undermine efforts to conserve biodiversity. Front Ecol Environ 17(5):259–264
Darwall WR, Holland RA, Smith KG, Allen D, Brooks EG, Katarya V, Pollock CM, Shi Y, Clausnitzer V, Cumberlidge N, Cuttelod A, Klaas-Douwe BD, Diop MD, García N, Seddon MB, Skelton PH, Snoeks J, Tweddle D, Vié JC (2011) Implications of bias in conservation research and investment for freshwater species. Conserv Lett 4(6):474–482
Darwall WR, Carrizo S, Numa C, Barrios V, Freyhof J, Smith K (2014) Freshwater key biodiversity areas in the Mediterranean Basin hotspot: informing species conservation and development planning in freshwater ecosystems. IUCN, Cambridge, UK, Malaga, Spain, x + 86pp
Dias MS, Tedesco PA, Hugueny B, Jézéquel C, Beauchard O, Brosse S, Oberdorff T (2017) Anthropogenic stressors and riverine fish extinctions. Ecol Ind 79:37–46
Di Marco M, Chapman S, Althor G, Kearney S, Besancon C, Butt N, Maina JM, Possingham HP, von Bieberstein KR, Venter O, Watson JE (2017a) Changing trends and persisting biases in three decades of conservation science. Global Ecol Conserv 10:32–42
Di Marco M, Watson JE, Possingham HP, Venter O (2017b) Limitations and trade-offs in the use of species distribution maps for protected area planning. J Appl Ecol 54(2):402–411
Dudgeon D, Arthington AH, Gessner MO, Kawabata ZI, Knowler DJ, Lévêque C, Naiman RJ, Prieur-Richard AH, Soto D, Stiassny ML, Sullivan CA (2006) Freshwater biodiversity: importance, threats, status and conservation challenges. Biol Rev 81(2):163–182
Dudley N, Boucher JL, Cuttelod A, Brooks TM, Langhammer PF (eds) (2014) Applications of key biodiversity areas: end-user consultations. IUCN, Cambridge, UK, Gland, Switzerland
Eken G, Isfendiyaroğlu S, Yeniyurt C, Erkol IL, Karataş A, Ataol M (2016) Identifying key biodiversity areas in Turkey: a multi-taxon approach. Int J Biodiver Sci Ecosyst Serv Manag 12(3):181–190
Föller K, Stelbrink B, Hauffe T, Albrecht C, Wilke T (2015) Constant diversification rates of endemic gastropods in ancient Lake Ohrid: ecosystem resilience likely buffers environmental fluctuations. Biogeosciences 12(23):7209–7222
Freyhof J, Bergner L, Ford M (2020) Threatened freshwater fishes of the Mediterranean Basin biodiversity hotspot: distribution, extinction risk and the impact of hydropower. EuroNatur and RiverWatch, i–viii + 1–348
Fricke R, Eschmeyer WN, van der Laan R (eds) (2023) Eschmeyer's catalog of fishes: genera, species, references. Retrieved 21 Dec 2023, from http://researcharchive.calacademy.org/research/ichthyology/catalog/fishcatmain.asp
Frogley MR, Preece RC (2004) A faunistic review of the modern and fossil molluscan fauna from lake Pamvotis, Ioannina, an ancient lake in NW Greece: implications for endemism in the Balkans. In: Griffiths HI, Kryštufek B, Reed JM (eds) Balkan biodiversity: pattern and process in the European hotspot. Springer Netherlands, Dordrecht, pp 243–260
Frogley MR, Griffiths HI, Heaton TH (2001) Historical biogeography and Late Quaternary environmental change of Lake Pamvotis, Ioannina (north-western Greece): evidence from ostracods. J Biogeogr 28(6):745–756
Grill G, Lehner B, Thieme M, Geenen B, Tickner D, Antonelli F, Babu S, Borrelli P, Cheng L, Crochetiere H, Ehalt Macedo H, Filgueiras R, Goichot M, Higgins J, Hogan Z, Lip B, McClain

ME, Meng J, Mulligan M, Nilsson C, Olden JD, Opperman JJ, Petry P, Reidy Liermann C, Sáenz L, Salinas-Rodríguez S, Schelle P, Schmitt RJP, Snider J, Tan F, Tockner K, Valdujo PH, van Soesbergen A, Zarfl C (2019) Mapping the world's free-flowing rivers. Nature 569(7755):215–221

Harrison I, Abell R, Darwall W, Thieme ML, Tickner D, Timboe I (2018) The freshwater biodiversity crisis. Science 362(6421):1369–1369

Heino J (2011) A macroecological perspective of diversity patterns in the freshwater realm. Freshw Biol 56:1703–1722

Holland RA, Darwall WRT, Smith KG (2012) Conservation priorities for freshwater biodiversity: the key biodiversity area approach refined and tested for continental Africa. Biol Cons 148(1):167–179

Hortal J, Nabout J, Calatayud JC, Carneiro FM, Padial A, Santos AMC, Siqueira T, Bokma F, Bini LM, Ventura M (2014) Perspectives on the use of lakes and ponds as model systems for macroecological research. J Limnol 46–60

IPBES (2019) Global assessment report of the Intergovernmental Science-Policy Platform on Biodiversity and Ecosystem Services. https://www.ipbes.net/global-assessment

IUCN (2016a) A global standard for the identification of key biodiversity areas, Version 1.0. Gland. IUCN, Switzerland

IUCN (2016b) Rules of procedure for IUCN red list assessments 2017–2020. Version 3.0. http://cmsdocs.s3.amazonaws.com/keydocuments/Rules_of_Procedure_for_Red_List_2017-2020.pdf

IUCN (2023) The IUCN red list of threatened species. Version 2022-2. Retrieved 2 Jan 2023, from https://www.iucnredlist.org

IUCN Red List Technical Working Group (2021) Mapping standards and data quality for the IUCN red list spatial data. Version 1.19. Prepared by the standards and petitions working group of the IUCN SSC red list committee. Downloadable from: https://www.iucnredlist.org/resources/mappingstandards

KBA Secretariat (2019) Key Biodiversity areas proposal process: guidance on proposing, reviewing, nominating and confirming sites. Version 1.0 March 2019, prepared by the KBA secretariat and KBA Committee of the KBA Partnership. Cambridge, UK. https://www.cepf.net/sites/default/files/kba-proposal-process.pdf.

KBA Standards and Appeals Committee of IUCN SSC/WCPA (2022) Guidelines for using a global standard for the identification of key biodiversity areas. Version 1.2. 462 IUCN, Gland, Switzerland

Law 5037/2023, Article 184, Key Biodiversity Areas article 33 law 4625/2020 amendment, Government Gazette of the Hellenic Republic (O.G.G. 78/A'/28-03-2023)

Lehner B, Grill G (2013) Global river hydrography and network routing: baseline data and new approaches to study the world's large river systems. Hydrol Process 27(15):2171–2186

Maasri A, Bonada N (2024) Ecology of Mediterranean freshwater ecosystems. In: Identification and ecology of freshwater arthropods in the Mediterranean Basin. Elsevier, pp 5–16

Máiz-Tomé L (2019) Donor report: freshwater key biodiversity areas in the lower Mekong River Basin informing species conservation and investment planning in freshwater ecosystems, June 2019

Máiz-Tomé L, Darwall W, Numa C, Barros V, Smith KG (2017) Freshwater key biodiversity areas in the north-western Mediterranean sub-region. IUCN, Gland, Switzerland, Cambridge, UK, Malaga, Spain, viii + 48pp. https://portals.iucn.org/library/sites/library/files/documents/SSC-OP-no.64.pdf

Máiz-Tomé L, Sayer C, Darwall W (2018) Freshwater key biodiversity areas in Madagascar. In: Máiz-Tomé L, Sayer C, Darwall W (eds) (2018) The status and distribution of freshwater biodiversity in Madagascar and the Indian Ocean islands hotspot. Gland, Switzerland, pp 98–114 (Chapter 9)

Mammides C (2020) A global assessment of the human pressure on the world's lakes. Glob Environ Chang 63:102084

Marr SM, Olden JD, Leprieur F, Arismendi I, Ćaleta M, Morgan DL, Šanda AMNR, Tarkan AS, García-Berthou E (2013) A global assessment of freshwater fish introductions in mediterranean-climate regions. Hydrobiologia 719:317–329
Myers N, Mittermeier RA, Mittermeier CG, Da Fonseca GA, Kent J (2000) Biodiversity hotspots for conservation priorities. Nature 403(6772):853–858
Oikonomou A, Leprieur F, Leonardos ID (2014) Biogeography of freshwater fishes of the Balkan Peninsula. Hydrobiologia 738:205–220
Pešić V, Miliša M, Milošević Đ (2022) Conclusions: small water bodies of the western balkans—values and threats. Springer International Publishing
Piria M, Simonović P, Kalogianni E, Vardakas L, Koutsikos N, Zanella D, Ristovska M, Apostolou A, Adrović A, Mrdak D, Tarkan AS, Milošević D, Zanella LN, Bakiu R, Ekmekçi FG, Povž M, Korro K, Nikolić V, Škrijelj R, Kostov V, Gregori A, Joy MK (2018) Alien freshwater fish species in the Balkans—vectors and pathways of introduction. Fish Fish 19(1):138–169
Plumptre AJ, Ayebare S, Behangana M, Forrest TG, Hatanga P, Kabuye C, Kirunda B, Kityo R, Mugabe H, Namaganda M, Nampindo S (2019) Conservation of vertebrates and plants in Uganda: identifying key biodiversity areas and other sites of national importance. Conserv Sci Pract 1(2):e7
Pyron M, Brown KM (2015) Introduction to Mollusca and the class Gastropoda, In: Thorp JH, Rogers DC (eds) Thorp and Covich's freshwater invertebrates, 4th edn, pp 383–421 (Chapter 18)
Ramsar Convention on Wetlands (2018) Global wetland outlook: state of the world's wetlands and their services to people. Ramsar Convention Secretariat, Gland, Switzerland. https://www.ramsar.org/sites/default/files/documents/library/gwo_e.pdf
Reid AJ, Carlson AK, Creed IF, Eliason EJ, Gell PA, Johnson PT, Kidd KA, MacCormack TJ, Olden JD, Ormerod SJ, Smol JP, Taylor WW, Tockner K, Vermaire JC, Dudgeon D, Cooke SJ (2019) Emerging threats and persistent conservation challenges for freshwater biodiversity. Biol Rev 94(3):849–873
Ribeiro F, Leunda PM (2012) Non-native fish impacts on Mediterranean freshwater ecosystems: current knowledge and research needs. Fish Manage Ecol 19(2):142–156
Rubec C, Alwash A, Bachmann A (2009) The key biodiversity areas project in Iraq: objectives and scope 2004–2008. BioRisk 3:39–53
Sayer CA, Palmer-Newton AF, Darwall WRT (2019) Conservation priorities for freshwater biodiversity in the Lake Malawi/Nyasa/Niassa catchment. International Union for Conservation of Nature, Cambridge, UK, Gland, Switzerland. https://portals.iucn.org/library/sites/library/files/documents/RL-2019-001-En.pdf
Sayer CA, Fernando E, Jimenez RR, Macfarlane NB, Rapacciuolo G, Böhm M, Brooks TM, Contreras-MacBeath T, Cox NA, Harrison A, Hoffmann M, Jenkins R, Smith KG, Vié JC, Abbott JC, Allen DJ, Allen GR, Barrios V, Boudot JP, Carrizo SF, Charvet P, Clausnitzer V, Congiu L, Crandall KA, Cumberlidge N, Cuttelod A, Dalton J, Daniels AG, De Grave S, De Kniff G, Dijkstra, BKD, Dow RA, Freyhof J, García N, García J, Getahun A, Gibson C, Gollock MJ, Grant MI, Groom AER, Hammer MP, Hammerson GA, Hilton-Taylor C, Hodgkinson L, Holland RA, Jabado RW, Bignoli DJ, Kalkman VJ, Karimov BK, Kipping J, Kottelat M, Lalèyè PA, Larson HK, Lintermans M, Lozano F, Ludwig A, Lyons TJ, Máiz-Tomé L, Molur S, Ng HH, Numa C, Palmer-Newton AF, Pike C, Pippard HE, Polaz CNM, Pollock C, Raghavan R, Rand PS, Ravelomanana T, Reis RE, Rigby CL, Scott JA, Skelton PH, Sloat MR, Snoeks J, Stiassny MLJ, Tan HH, Taniguchi Y, Thorstad EB, Tognelli MF, Torres AG, Torres Y, Tweddle D, Watanabe K, Westrip JRS, Wright EGE, Zhang E, Darwall WR (2025) One-quarter of freshwater fauna threatened with extinction. Nature 1–8
Skoulikidis N, Economou AN, Gritzalis K, Zogaris S (2009) Rivers of the Balkans. In: Tockner K, Uehlinger U, Robinson CC (eds) Rivers of Europe. Academic Press, Elsevier, Amsterdam, pp 421–466
Smith KG, Allen DA, Carrizo S (2012) Synthesis for all taxa. In: Smith KG, Barrios V, Darwall WRT, Numa C (eds) The status and distribution of freshwater biodiversity in Indo-Burma. Cambridge, UK, Gland, Switzerland, pp 89–110 (Chapter 8)

Solheim AL, Globevnik L, Austnes K, Kristensen P, Moe SJ, Persson J, Phillips G, Poikane S, van de Bund W, Birk S (2019) A new broad typology for rivers and lakes in Europe: development and application for large-scale environmental assessments. Sci Total Environ 697:134043
Spiliopoulou K, Starnes T (2021) Key biodiversity areas. In: Starnes T, Darwall WRT (eds) Identification and validation of Western African freshwater key biodiversity areas. IUCN, Gland, Switzerland, pp102–122 (Chapter 9)
Spiliopoulou K, Brooks TM, Dimitrakopoulos PG, Oikonomou A, Karavatsou F, Stoumboudi MT, Triantis KA (2023) Protected areas and the ranges of threatened species: towards the EU biodiversity strategy 2030. Biol Cons 284:110166
Spiliopoulou K, Rigal F, Plumptre AJ, Trigas P, Paragamian K, Hochkirch A, Lymberakis P, Portolou D, Soumboudi MT, Triantis KA (2024) KBAscope: key biodiversity area identification in R. Ecography 2024(9):e07061
Stelbrink B, Wilke T, Albrecht C (2020) Ecological opportunity enabled invertebrate radiations in ancient Lake Ohrid. J Great Lakes Res 46(5):1156–1161
Stefanidis K, Oikonomou A, Stoumboudi M, Dimitriou E, Skoulikidis NT (2021) Do water bodies show better ecological status in Natura 2000 protected areas than non-protected ones?—The case of Greece. Water 13(21):3007
Thieme M, Lehner B, Abell R, Hamilton SK, Kellndorfer J, Powell G, Riveros JC (2007) Freshwater conservation planning in data-poor areas: an example from a remote Amazonian basin (Madre de Dios River, Peru and Bolivia). Biol Cons 135(4):484–501
Thieme ML, Khrystenko D, Qin S, Golden Kroner RE, Lehner B, Pack S, Tockner K, Zarfl C, Shahbol N, Mascia MB (2020) Dams and protected areas: quantifying the spatial and temporal extent of global dam construction within protected areas. Conserv Lett 13(4):e12719
Tierno de Figueroa JM, López-Rodríguez MJ, Fenoglio S, Sánchez-Castillo P, Fochetti R (2013) Freshwater biodiversity in the rivers of the Mediterranean Basin. Hydrobiologia 719:137–186
Tognelli MF, Máiz-Tomé L, Kraus D, Lepitzki D, Mackie G, Morris T, Carney J, Alfonso N, Tonn B, Cox NA, Smith KG (2017) Freshwater key biodiversity areas in Canada. Informing species conservation and development planning in freshwater ecosystems. IUCN, Gland, Switzerland, Cambridge, UK, Arlington, USA
Tordoff AW, Baltzer MC, Fellowes JR, Pilgrim JD, Langhammer PF (2012) Key biodiversity areas in the Indo-Burma hotspot: process, progress and future directions. J Threat Taxa, pp 2779–2787
UNEP-WCMC, & IUCN (2023) Protected planet: the world database on protected areas (WDPA) and world database on other effective area-based conservation measures (WD-OECM). Retrieved Feb 2023, from: www.protectedplanet.net
United Nations (2016) The sustainable development goals report 2016. https://unstats.un.org/sdgs/report/2016/the%20sustainable%20development%20goals%20report%202016.pdf
WWF (2021) The world's forgotten fishes. https://wwfint.awsassets.panda.org/downloads/world_s_forgotten_fishes__final_april9_.pdf
WWF (2022) Living planet report 2022—building a nature positive society. WWF, Gland, Switzerland. https://wwfint.awsassets.panda.org/downloads/embargo_13_10_2022_lpr_2022_full_report_single_page_1.pdf

Freshwater KBAs in Türkiye's Mediterranean Region: Challenges and Conservation Strategies for Fish Fauna

Fatih Mangıt, Mustafa Korkmaz, Mehmet Arda Çolak, and Erik Jeppesen

Abstract The Mediterranean region is recognised as a global biodiversity hotspot, with Türkiye playing a key role in the conservation of freshwater ecosystems. This chapter provides a comprehensive assessment of the ecological status, prevailing threats and conservation priorities for the important freshwater Key Biodiversity Areas (KBAs) in the Mediterranean region of Türkiye, with a particular focus on ten ecologically significant lakes. These lakes serve as critical habitats for endemic and globally threatened fish species, but are increasingly affected by anthropogenic pressures, including habitat degradation, pollution, invasive species and unsustainable water management practices. Climate projections based on Shared Socioeconomic Pathways scenarios indicate an increased risk of drought, reduced precipitation and rising temperatures, exacerbating the vulnerability of aquatic biodiversity. Long-term analyses of lake surface dynamics, using remote sensing and historical datasets, reveal significant declines due to excessive water abstraction and hydrological alterations. The results highlight the urgent need for integrated conservation strategies, strengthened water governance and adaptive management approaches to mitigate these threats and safeguard the ecological integrity of Türkiye's Mediterranean freshwater KBAs.

F. Mangıt
Department of Biology, Hacettepe University, Ankara, Türkiye

M. Korkmaz (✉) · E. Jeppesen
Centre for Ecosystem Research (EKOSAM), Middle East Technical University, Ankara, Türkiye
e-mail: korkmazm@metu.edu.tr

Department of Biological Sciences, Middle East Technical University, Ankara, Türkiye

M. A. Çolak
Department of Geomatics Engineering, Hacettepe University, Ankara, Türkiye

E. Jeppesen
Department of Ecoscience and WATEC, Aarhus University, Aarhus, Denmark

Sino-Danish Centre for Education and Research (SDC), Beijing, China

Institute for Ecological Research and Pollution Control of Plateau Lakes, School of Ecology and Environmental Science, Yunnan University, Kunming, China

K. Stefanidis et al. (eds.), *Lakes of the Eastern Mediterranean and Balkans*,
https://doi.org/10.1007/978-3-032-02168-7_2

Keywords Freshwater KBAs · Endemic fish · Habitat degradation · Climate change

1 Introduction

The Mediterranean region is a globally recognized biodiversity hotspot, known for its rich ecosystems and endemic species (Médail and Quezel 1999; Blondel et al. 2010). Together with the North Eastern European region, it has been identified as a primary hotspot for climate change impacts (Giorgi 2006; Lionello and Scarascia 2018). Within this region, Türkiye plays a vital role due to its significant number of Key Biodiversity Areas (KBAs), serving as critical habitats for endemic and native species. KBAs are defined as locations that make substantial contributions to the global persistence of biodiversity and are classified based on five criteria: biological processes, ecological integrity, threatened biodiversity, geographically restricted biodiversity, and irreplaceability (IUCN 2016).

Türkiye hosts a total of 324 KBAs, 120 of which are located in the Mediterranean region (Eken et al. 2006; KBAP 2025). More than 60 lakes in Türkiye, including 17 lakes in the Mediterranean hotspot, are designated as KBAs due to their ecological importance and the presence of globally threatened and endemic species (Table 1). These lakes not only play a crucial role in preserving biodiversity but also provide essential ecosystem services. However, escalating pressures from climate change and human activities, including habitat degradation and water abstraction, jeopardize their ecological balance and long-term sustainability, highlighting the urgent need for conservation and adaptive management strategies.

Türkiye's freshwater ecosystems, particularly those designated as KBAs, are under severe stress from invasive species, habitat degradation, pollution, and water abstraction. Invasive species, such as *Carassius gibelio* (Bloch 1782), *Gambusia holbrooki* Girard 1859, *Lepomis gibbosus* (Linnaeus 1758), *Pseudorasbora parva* (Temminck and Schlegel 1846), and *Atherina boyeri* Risso 1810, have led to significant ecological shifts, displacing native species and contributing to the decline of the endemic species and extinction like that of *Pseudophoxinus handlirschi* (Pietschmann 1933) in Eğirdir Lake (Küçük 2012) and *Alburnus nicaeensis* Battalgil 1941 in İznik Lake (Freyhof 2014a, b), see Appendix 1. Anthropogenic activities, such as dam construction and agricultural runoff, further fragment habitats and degrade water quality, endangering the survival of fish species uniquely adapted to these ecosystems (Freyhof et al. 2020).

Agricultural water abstraction from rivers, lakes, and groundwater sources, exacerbates the threats to Mediterranean lakes. Lakes in the Burdur Closed Basin and Konya Closed Basin exemplify the severe impacts of irrigation-dependent agriculture (Yılmaz et al. 2021; Çolak et al. 2022). Here, excessive water withdrawal, often unregulated, and dam constructions have significantly reduced inflows, leading to shrinkage of lakes like Burdur, Eğirdir, and Acıgöl. This has resulted in loss of critical habitats for endemic species such as *Anatolichthys transgrediens* (Ermin 1946),

Table 1 Mediterranean lake KBAs of Türkiye

KBA code	Name	This study	Latitude	Longitude	Area (km^2)
759	Bafa Lake	+	37.5333	27.4333	176.90
766	Burdur Lake	+	37.7333	30.2167	251.43
767	Eğirdir Lake	+	38.0467	30.8863	627.44
790	Işıklı Lake	+	38.2333	29.9333	97.47
777	İznik Lake[a]	+	40.4333	29.5167	380.19
801	Köyceğiz Lake	+	36.8667	28.6333	399.33
763	Lake Acıgöl	+	37.8276	29.8825	328.00
783	Manyas Lake	+	40.1833	27.95	227.14
770	Salda Lake	+	37.55	29.6833	62.35
782	Uluabat Lake	+	40.1667	28.5833	245.43
24056	Çığlıkara Forests (Avlan Lake)	NE	36.5796	29.9421	494.74
793	Çorak Lake	NE	37.6667	29.7667	19.34
798	Gavur Lake	NE	37.3009	36.8333	66.64
9603	Girdev Lake and Akdağlar	NE	36.5833	29.6167	751.05
31611	Gölcük Lake	NE	37.7307	30.4953	4.34
769	Karataş Lake	NE	37.3833	29.9667	24.32
762	Marmara Lake	NE	38.6167	28.0167	69.26
771	Yarışlı Lake	NE	37.5667	29.9667	26.26

NE Not evaluated in this study
[a]According to the original delineation, the Mediterranean region does not include İznik Lake. However, the authors included it due to its significance

Anatolichthys sureyanus (Neu 1937), and *Pseudophoxinus burduricus* (Küçük et al. 2013; Freyhof et al. 2020).

The impacts of climate change compound the challenges faced by KBAs. Projections under Shared Socioeconomic Pathways (SSP) scenarios indicate that significant changes will occur in precipitation and temperature patterns in Türkiye's Mediterranean region, resulting in reduced lake levels, increased salinity, and altered ecological dynamics. Historical trends over the past 40 years already show variability in rainfall and rising temperatures, affecting water availability and habitat suitability for aquatic species. These pressures accelerate the degradation of Türkiye's lakes and heightens the vulnerability of their endemic species.

This chapter provides an assessment of the current state and future prospect of the lakes in Türkiye's Mediterranean KBAs. We focus on ten selected lakes that are of particular significance due to their fish fauna and ecological importance. We examine the historical changes in fish assemblages and evaluate the impacts of threats such as pollution, irrigation, invasive species, and climate change. We conduct climate projections for the lakes as climate change is a key ongoing pressure with a likely high impact on the fish communities. Furthermore, we highlight the urgent need for integrated conservation.

2 Materials and Methods

2.1 Lakes

The Mediterranean region of Türkiye is home to 17 designated lake KBAs with ecological and biodiversity significance (see Table 1 for details). The lakes are vital habitats supporting endemic and globally threatened species and provide ecosystem services such as water supply, fisheries, and recreation. To illustrate the risks presented by climate change to these ecosystems, ten lakes were chosen for more detailed examination: İznik, Uluabat, Manyas, Bafa, Köyceğiz, Eğirdir, Burdur, Salda, Işıklı, and Acı (Figs. 1 and 2).

2.2 Land Cover Analysis

Data for the Mediterranean region were retrieved from the Key Biodiversity Areas website (KBAP 2025) and sub-basin data for the selected lakes were extracted from the HydroBASINS (Level 10) v1.c (Lehner and Grill 2013) dataset. In order to delineate basin of the lake as accurate as possible, river catchments connected to the lakes were combined, but connections were terminated if a river was cut off by a reservoir. Lake outflows, if any, were manually removed based on the elevation layer.

Land cover data from Corine Land Cover 1990 (European Environment Agency 2020a) and Corine Land Cover 2018 (European Environment Agency 2020b) were obtained from Copernicus. A preliminary analysis revealed inconsistencies leading to an underestimation of artificial surfaces. Therefore, level 1 areas of the CLC1990 were merged with the CLC2018 layer, which attributes small rural areas to agricultural areas, to obtain accurate, comparable results. Land cover data for subsequent years were obtained from ESRI, Sentinel-2 Land Cover Explorer (2018 and 2022) (Karra et al. 2021). Due to incompatibility between Corine and ESRI datasets, percentage changes between the two periods, 1990–2018 and 2018–2023, are reported separately.

2.3 Lake Surface Area

Information of water area the last 40 years can be obtained for using optical sensors on satellites that detect different wavelengths. Water has low reflectance characteristics in the Near-infrared part of the electromagnetic spectrum, enabling development of indices to detect water. The normalized difference between the Green and the Shortwave Infrared (SWIR) parts of the spectrum enhances water observability. A Modified Normalized Difference Water Index (MNDWI) (Eq. 1) was developed for water observation by remote sensing (Xu 2006).

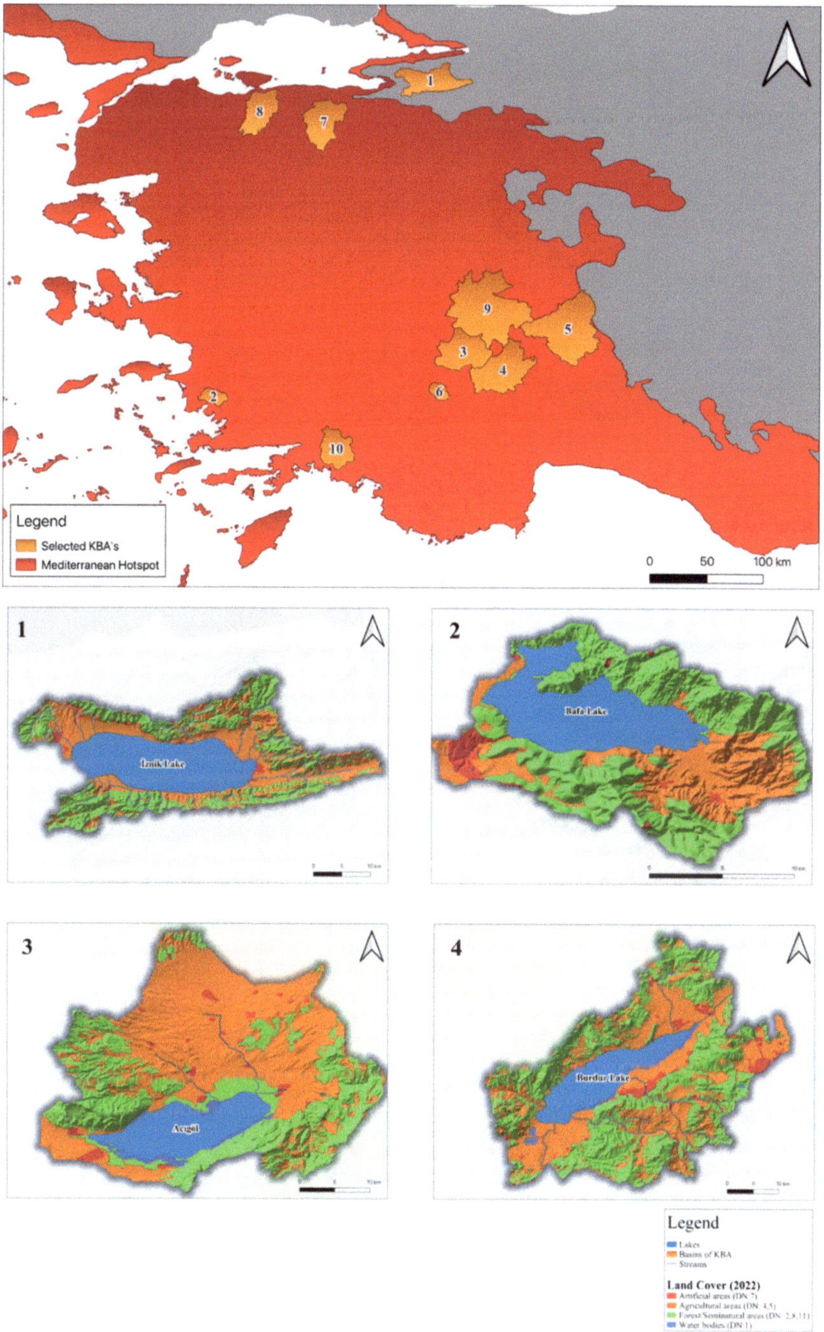

Fig. 1 Selected Mediterranean lake KBAs (1: İznik Lake, 2: Bafa Lake, 3: Lake Acı, 4: Burdur Lake)

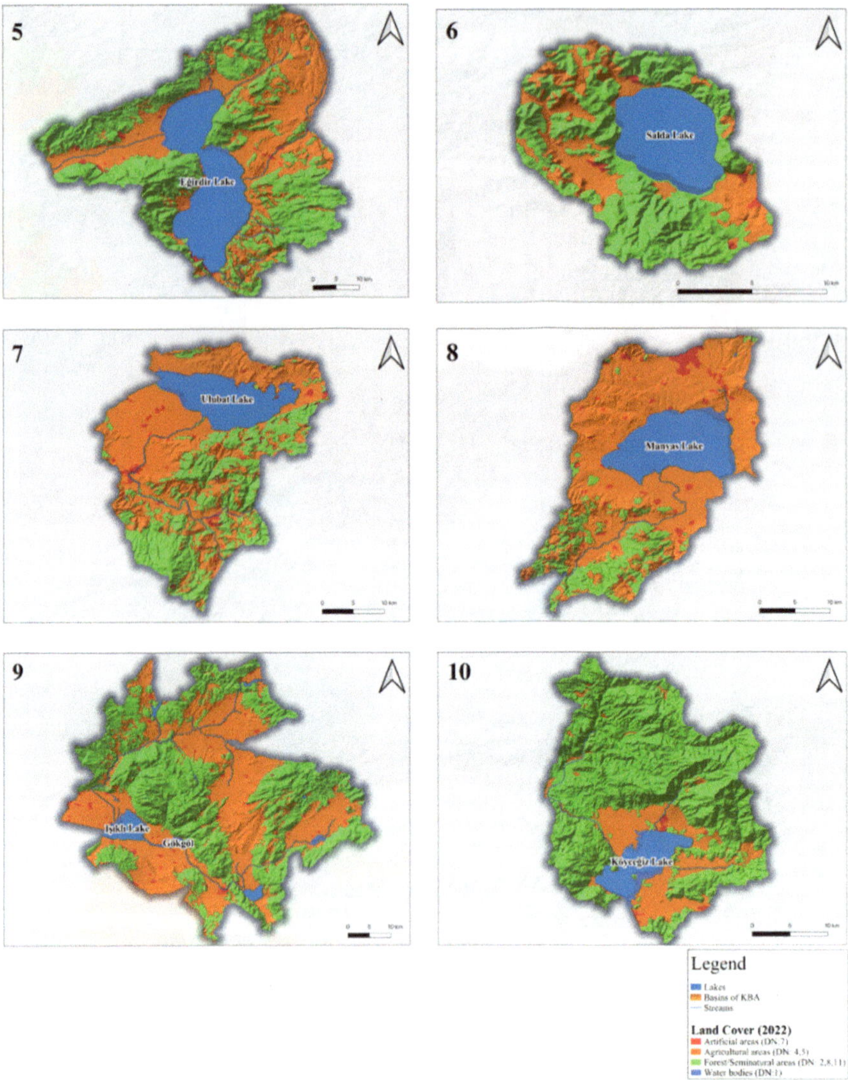

Fig. 2 Selected Mediterranean lake KBAs (5: Eğirdir Lake, 6: Salda Lake, 7: Uluabat Lake, 8: Manyas Lake, 9: Işıklı Lake, 10: Köyceğiz Lake)

$$MNDWI = \frac{Green - SWIR}{Green + SWIR} \quad (1)$$

Landsat, the longest serving earth observation program, has provided continuous data since 1972. Landsat-5 Thematic Mapper (TM) and Landsat-8 Operational Land Imager (OLI) Level-2 data were used for the years between 1985 and 2011 and 2013–2024 using Google Earth Engine (GEE) (Gorelick et al. 2017). Images with cloud cover < 2% were selected for this study. Positive indexed pixels were identified

Table 2 SPEI classes and descriptions

SPEI class	Range	Description
Extremely wet	> 2.00	Unusually wet conditions
Very wet	1.50–1.99	Very wet conditions
Moderately wet	1.00–1.49	Moderately wet conditions
Near normal	– 0.99 to 0.99	Normal conditions, within the average range
Moderately dry	– 1.00 to – 1.49	Moderately dry conditions
Very dry	– 1.50 to – 1.99	Very dry conditions
Extremely dry	< – 2.00	Extreme drought

as water, and binary water-land images were created. These images were smoothed with a 50 × 50 median filter for large lake surfaces to better fit the water body. The smoothed images were vectorized, and surface areas were calculated by time-series analysis. To address misdetection issues, a Hampel filter was applied to detect outliers in lake surface area time-series (Hampel 1974).

2.4 *SPEI*

The Standardized Precipitation Evapotranspiration Index (SPEI) was used to evaluate variations in drought conditions (Bohn and Piccolo 2018; Ye et al. 2016). SPEI is a widely used metric that integrates precipitation and potential evapotranspiration to assess the climatic water balance, providing a comprehensive measure of drought intensity, duration, and frequency. SPEI data enable a more nuanced analysis of hydrological and climatic stressors under the projected climate scenarios.

ERA-5 data were used with a 24-month scale to calculate the monthly water balance (WB) for the historical period 1970–2024 using GEE ERA5 monthly data (Hersbach et al. 2020). The dataset includes "Precipitation" (P) and "Potential Evaporation" (PET) bands. WB was used for SPEI calculations, WB = P – PET, using the R-SPEI package (R Package: "SPEI", Vicente-Serrano et al. 2010). SPEI classes provide information about the wet/dry status of an area, as shown in Table 2.

2.5 *Trend Analysis for Lake Surface Area and SPEI*

Seasonal Mann–Kendall (SMK) was used for the trend analysis for the lake surface area and SPEI data (Hirsch et al. 1982). For the analysis, we used the R software (R Package: "Kendall"). The parameter "Tau" indicates whether the trend is decreasing or increasing. The threshold for significance was set to $p < 0.005$.

2.6 *Present and Future Precipitation and Temperature Data*

Precipitation and temperature are two key variables related to climatological changes (Masson-Delmotte et al. 2021). We used the WorldClim dataset, a widely used high-resolution global climate database, to assess projected changes in these variables (Fick and Hijmans 2017). The 2.5 arcminute ($\approx$ 1 km^2) resolution dataset was employed, providing detailed spatial climate data essential for localized analysis. Annual Mean Temperature (Band 1) and Annual Precipitation (Band 12) were used as bioclimatic variables. Future climate projections were derived from the CMIP6 (Coupled Model Intercomparison Project Phase 6), a framework for simulating the Earth's climate system under various socio-economic and greenhouse gas concentration scenarios (Eyring et al. 2016). HadGEM3-GC3.1 climate model was selected as the future climate projection in CMIP6 since it provides 2.5 arcminute projected data (Andrews et al. 2020).

Two shared socioeconomic pathways (SSPs), SSP2-4.5 and SSP5-8.5, were selected. SSPs represent integrated scenarios of future social development and greenhouse gas emissions, with SSP2-4.5 representing a "middle-of-the-road" scenario and SSP5-8.5 a high emissions trajectory (Riahi et al. 2017). Climate projections were examined for two-time periods, 2021–2040 and 2081–2100, to highlight near- and long-term trends.

3 Lakes

3.1 *İznik Lake*

3.1.1 Introduction

İznik Lake, a deep, tectonic lake in Marmara Basin, is the largest lake in the region and the fifth largest lake in Türkiye. The lake's small sub-basin includes a few permanent streams (Sölöz, Kırandere, Çakırca, Kuru, and Orhangazi) and an outlet to the Sea of Marmara, Karsak stream (Özuluğ et al. 2005). Our analysis of Landsat data from 1985 onward revealed that the lake's area has fluctuated from 299 to 307 km^2 (Fig. 3). The lake sub-basin has historically been important for agriculture, and the lake level has always been controlled (Weissova and Polla 2019). The wetland area in the western part dried up after a barrier was built in 1963, and the lake level is now controlled by the Karsak outlet at 83 m. The lake was declared a "Wetland of National Importance" in 2018, and a total area of 61,600 ha is protected (MoAF 2025).

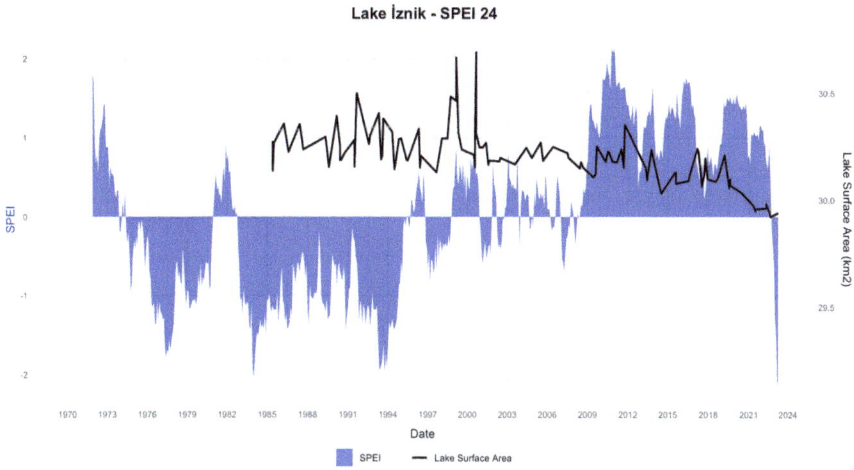

Fig. 3 SPEI and lake surface area for İznik Lake

3.1.2 Pressures

Water management practices at İznik Lake date back to the Greek and late Roman periods to protect the agricultural and urban areas of the İznik Plain (Weissova and Polla 2019). From the second half of the twentieth century, flood control measures were implemented in the lake's tributaries, including the Çakırca (in 1986) and Kırandere (in 2016) streams. In 1963, wetlands in the western part of İznik Lake were drained by damming, leading to significant habitat modifications (MoEUCC 2017). This intervention, serving also water retention purposes, has partially transformed the lake into a reservoir, with approximately 9000 ha of agricultural land now being irrigated with lake water (MoEUCC 2017). Furthermore, the ongoing construction of İznik Reservoir (36.8 million m^3), as well as other planned reservoirs, poses a growing threat by further altering riverine habitats and reducing water inflows to the lake.

The previous KBA assessment in 2004 identified three main pressures on İznik Lake (KBAP 2025), see Appendix 2. Pollution remains a major concern, with agricultural runoff containing herbicides and pesticides long being recognized as a significant contamination source (Ayaz et al. 2012; KBAP 2025). The lake's water quality has been considered acceptable (Akkoyunlu 2003), but recent studies indicate deteriorations in both the lake (Mercan and Arslan 2023) and its inflowing streams (Ayaz et al. 2012). Additionally, newly identified heavy metal pollution from industrial discharges has come to presents an ecological risk. Thus, arsenic and mercury have been detected in the water and arsenic, chromium, copper, and nickel in the sediment (Bulut 2023). The contaminants have also been found in the flora and fauna (Köker 2022; Bulut 2023), highlighting widespread ecosystem contamination, potentially threatening the aquatic organisms and the broader ecological integrity of the lake.

Table 3 Land cover changes in the İznik Lake sub-basin over the past three decades

İznik Lake	Corine		% Change	Sentinel-2		% Change
	1990	2018	1990–2018	2018	2023	2018–2023
Artificial areas	0.87	1.65	89	3.13	3.94	26
Agricultural areas	33.07	35.81	8	20.83	20.35	− 2
Forest/seminatural areas	38.67	35.14	− 9	48.83	48.68	0
Water	27.39	27.40	0	27.19	27.01	− 1

Land cover analysis (Table 3) further highlights increasing anthropogenic pressures. Between 1990 and 2018, artificial areas increased by 89% and agricultural areas by 8%, contributing to habitat degradation. More recent data from 2018 to 2023 indicate continued expansion of artificial areas by 26%, while forest and seminatural areas remain stable. The lake's water surface area exhibited a slight decline (− 1%), likely influenced by cumulative effects of water abstraction and climate and land use changes.

3.1.3 Fish Fauna

A total of 20 fish species have been recorded in İznik Lake (Özuluğ et al. 2005; Yağcı et al. 2019; Giannetto and İnnal 2021). The native species are *Cobitis taenia* Linnaeus 1758, *Cyprinus carpio* Linnaeus 1758, *Gasterosteus aculeatus* Linnaeus 1758, *Knipowitschia caucasica* (Berg 1916), *Proterorhinus semilunaris* (Heckel 1837), *Rutilus rutilus* (Linnaeus 1758), *Rutilus frisii* (Nordmann 1840), *Salariopsis fluviatilis* (Asso y del Rio 1801), *Silurus glanis* Linnaeus 1758, *Squalius cii* (Richardson 1857), *Tinca tinca* (Linnaeus 1758), and *Vimba vimba* (Linnaeus 1758) that are all listed as 'Least Concern—(LC)' by the IUCN (IUCN 2024).

The lake is home to four endemic species of great ecological and conservation importance due to their restricted geographical range. Among these, *A. nicaeensis* has been reported extinct (EX) (IUCN 2024). This species, known only from the İznik Lake sub-basin, has not been recorded since the late twentieth century, despite several intensive studies between 1996 and 2003 (Freyhof 2014a). Its extinction is most likely due to the introduction of the invasive species *A. boyeri*, which was stocked into the lake to enhance fishery yields, leading to an invasion (Freyhof 2014a; IUCN 2024). The other three endemic species, *Barbus oligolepis* Battalgil 1941, *Capoeta tinca* (Heckel 1843), and *Oxynoemacheilus angorae* (Steindachner 1897), are endemic to Anatolia and listed as 'Least Concern' by the IUCN (IUCN 2024). However, due to increasing anthropogenic pressures such as habitat modification (MoEUCC 2017), pollution (Ayaz et al. 2012), and limited distribution, these species require ongoing monitoring and conservation efforts to prevent further decline. Invasive species, including *G. holbrooki*, *L. gibbosus*, *A. boyeri*, and *C. gibelio*, having invaded İznik Lake, pose a significant threat to its biodiversity (Yağcı et al. 2019; Giannetto and İnnal (2021).

Endemic species in the lake are highly vulnerable due to their restricted distribution and habitat specialization, making them susceptible to extinction. The loss of *A. nicaeensis* and the decline in abundance of other endemic species highlight the urgent need for targeted conservation measures.

3.1.4 Future

From the 1970s to 1995, the lake sub-basin was moderately dry, followed by a normal period for 15 years (Fig. 3). However, after 2009, it was moderately wet until an extreme drought suddenly occurred in 2023. The lake's surface area decreased by 1% after 2018, which is consistent with data from the Corine and MNDVI sensors. (Table 3; Fig. 3). Under the moderately optimistic SSP2-4.5 scenario, which assumes moderate socioeconomic development and gradual climate mitigation, precipitation is projected to decline substantially between 2021 and 2040 (see Appendix 3). This decline, coupled with a modest rise in temperature, suggests an elevated probability of drought in the sub-basin. Under the highly pessimistic SSP5-8.5 scenario, similar conditions are projected, with a substantial escalation in severity by 2081.

3.2 Bafa Lake

3.2.1 Introduction

Bafa Lake, located near the mouth of the Büyük Menderes River in the Aegean region, is a brackish lake of historical, ecological, and economic importance. Our analysis of Landsat data from 1985 onward showed that the lake area fluctuated between 53 and 81 km^2, averaging 67 km^2 in 2023 (Fig. 4). Historically, the lake was part of the Aegean Sea but formed Latmos Bay before being separated by alluvial deposits from the Büyük Menderes River between 50 and 300 AD (Kasparek 1988), becoming a lake by the fifteenth century (Brückner et al. 2017). Livestock farming, olive cultivation, and fishing are the primary economic activities in the surrounding areas (Tuna 2015). Its unique ecosystem and historical significance make Bafa Lake a vital natural and cultural resource for the region. The lake was declared a "Nature Park" in 1994, and a total area of 12,281 ha is protected (MoAF 2025).

3.2.2 Pressures

The previous KBA assessment in 2004 identified aquaculture, tourism and ecosystem modifications as main threats for the lake (KBAP 2025), see Appendix 2. Big aquaculture companies continue their operations in the basin and touristic tours seem to continue. In addition to these threats, pollution and natural system modification are key threats.

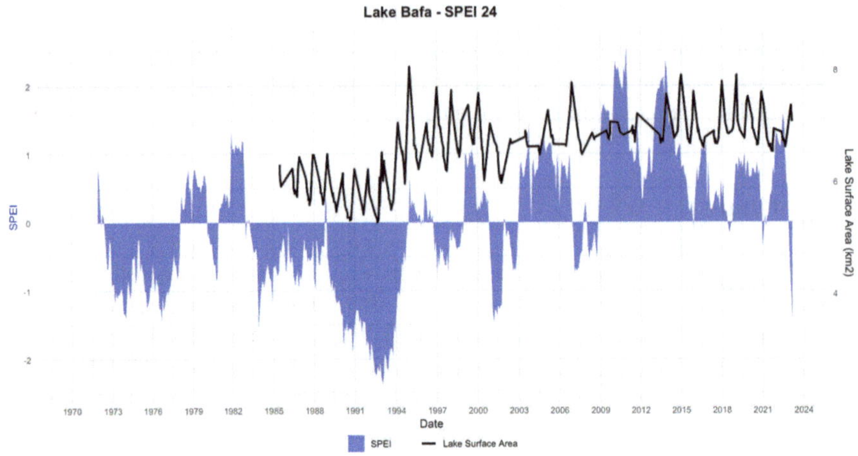

Fig. 4 SPEI and lake surface area for Bafa Lake

In 1985, the State Hydraulic Works agency built two regulators near the river to control the lake water level. Following a major fish kill in 2006, the impact of these regulators was highlighted, along with hydrochemical anomalies such as high organic pollution (nitrogen, phosphorus, etc.), increased salinity, and heavy metal pollution (Erdoğan 2011). Despite efforts by the State Hydraulic Works, sedimentation due to Büyük Menderes River inflow continues (Üzücck 2005), as evidenced by high organic input (Koçak et al. 2017) and high chlorophyll-a levels (Sukatar et al. 2021). Sediment analyses show significant heavy metal pollution, including cadmium, chromium, cobalt, aluminum, and especially nickel (Algül and Beyhan 2020; Yozukmaz and Yabanlı 2023).

Land cover analysis was conducted for the lake's natural sub-basin, without considering the artificial connections from Büyük Menderes River. It revealed significant and ongoing expansion of artificial areas, possibly due to conversion from agricultural and forest/semi natural areas (Table 4).

Table 4 Land cover changes in the Bafa Lake sub-basin over the past three decades

Bafa Lake	Corine		% Change	Sentinel-2		% Change
	1990	2018	1990–2018	2018	2023	2018–2023
Artificial areas	1.01	3.65	263	2.37	3.31	40
Agricultural areas	27.42	26.09	− 5	3.63	3.44	− 5
Forest/seminatural areas	50.08	47.87	− 4	70.70	70.10	− 1
Water	21.49	22.39	4	23.27	23.13	− 1

3.2.3 Fish Fauna

The lake hosts a diverse fish fauna, including 21 species (Sarı et al. 1999; Demir 2015; Çiçek et al. 2023; Kurtul et al. 2024). Native species are *Anguilla anguilla* (Linnaeus 1758) (CR), *Aphanius fasciatus* (Valenciennes 1821) (LC), *Chelon auratus* (Risso 1810) (NT), *Chelon ramada* (Risso 1827) (NT), *Dicentrarchus labrax* (Linnaeus 1758) (NT), *Gobius niger* Linnaeus 1758 (LC), *Mugil cephalus* Linnaeus 1758 (LC), *Salaria pavo* (Risso 1810) (LC), *S. fluviatilis* (LC), *S. glanis* (LC), and *Syngnathus abaster* Risso 1827 (LC). Endemic species include *Chondrostoma meandrense* Elvira 1987 (VU), *Luciobarbus kottelati* Turan, Ekmekçi, İlhan and Engin, 2008 (VU), *Oxynoemacheilus germencicus* (Erk'akan et al. 2007) (VU), and *Vimba mirabilis* (Ladiges 1960) (LC). Introduced species encompass *C. carpio* (LC) and *K. caucasica* (LC), while invasive species include *C. gibelio*, *G. holbrooki*, *A. boyeri*, and *L. gibbosus*.

Endemic species in the lake are of crucial importance due to their limited distribution and critical ecological role (Çiçek et al. 2023). *C. meandrense*, endemic to the Büyük Menderes drainage, faces severe threats from pollution, groundwater abstraction, and industrial effluents in the floodplain (Korkmaz et al. 2017; Freyhof et al. 2020). Habitat fragmentation from dam construction and reduced rainfall due to climate change further limit its chance of survival (Freyhof 2014c). Accordingly, a significant decline in *C. meandrense* abundance has occurred, making conservation efforts important. Similarly, *L. kottelati*, belonging to the same drainage system, is threatened by pollution, reduced water flow, and habitat degradation, risking local extinction (Freyhof et al. 2020). *V. mirabilis*, once present, is now extirpated from the lake but persists in reservoirs. Its riverine populations have sharply declined due to habitat destruction and hydrological changes, highlighting the urgent need for habitat restoration to ensure its long-term viability (Freyhof 2014d).

Anthropogenic activities, including water diversion and agricultural practices, have significantly changed the lake (Erdoğan 2011; Kurtul et al. 2024). Salinity has increased from 3.5 to 12.4‰ since the first measurement. Damming has reduced the freshwater inflow, creating a brackish environment that favors invasive species like *A. boyeri*, which now dominates the fish populations (Kurtul et al. 2024). These changes have led to reduced abundance of native and endemic species, highlighting the need for effective management and conservation strategies to protect its endemic and threatened species.

3.2.4 Future

From 1976 to 1999, the lake surface area increased slightly at moderately dry conditions. Following a brief period of dry conditions, the sub-basin experienced a moderate increase in wetness, which continues today (Fig. 4). Water diversion from the sea likely prevents water level reduction but alters lake salinity. Projections under the SSP2-4.5 and SSP5-8.5 scenarios suggest a modest decline in precipitation and a significant temperature rise in the highly pessimistic scenario, potentially leading to manageable droughts (see Appendix 3). However, ineffective management strategies could pose significant challenges by 2081–2100.

Local NGO's have highlighted biodiversity and the lake's future, prompting an on-site inspection by a delegation, including members of parliament, influenced by civil society organizations. This initiative looks promising for the lake's future.

3.3 *Lake Acıgöl*

3.3.1 Introduction

Lake Acıgöl, a shallow hypersaline lake located in the Burdur Closed Basin, has a maximum salinity of 211 g/L (Karami et al. 2019). In Turkish "Acı" means bitter and hydrobiologically refers to brackish water systems, and Acıgöl is therefore a common lake name in Türkiye. The lake's main water source is groundwater; magnesium carbonate-rich discharge water, and sodium sulphate-rich spring water (Mutlu et al. 1999). Our analysis of Landsat data from 1985 onward showed that the lake area gradually decreased from 194 to 54 km^2 by 2021 (Fig. 5). The lake is an important reserve for economically significant sodium sulphate (Gündoğan et al. 1995; Uca Avcı et al. 2014), with production pools on the north side containing 211 g/L (Karami et al. 2019). The lake was declared a "Wetland of National Importance" in 2015, and a total area of 55,095 ha is protected (MoAF 2025).

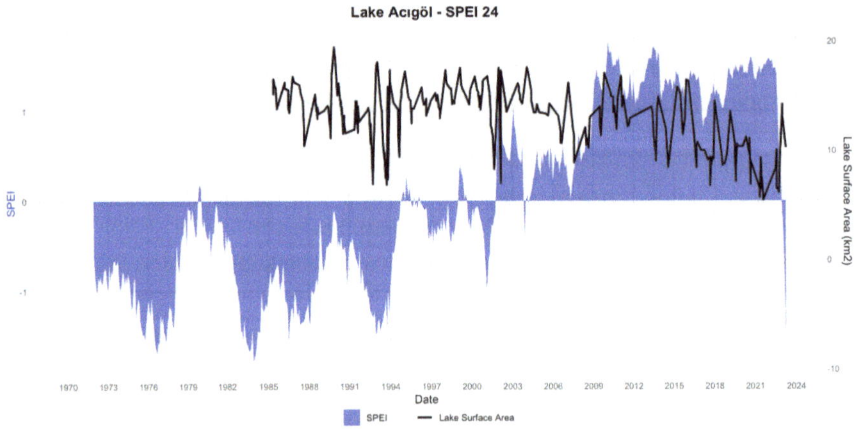

Fig. 5 SPEI and lake surface area for Lake Acıgöl

Table 5 Land cover changes in the Lake Acıgöl sub-basin over the past three decades

Lake Acıgöl	Corine		% Change	Sentinel-2		% Change
	1990	2018	1990–2018	2018	2023	2018–2023
Artificial areas	1.93	2.26	17	2.52	2.76	10
Agricultural areas	50.43	51.04	1	25.07	20.26	– 19
Forest/seminatural areas	43.84	40.60	– 7	68.83	73.55	7
Water	3.80	6.10	60	3.57	3.43	– 4

3.3.2 Pressures

In the 2004 KBA assessment, surface water abstraction, mining activities, and terrestrial animal hunting were identified within the vicinity of the lake (KBAP 2025), see Appendix 2.

Agricultural fertilizers have caused nitrate contamination (up to 487 mg/L) of groundwater (Karaman et al. 2017). Data from 2016 show that the lake is seriously contaminated by heavy metals (Zn, Fe, Cu, Co, Ni, Al, As, V) (Oruçoğlu and Beyhan 2019).

Although not reflected in the Corine data (Table 5), the lake area shrunk dramatically (67%) from 1985 to 2015 (Temiz and Durduran 2016), and this shrinkage is also evident in our lake surface area (MNDWI) analysis (Fig. 5). This inconsistency in the available data (Table 5) makes it impossible to provide solid information on land cover change.

3.3.3 Fish Fauna

Being a hypersaline soda lake, Lake Acıgöl has unique ecological characteristics (Karami et al. 2019) that make it unsuitable for most freshwater fish. However, it is an important habitat for the critically endangered (CR) *Anatolichthys transgrediens*, an endemic species that relies on the springs around the lake (Yoğurtçuoğlu and Ekmekçi 2014), which produce fresh to lightly brackish water and serve as essential migration and wintering areas (Güçlü and Küçük 2012). *A. transgrediens* generally avoids the lake itself due to its high salinity and periodical summer dry-out, but heavy rainfall temporarily lowers the lake's salinity, enabling migration from the springs to the lake (Güçlü and Küçük 2012).

The survival of *A. transgrediens* is threatened by over-extraction of water by nearby sodium sulfate factories, which significantly reduces water levels, increases salinity, and desiccates vital springs around the lake (Güçlü and Küçük 2012; Freyhof 2014e). Groundwater abstraction for irrigation exacerbates these challenges, causing habitat loss in coastal regions, and prolonged droughts and water resource depletion have severely reduced the available habitat and restricted *A. transgrediens* to a few spring habitats (Yoğurtçuoğlu and Ekmekçi 2014). The invasive *G. holbrooki*, introduced in the late 1990s, competes for resources and preys on juveniles (Wildekamp et al. 1999; Pyke 2008), further stressing *A. transgrediens*. The presence of *G: holbrooki* has drastically reduced the range of *A. transgrediens* (Freyhof 2014e).

3.3.4 Future

According to SPEI results, the sub-basin's climate was arid, and substantial changes in lake area occurred up until 2002. However, despite the severity of these changes, the lake surface area remained relatively stable. After 2002, the climate became wetter and more humid (positive SPEI values), and the lake area decreased substantially (Fig. 5). However, it should be noted that the CORINE Land Cover data on lake surface area are not reliable. From 1990 to 2018, the CLC projected a 60% increase in water areas, NDVI analysis showed that this was not the case (Fig. 5). Both the SSP2-4.5 and SSP5-8.5 scenarios project a precipitation decline and a modest temperature increase during 2021–2040. However, the SSP5-8.5 scenario anticipates a substantial decline in precipitation and a notable temperature increase. Both scenarios foresee significant temperature fluctuations during 2081–2100 (see Appendix 3).

3.4 *Burdur Lake*

3.4.1 Introduction

Burdur Lake, located in the Burdur Closed Basin, is a tectonic, highly alkaline, warm monomictic saline lake (Girgin et al. 2004), which holds significant ecological importance as a Ramsar Convention-designated wetland. The lake's sub-basin is approximately 3185 km^2 with an average lake depth of 40 m in the past (Çolak et al. 2022). The lake is replenished by water from seasonal and perennial streams, groundwater sources, and rainfall. Our analysis of Landsat data from 1985 onward shows that the lake area has decreased from 209 to 119 km^2 over four decades (Fig. 6).

Burdur Lake plays a vital role as a wintering habitat for the globally endangered white-headed duck (*Oxyura leucocephala*), whose population has recently declined (Çolak et al. 2022). Moreover, the lake supports wintering waterfowl species and is home to the endemic Burdur killifish (*Anatolichthys sureyanus*). The lake was declared a "Ramsar site" in 1994, and a total area of 25,096 ha is protected. In addition, the lake was declared a "Wildlife Development Area" for waterfowl in 2006 (MoAF 2025).

3.4.2 Pressures

A previous KBA assessment identified surface water abstraction for agricultural use as a threat Burdur Lake, and this is still the case (KBAP 2025), see Appendix 2. Since 1974, about 20 dams/ponds have been revived in the basin, which increasingly affects

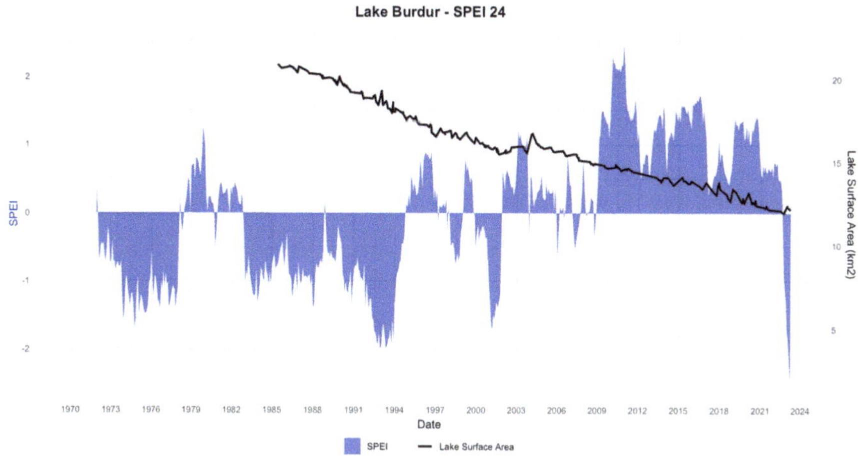

Fig. 6 SPEI and lake surface area for Burdur Lake

the lake's area and volume (Davraz et al. 2019a). Excessive extraction of underground water from numerous wells, coupled with inefficient agricultural irrigation practices, exacerbates the situation (Ataol 2010). Consequently, the water level has plummeted by over 12 m, reducing the lake's surface area to 153 km^2 (Şener 2016), which is significantly smaller than its original size.

Nutrient loads, salinization, and added abstraction of ground water for agricultural use are the key threats to Burdur Lake (see Appendix 2), which is alkaline with high salinity and elevated ion concentrations and absence of outflows. Projected increases in evaporation rates, unmitigated by precipitation and anticipated rises in water consumption, will likely escalate salt and ion concentrations, and the subsequent salinity surge will expectedly affect ecosystem services and functions, and not least biodiversity (Çolak et al. 2022). Besides salinization, the lake's chlorophyll-a levels are steadily increasing, and the lake is therefore classified as hypertrophic (Tuygun et al. 2023).

Land cover analysis revealed a significant expansion of artificial areas; however, the water level change does not reflect the change in lake surface area, which was found to be significantly higher than expected from the Corine data (Table 6) and Çolak et al. (2022).

3.4.3 Fish Fauna

Burdur killifish *A sureyanus* is classified as Endangered (EN) by the IUCN and only found in Burdur Lake where it is threatened by the rapid water level decline that has led to elevated salinity and pollution (Gülle et al. 2016; IUCN 2024). In the assessment of Burdur Lake's fish fauna, only the lake itself was considered due to its unique saline and alkaline conditions, while the fish fauna of the freshwater springs

Table 6 Land cover changes in the Burdur Lake sub-basin over the past three decades

Burdur Lake	Corine		% Change	Sentinel-2		% Change
	1990	2018	1990–2018	2018	2023	2018–2023
Artificial areas	1.92	3.82	99	6.21	7.40	19
Agricultural areas	38.51	39.50	3	21.11	18.09	– 14
Forest/seminatural areas	48.09	45.57	– 5	65.34	67.62	3
Water	11.48	11.11	– 3	7.33	6.89	– 6

feeding Burdur Lake was not included. Although *A. sureyanus* is adapted to the lake's saline and alkaline conditions (Gülle et al. 2016), it faces significant threats due to the rapid water level decline, which is primarily driven by excessive water abstraction, dam construction within the lake's catchment, and climate change-induced rainfall reduction (Freyhof et al. 2020; Çiçek et al. 2023; IUCN 2024).

Stocking of economically important fish species into Burdur Lake began in the 1950s with Tilapia and eel, but the efforts were unsuccessful due to the lake's physical and chemical conditions (Akşıray 1982). In 1969, the endemic fish species *Alburnus tarichi* (Güldenstädt 1814) from Lake Van was introduced to Burdur Lake, based on presumed compatibility with the lake's physiochemical conditions. Although the species adapted well to the environment, it failed to establish a permanent population due to pollution discharged into the lake and overfishing during its breeding migration to streams (Akşıray 1982). The Burdur killifish remains a key indicator of the lake's ecological health, and its continued survival depends on urgent conservation actions (Gülle et al. 2016; Çiçek et al. 2023).

3.4.4 Future

Until 2009, the Burdur Lake sub-basin experienced average climatic conditions, followed by a relatively wet period lasting 10 years. However, drought conditions have significantly increased in recent years. The lake's surface area has steadily decreased to half its original size, likely due to the construction of dams in the sub-basin (Fig. 6). Precipitation is expected to decrease in both scenarios, with a significant temperature increase projected for the period 2081–2100 (see Appendix 3). Without substantial management actions, the lake's water level will be seriously threatened.

Recent modernization of the water treatment plant near the lake may improve the lake water quality. An action plan was implemented in 2015 to control the use of water resources, but the water level continues to decline.

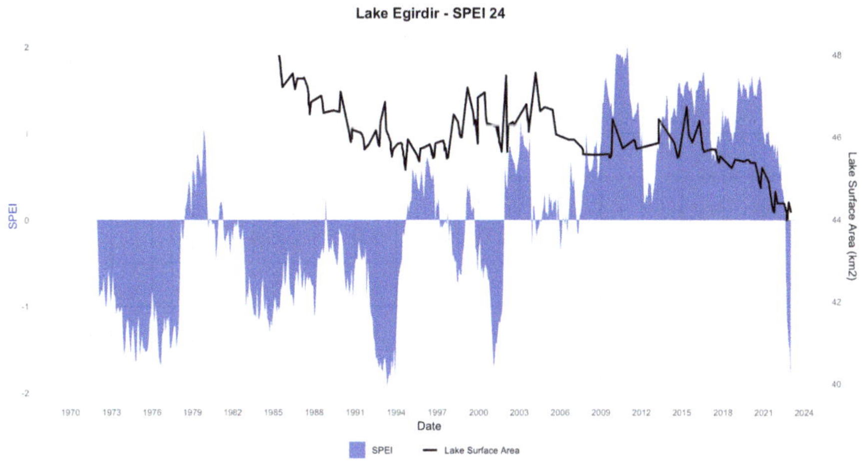

Fig. 7 SPEI and lake surface area for Eğirdir Lake sub-basin

3.5 *Eğirdir Lake*

3.5.1 Introduction

Eğirdir Lake, the fourth largest lake in Türkiye, is of tectonic-karstic origin and connected to other lakes and the Mediterranean Sea via underground river systems (Atalay 2003). The sub-basin contains a single, small reservoir, Sücüllü. Main surface flows to the lake include Pupa, Hoyran, Yalvaç (Gelendost), and Çay streams. Historically, Lake Eğirdir drained into Lake Kovada through a natural stream, which was canalized in the 1950s and its flow regulated after 1968 (Kesici and Kesici 2006). Our analysis of Landsat data from 1985 onward showed that the lake area has fluctuated between 435 and 480 km^2 (Fig. 7). Since 1994, Eğirdir Lake has served as a source of drinking water for Isparta. Additionally, most of the surrounding agricultural land in the Isparta region is irrigated using water from Eğirdir Lake (Özseven et al. 2020). The lake is considered a "Sensitive Area" for drinking water use.

3.5.2 Pressures

Invasive species and tourism pressures, identified in 2004 (KBAP 2025), remain ongoing threats to the lake. Additional pressures are water abstraction and coastal urbanization, see Appendix 2, which Kesici and Kesici (2006) also identified as the main threats to the. Fishing, an important sector, has significantly impacted the fish community due to management decisions favoring high-priced species, resulting in the loss of native species and a sharp decline in fishery-related species (Yerli et al. 2013).

Table 7 Land cover changes in the Eğirdir Lake sub-basin over the past three decades

Eğirdir Lake	Corine		% Change	Sentinel-2		% Change
	1990	2018	1990–2018	2018	2023	2018–2023
Artificial areas	0.76	1.15	51	1.57	1.85	18
Agricultural areas	34.26	34.50	1	20.62	20.05	− 3
Forest/seminatural areas	47.27	46.63	− 1	60.70	61.41	1
Water	17.71	17.73	0	17.10	16.67	− 3

In addition, nutrient loads, coupled with increased heavy metal and pesticide contamination, are polluting the lake. In 2016, only aluminium and chromium levels exceeded quality standards, although heavy metal levels were higher in downstream Kovada Lake (Oruçoğlu and Beyhan 2019). Pyridaben, chlorpyrifos, acetamiprid, methidathion, and cyprodinil were also detected in sediment samples (Kılıç et al. 2023). Water quality assessments have yielded conflicting results, requiring further detailed analysis. Data from 2011 indicated eutrophy for nitrogen and hypertrophy for total phosphate (Beyhan and Kaçıkoç 2014), but by 2014, the overall trophic state was mesotrophic (Bulut and Kubilay 2018). Eğirdir Lake has a high-water renewal rate (Karagüzel et al. 1995).

According to land cover analysis, artificial areas in the sub-basin expanded by 50% from 1990 to 2018 and have continued to grow since. Other land cover classes showed only minor changes (Table 7).

3.5.3 Fish Fauna

Eğirdir Lake hosts a diverse fish fauna of 16 recorded species, including endemic, native, and invasive species. Over the past century, human activities, such as the introduction of invasive species, overfishing, and habitat alteration, have profoundly changed the fish fauna (Yerli et al. 2013). Historically, Eğirdir Lake was home to nine native and endemic species (Kosswig and Geldiay 1952; Yerli et al. 2013). The introduction of pike-perch *Sander lucioperca* (Linnaeus 1758) in 1955 marked a critical turning point (Küçük et al. 2009). While pike-perch initially boosted fishery yields to 2000 tons per year by the 1970s, it caused significant declines in native populations (Yerli et al. 2013). By the 1980s, several species, including *P. handlirschi* and *Egirdira nigra* (Kosswig and Geldiay 1952), were driven to extinction (Küçük et al. 2009). In subsequent decades, invasive and introduced species such as *S. lucioperca*, *C. gibelio*, *A. boyeri*, *P. parva*, *K. caucasica*, and *G. holbrooki* further disrupted the ecosystem, increasing competition for resources and degrading habitat quality. *A. boyeri* and *C. gibelio* became dominant species, exerting significant ecological pressure on native and endemic fish populations (Yerli et al. 2013; Yağcı et al. 2021).

Today, the lake is home to seven endemic species: *Anatolichthys iconii* (Akşiray 1948) (NE), *Capoeta pestai* (Pietschmann 1933) (CR), *Cobitis turcica* Hankó 1925

(EN), *E. nigra* (EN), *Garra klatti* (Kosswig 1950) (EN), *Oxynoemacheilus mediterraneus* (Erk'akan, Nalbant and Özeren 2007) (LC), and *Seminemacheilus ispartensis* Erk'akan, Nalbant and Özeren 2007 (VU) (Küçük et al. 2009; Yağcı et al. 2022; Çiçek et al. 2023; IUCN 2024). *P. handlirschi*, once endemic to the lake, is now extinct (EX) due to predation by pike-perch (Freyhof 2014b). *E. nigra*, thought to be extinct, has been rediscovered in restricted habitats (Freyhof 2014f). Similarly, *C. pestai* now depends on a single tributary affected by damming and unregulated water abstraction. Meanwhile, *S. ispartensis* remains resilient but faces threats from habitat alteration and water mismanagement (Freyhof 2014g). The lake also supports two native species, *C. carpio* and *V. vimba*, which remain relatively stable (Yerli et al. 2013; Yağcı et al. 2021).

While the lake hosts a unique assemblage of endemic and native fish species, their populations have significantly declined (Yerli et al. 2013; Freyhof et al. 2020). Endemic species such as *C. pestai* and *E. nigra* now occur in extremely limited habitats, relying on small tributaries or springs for survival (Freyhof et al. 2020; IUCN 2024). Once abundant, these species are now critically endangered and face increasing threats from habitat alteration and competition with invasive species (Yerli et al. 2013; Yağcı et al. 2021).

3.5.4 Future

Regardless of changes in precipitation and temperature, the lake surface area has decreased by 10% since 1985. Water levels dropped significantly after the 2000s, despite generally wet conditions, as indicated by the SPEI index values above one. The index changed sign in 2023, and expected changes of precipitation and temperature are likely to exacerbate this effect (Fig. 7). The lake sub-basin is also expected to experience aridity under both scenarios (see Appendix 3).

The Ministry of Environment, Urbanization, and Climate Change announced an action plan to restore Eğirdir Lake. The plan includes sediment removal, increasing oxygen levels, removing bottom silt, and replenishing water. As the lake is one of Türkiye's largest and economically important lakes, ongoing action plans are essential. However, their effectiveness needs careful monitoring, especially regarding water transport from adjacent areas.

3.6 *Salda Lake*

3.6.1 Introduction

Salda Lake, located in the Burdur Closed Basin in southwest Türkiye, is a unique alkaline soda lake characterized by its deep waters and the remarkable hydromagnesite stromatolites along its shores. Formed through tectonic activity at the end of the Neogene, the lake reaches a depth of 184 m, making it the third deepest lake

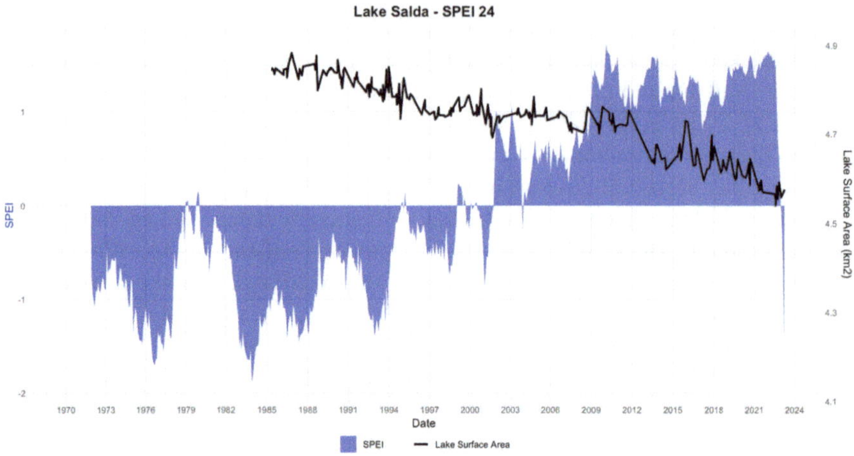

Fig. 8 SPEI and lake surface area for Salda Lake

in Türkiye (Davraz and Aksever 2023). Its extreme conditions, including high pH and magnesium levels, contribute to its distinctive limnological characteristics and support a limited but specialized biodiversity (Kazanci et al. 2004). The lake primarily receives water from groundwater, both directly and through springs, as well as from the Değirmendere and Zahra streams (Davraz and Aksever 2023). Our analysis of Landsat data from 1985 onward showed that the lake area fluctuated has fluctuated between 45 and 49 km^2 (Fig. 8). Mineralogical and geological similarities to the Jezero Crater on Mars excited NASA and international scientists to conduct detailed research in the area (Karaman 2021). The lake was declared a "Nature Park" in 2011, and a total area of 4451 ha is protected. In addition, the lake and its surrounding area were declared a "Specially Protected Area" in 2019 by the Ministry of Environment, Urbanization and Climate Change, increasing the size of the protected area (MoEUCC 2025).

3.6.2 Pressures

Abstraction of groundwater for agricultural use was the only recognized threat to the lake by 2004 (KBAP 2025). However, recent studies indicate the existence of more threats such as heavy metal pollution (Çaldırak et al. 2017) and introduction of invasive species (Freyhof 2020), see Appendix 2. Geogenic and anthropogenic (agricultural) arsenic pollution and analysis of fertilizers and trace elements have shown that the water is not suitable for drinking, and gradual deterioration of the lake water is expected (Varol et al. 2021). High levels of lead, nickel, and chromium have also been recorded (Çaldırak et al. 2017).

A decrease in water flow is anticipated due to climatic and anthropogenic effects such as groundwater abstraction from wells, ponding of streams, and alteration of

Table 8 Land cover changes in the Salda Lake sub-basin over the past three decades

Salda Lake	Corine		% Change	Sentinel-2		% Change
	1990	2018	1990–2018	2018	2023	2018–2023
Artificial areas	1.04	1.08	4	1.91	2.02	6
Agricultural areas	22.87	24.68	8	8.87	9.89	11
Forest/seminatural areas	55.72	54.62	− 2	68.96	68.11	− 1
Water	20.37	19.62	− 4	20.24	19.95	− 1

springs (Davraz and Aksever 2023). In recent years, in addition to agricultural activities and sewage from nearby settlements, tourism has begun to affect the lake (Ceylan and Bulut 2019). As of 2015, the lake was relatively clean and classified as Class I (Davraz et al. 2019b), with similar results in 2018 (Hanedar and Tanik 2023). However, the Burdur Basin Management Plan project results indicate that biological quality elements are 'moderate', and the physico-chemical status is also 'moderate'.

Since 1990, the area of forest and seminatural zones, as well as water bodies, has gradually been replaced by artificial and agricultural areas (Table 8).

3.6.3 Fish Fauna

The lake hosts four endemic species, one introduced, and two invasive species. The endemic species are *Anatolichthys saldae* (Akşiray 1955) (NE), *Cobitis phrygica* Battalgil 1944 (EN), *P. burduricus* (EN), and *Oxynoemacheilus anatolicus* Erk'akan, Özeren and Nalbant 2008 (EN), found only in the sub-basin of Salda Lake and Burdur Closed Basin. *C. carpio* (LC) is an introduced species, while *A. boyeri* and *G. holbrooki* are invasive (Erkakan et al. 1999; Küçük et al. 2013; Gülle et al. 2016; Korkmaz et al. 2023).

The endemic species in the lake are threatened by introduction of invasive species, water abstraction, pollution, and dam construction on its tributaries. *A. saldae,* an endemic species found exclusively in Salda Lake and its surrounding river, highlights the lake's global conservation importance (Gülle et al. 2016; Freyhof et al. 2020). However, *A. saldae* is increasingly threatened by habitat degradation, invasive species, and water management issues (Freyhof et al. 2020; Çiçek et al. 2023). Dams, water abstraction, and reduced environmental flows are drying up streams, severely impacting its habitat, and invasive species such as *A. boyeri* and *G. holbrooki* are threatening its survival (Freyhof et al. 2020). *C. phrygica,* although restricted to small areas in Burdur, Salda, and Tefenni, is often abundant in its habitats (Freyhof 2014h). *P. burduricus,* endemic to springs and streams in the Salda and Burdur basins, is threatened by water abstraction, pollution, and climate change. Many dams in the area lack environmental flow provisions, causing streams to dry up in summer, further impacting *P. burduricus* (Freyhof 2014j).

Beyond its fish fauna, Salda Lake's ecological and geological uniqueness further enhances its conservation value. Urgent conservation efforts are essential to mitigate the pressures, protect endemic species, and preserve this globally significant ecosystem.

3.6.4 Future

Year 2000 marked a pivot point in the sub-basin's climatic conditions that shifted from relatively dry to moderately wet. Although the lake area has steadily decreased, the change is relatively small (Fig. 8). The SSP2-4.5 scenario, assuming moderate socioeconomic development and gradual climate mitigation, projects a substantial decline in precipitation. The decline is less significant in the highly pessimistic SSP5-8.5 scenario. This relationship reverses for the 2081–2100 period. The projected changes in annual mean temperature are negligible in the near future but show a marked increase for 2081–2100 (see Appendix 3).

In 1989, Salda Lake was designated as a protected area, and by 2006 the lake and its surrounding area were declared a tourism centre. However, by 2019, the situation was reversed, and the area was declared a "Specially Protected Area" to safeguard the lake. The protected area was later expanded in size. While protection management has increased in the area, intense tourist activity persists at the lake.

3.7 *Uluabat Lake*

3.7.1 Introduction

Uluabat Lake, also referred to as Apolyont Lake, is a shallow lake in the Susurluk Basin in the Marmara Region of Türkiye. The lake, formed as an alluvial barrier lake under tectonic control, has an average depth of 2.5 m, with seasonal variations ranging from 0.5–1.5 to 3.5–4.5 m during dry and rainy months, respectively (İleri et al. 2014). It is primarily fed by the Mustafakemalpaşa Stream, which originates from the Emet and Orhaneli streams (Sarmaşık 2012), although the flow of these tributaries has been significantly reduced by the establishment of the Devecikonağı and Çınarcık reservoirs. Our analysis of Landsat data from 1985 onward showed that the lake area fluctuated between 110 and 185 km^2, averaging 116 km^2 in 2023 (Fig. 9). As a vital wetland with rich flora and fauna, Uluabat Lake plays a crucial role in the migration route of birds entering Anatolia from the northwest. Recognized for its ecological importance, the lake was designated as a Ramsar Site in 1998, and a total area of 19,900 hectares are now under protection (MoAF 2025).

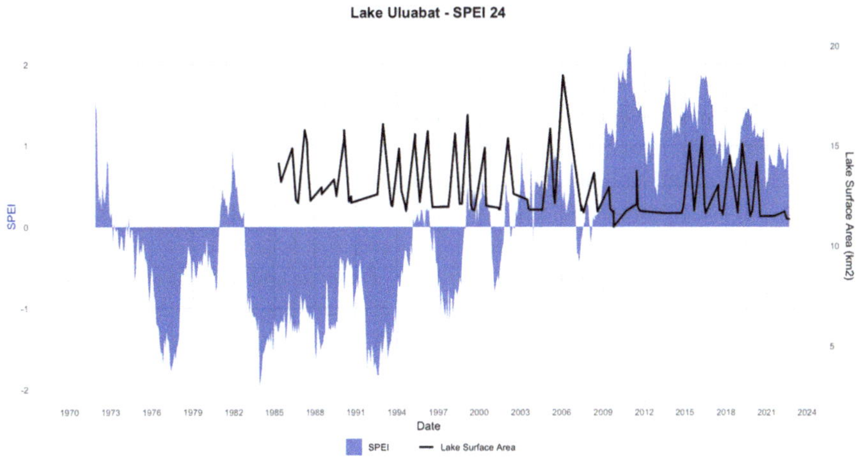

Fig. 9 SPEI and lake surface area for Uluabat Lake

3.7.2 Pressures

Only roads and a railroad were identified as threats to the lake in the 2004 assessment (KBAP 2025), but heavy metal pollution and nutrient loading to the lake have become evident, see Appendix 2. In 2009, chromium and nickel levels in the lake were higher than the "severe effect level" (Katip et al. 2012). By 2017, arsenic concentrations in the water and sediment exceeded recommended limits (Şener et al. 2023). Spatial analysis indicates that the Mustafakemalpaşa stream is a significant source of heavy metal (Şener et al. 2023) and phosphorus pollution (Bulut and Aksoy 2008).

In 2023, the lake was classified as eutrophic based on Secchi disk and chlorophyll-a measurements and as hypereutrophic based on total phosphorus and nitrogen concentrations (Elmaci et al. 2009). This trend towards hypertrophy has continued (Katip et al. 2015; Kesmez and Dalkıran 2023). Lake area has decreased, and the main change from 1990 to 2023 has been the expansion of the artificial areas (Table 9).

Table 9 Land cover changes in the Uluabat Lake sub-basin over the past three decades

Uluabat Lake	Corine		% Change	Sentinel-2		% Change
	1990	2018	1990–2018	2018	2023	2018–2023
Artificial areas	1.19	2.66	124	4.11	5.02	22
Agricultural areas	52.63	52.80	0	44.80	43.40	– 3
Forest/seminatural areas	33.21	31.66	– 5	41.16	41.78	2
Water	12.98	12.88	– 1	9.92	9.79	– 1

3.7.3 Fish Fauna

Faunistic studies show that Uluabat Lake hosts a total of 29 fish, including native, endemic, and invasive species. The native species include *Alosa maeotica* (Grimm 1901), *A. anguilla*, *Blicca bjoerkna* (Linnaeus 1758), *Cobitis puncticulata* Erk'akan, Atalay-Ekmekçi and Nalbant 1998, *C. carpio*, *Esox lucius* Linnaeus 1758, *K. caucasica*, *M. cephalus*, *Petroleuciscus borysthenicus* (Kessler 1859), *P. semilunaris*, *Rhodeus amarus* (Bloch 1782), *R. rutilus*, *S. fluviatilis*, *Scardinius erythrophthalmus* (Linnaeus 1758), *S. glanis*, *S. cii*, *S. abaster*, *T. tinca*, and *V. vimba* (Berber et al. 2011; Çınar et al. 2013; Çiçek et al., 2023). All native species are listed as Least Concern (LC) by the IUCN (IUCN 2024), with the exception of *A. anguilla*, which is classified as Critically Endangered (CR) due to its high risk of global extinction.

The lake and its tributaries are also home to five endemic species, which include *Alburnus carinatus* Battalgil 1941 (EN), *B. oligolepis* (LC), *Barbus niluferensis* Turan, Kottelat and Ekmekçi 200 (NT), *Clupeonella muhlisi* Neu 1934 (EN), and *Oxynoemacheilus marmaraensis* Turan, Bayçelebi and Kalayci 2023 (NE) (IUCN 2024). Of these, *A. carinatus* and *C. muhlisi* are of particular concern. *A. carinatus* is endemic to Uluabat Lake and Manyas Lake and faces significant environmental pressure due to pollution, water abstraction, and habitat degradation (Freyhof 2014i). Although resilient, its populations have declined significantly. Similarly, *C. muhlisi*, endemic to the lake, is highly vulnerable due to its restricted range, making it vulnerable to threats (Di Dario 2018). It has suffered severe population declines from targeted commercial fishing, bycatch, and pollution with industrial and domestic wastewater from its catchments (Di Dario 2018; Çiçek et al. 2023). Eutrophication and water abstraction for irrigation have further degraded its habitat quality (Altınsaçlı and Griffiths 2001; Fricke et al. 2007).

The lake faces significant ecological challenges due to the presence of invasive species, including *A. boyeri*, *C. gibelio*, *G. holbrooki*, *L. gibbosus*, and *P. parva* (Çınar et al. 2013; Çiçek et al., 2023). Habitat degradation, pollution, invasive species, and climate change may intensify the pressures on the lake's biodiversity. The decline of species like *A. carinatus* and *C. muhlisi* underscores the urgent need for targeted conservation efforts to protect these vulnerable populations and preserve the lake's ecological integrity.

3.7.4 Future

The moderately dry period that ended in 1995 was followed by a 15-year normal period, after which the sub-basin became moderately wet (Fig. 9). The lake, with its large sub-basin, shows a slight trend of decreasing surface area and reduced precipitation. According to the model, precipitation is expected to decrease, especially in the optimistic scenario. However, under a more optimistic scenario, precipitation is predicted to increase, with an even greater rise under the SSP5-8.5 scenario by 2081–2100. Additionally, a noticeable rise in the annual mean temperature is anticipated from 2021 to 2040, with a particularly pronounced increase projected for 2081–2100 (see Appendix 3).

The lake's water quality is compromised by pollution, with limited literature addressing its management. According to the model, climate effects on the sub-basin are expected to be substantial. Coupled with high pollution levels, the potential consequences for the lake could be severe. A wetland management plan for the lake is currently in effect; however, implementing a new plan or a detailed re-analysis may be necessary.

3.8 Manyas Lake

3.8.1 Introduction

Manyas Lake, also known as "Kuş Cenneti" (Bird Paradise) Lake, is a shallow lake located in the Susurluk Basin in Marmara Region of Türkiye, with a maximum depth of 3.4 m and a surface area of 159 km^2 (Dorak et al. 2017; Özbayram et al. 2021), is primarily fed by the Kocaçay Stream along with the Sığırcı and Mürvetler streams and groundwater (Saç et al. 2020). Its outlet, the Karadere Stream, connects it to the Susurluk River. Seismic activity is believed to have played a role in the lake's formation during the late Holocene by damming a meandering stream system (Leroy et al. 2002). Our analysis of Landsat data from 1985 onward showed that the lake area fluctuated between 147 and 178 km^2, with the minimum value recorded in 2023 (Fig. 10). The lake is a critical wetland on the main bird migration route through Türkiye (Arı 2019) and serves as a vital breeding and wintering area for numerous bird species. Due to its ecological significance, the lake has been designated as a Ramsar Site (1994) and has held National Park status since 1959, making it one of the first protected areas of its kind in Türkiye (MoAF 2025).

3.8.2 Pressures

The first KBA assessment identified herbicide and pesticide pollution as a threat to the lake (KBAP 2025). By late 1990's, organochlorine pesticides and their residues were present in the water, sediments, and tributary streams (Erkmen et al. 2013). Recent data also indicate high nutrient loading, and the lake is classified as hypertrophic according to Carlson's Trophic State Index (Özbayram et al. 2021). Organic pollution is evident from the macroinvertebrate composition as well as dissolved oxygen and biological oxygen demand levels (Mercan and Arslan 2023).

Land cover analysis shows a steady increase in artificial areas and a significant decrease in water areas (Table 10). Water area loss is also evident in our LSA analysis (Fig. 10).

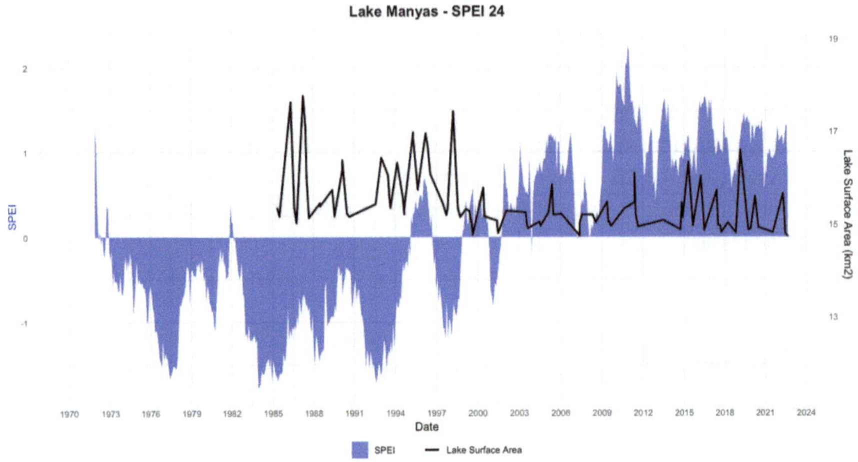

Fig. 10 SPEI and lake surface area for Manyas Lake

Table 10 Land cover changes in the Manyas Lake sub-basin over the past three decades

Manyas Lake	Corine		% Change	Sentinel-2		% Change
	1990	2018	1990–2018	2018	2023	2018–2023
Artificial areas	2.00	3.81	90	4.48	4.99	11
Agricultural areas	66.21	65.43	− 1	61.74	60.80	− 2
Forest/seminatural areas	13.78	13.92	1	18.55	18.98	2
Water	18.01	16.84	− 7	15.22	15.21	0

3.8.3 Fish Fauna

The lake hosts 25 fish species, including native, endemic, and invasive species (Balaban 2010; Arı 2019; Saç et al. 2020; Çiçek et al. 2023). The native species are *A. maeotica, C. puncticulata, E. lucius, C. carpio, S. erythrophthalmus, B. bjoerkna, P. borysthenicus, R. rutilus, S. glanis, Neogobius fluviatilis* (Pallas 1814), *R. amarus, K. caucasica, P. semilunaris*, and *V. vimba,* all classified as “Least Concern” (LC) by the IUCN (IUCN 2024), except *A. anguilla* (CR). Manyas Lake is also home to six endemic species that include *A. carinatus* (EN), *Alburnoides manyasensis* Turan, Ekmekçi, Kaya and Güçlü 2013 (LC), *S. cii* (LC), *B. oligolepis* (LC), *B. niluferensis* (NT), and *O. marmaraensis* (NE).

Among the endemic species, *A. carinatus* and *B. niluferensis* are of particular concern. *A. carinatus*, also found in Uluabat Lake, has experienced significant population declines due to habitat degradation and pollution (Freyhof 2014i). Although commercially important for fisheries, its habitat has been adversely affected by water abstraction, eutrophication, and pollution, leading to reduced population size (Freyhof 2014i). *B. niluferensis*, endemic to the limited drainages of the Susurluk

Basin in Türkiye, has a relatively restricted distribution (Korkmaz 2017). These populations are treated as distinct locations based on the primary threat of pollution. Despite occurring in multiple locations, pollution remains the primary threat to its survival, significantly affecting its habitat (Freyhof 2014k).

The introduction of invasive species has further exacerbated the lake's ecological challenges. Recorded invasive species include *C. gibelio*, *G. holbrooki*, *L. gibbosus*, and *P. parva* (Arı 2019; Saç et al. 2020). *C. gibelio* was first reported in the 2000s and quickly became one of the most intensively fished species in Manyas Lake due to its rapid population growth. However, excessive fishing pressure in recent years has led to a decrease in its population density (Saç et al. 2020). Despite this, *C. gibelio* remains the dominant species in the lake today (Arı 2019). The more recently (in 2020) detected invasive species, *G. holbrooki* and *P. parva*, pose additional threats to the lake's native biodiversity (Saç et al. 2020).

Manyas Lake faces numerous threats to its fish fauna and overall ecosystem health (Özbayram et al. 2021; Arı 2019; Saç et al. 2020). Eutrophication, resulting from agricultural runoff, industrial effluents, and domestic wastewater, has severely degraded the lake's water quality (Karafistan and Arik-Çolakoğlu 2005). The combined pressures of pollution, habitat degradation, and invasive species underscore the urgent need for conservation measures.

3.8.4 Future

The Susurluk Basin significantly impacts the climatic conditions and water levels of its lakes. While both lakes exhibit similar patterns, Uluabat Lake shows more pronounced seasonal water level changes than Manyas Lake. The climatic drought pattern (SPEI) is similar for both lakes, but the expected change in precipitation is moderately lower for Manyas Lake (Figs. 9 and 10, Appendix 3) Under the SSP2-4.5 scenario, unfavorable conditions are projected for the near future, however an increase is expected in precipitation between 2081 and 2100 (see Appendix 3). In the highly pessimistic SSP5-8.5 scenario, precipitation decline is less pronounced for Manyas Lake compared to Uluabat Lake, and the temperature increase patterns are similar (Fig. 10).

Manyas Lake is protected under multiple protection statuses. However, the lake is significantly polluted, and measures for its protection are discussed. A wetland management plan for the lake is in effect, but a new plan or a detailed re-analysis may be necessary.

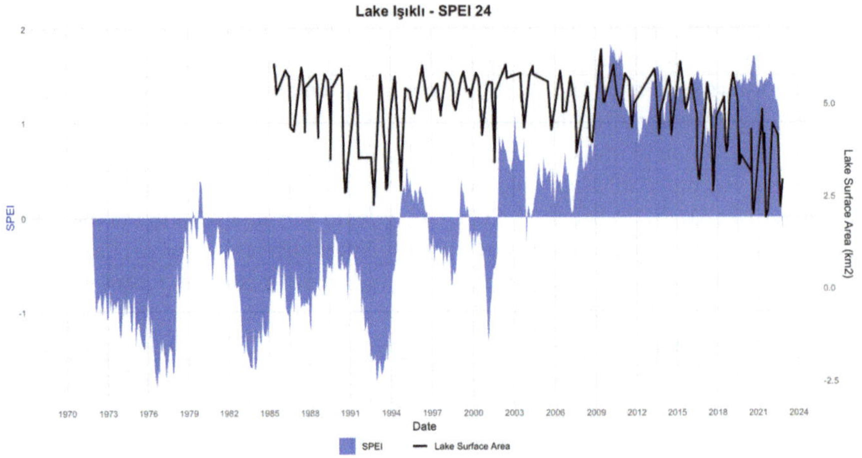

Fig. 11 SPEI and lake surface area for Işıklı Lake

3.9 *Işıklı Lake*

3.9.1 Introduction

Işıklı Lake is located in the headwater region of the Büyük Menderes River and is fed by significant water sources such as freshwater springs and two main streams: Dinar and Kufi. Our analysis of Landsat data from 1985 to 2021 onward showed that the lake area decreased from 65 to 19 km^2 by 2021 (Fig. 11). The maximum lake depth is < 7 m (Akbaba and Boyacı 2015). The Dinar Stream initially flows into shallow Lake Gökgöl, which is connected to Işıklı Lake. The two lakes and the Karakuyu Marshes located within the lake's drainage area are protected as 'Wetland of National Importance'.

Between 1949 and 1968, a substantial area of the lake was enclosed with barriers to manage flooding and facilitate irrigation. Originally a natural lake, it was converted into a reservoir by the State Hydraulic Works through dam construction (Akbaba and Boyacı 2015). Moreover, the lake holds wetland status, granting it international recognition.

3.9.2 Pressures

The KBA assessment in 2004 identified terrestrial animal persecution and control activities, but the report failed to comment on the terrestrial ecosystem threats (KBAP 2025). Additionally invasive species and water management threats have been identified, see Appendix 2.

The lake sub-basin contains extensive agricultural land (Çelik and Gülersoy 2013), supported by two large reservoirs, Örenler (1969) and Yavaşlar (2018), dams, and

Table 11 Land cover changes in the Işıklı Lake sub-basin over the past three decades

Işıklı Lake	Corine		% Change	Sentinel-2		% Change
	1990	2018	1990–2018	2018	2023	2018–2023
Artificial areas	1.42	1.75	23	2.21	2.54	15
Agricultural areas	43.70	43.90	0	38.99	36.65	– 6
Forest/seminatural areas	52.02	51.41	– 1	57.59	59.77	4
Water	2.86	2.94	3	1.20	1.04	– 13

numerous small ponds. Land cover data indicate that water levels remain stable, and lake water was previously classified as "very good quality waters" (Aksever and Büyükşahin 2017). In 2013, no heavy metal contamination was detected (Gülcü-Gür 2017). However, with the expansion of agricultural and artificial areas and the Yavaşlar dam now retaining water, reassessment of the lake's water quality is necessary.

Land cover data from 1990 to 2018 (Table 11) suggest an increase in water areas within the sub-basin. However, lake surface area analysis (Fig. 11) indicates an overall decrease. The observed increase between 1990 and 2018 may be attributed to expanded reservoir capacity during this period.

3.9.3 Fish Fauna

Işıklı Lake hosts 14 fish species, including native, endemic, introduced, and invasive species (Sarı et al. 2008; Uysal et al. 2008; Freyhof et al. 2020; Giannetto and İnnal 2021; Çiçek et al. 2023). The native species are *C. carpio* and *E. lucius*, all classified as "Least Concern" (LC) by the IUCN (IUCN, 2024). The lake's conservation significance lies in its seven endemic species. These include *Anatolichthys anatoliae* (Leidenfrost 1912) (NT), *C. meandrense* (VU), *G. klatti* (EN), *Gobio maeandricus* Naseka, Erk'akan and Küçük 2006 (EN), *Oxynoemacheilus mesudae* Erk'akan 2012 (EN), and *Squalius carinus* Özuluğ and Freyhof 2011 (EN). The lake is a critical refuge for these endemic species, but their survival is under threat from water abstraction, pollution, habitat fragmentation, and climate change (Freyhof et al. 2020; Çiçek et al. 2023). The introduction of non-native introduced species, *T. tinca* (LC) and *K. caucasica* (LC), along with the invasive species *C. gibelio*, *G. holbrooki*, *S. lucioperca* and *Oncorhynchus mykiss* (Walbaum 1792), complicates the situation by altering food web dynamics and degrading habitat quality (Freyhof et al. 2020; Giannetto and İnnal 2021).

The endemic species are highly specialized and particularly vulnerable to threats. *A. anatoliae* has suffered habitat loss due to water abstraction, desiccation of small streams, and competition from invasive *G. holbrooki*, leading to loss of many populations (Freyhof 2014l). *G. maeandricus* and *G. klatti*, inhabiting springs and streams like Işıklı Spring, Karadirek Stream, and Işıklı Lake, are severely impacted by water abstraction for agriculture, pollution, and climate change, reducing their available habitats (Freyhof 2014m). *C. meandrense* faces challenges from agricultural runoff, industrial pollution, and dam construction in the Büyük Menderes Basin. Meanwhile, *S. carinus*, which relies on both the lake and its streams for spawning, is threatened by habitat fragmentation and reduced water availability (Freyhof et al. 2014n, 2020; IUCN 2024). Protecting Işıklı Lake and its catchments is essential for maintaining the ecological integrity of the Büyük Menderes River Basin and preserving the lake's role as a key habitat for these endemic and endangered species.

3.9.4 Future

Işıklı Lake experiences significant annual fluctuations in surface area, at times losing up to half its extent (Fig. 11). These variations appear to be independent of climatic conditions and most likely driven by intensive agricultural activities in the sub-basin. Both precipitation and temperature are expected to increase slightly during 2021–2040, and substantial changes are projected for 2081–2100 under the highly pessimistic SSP5-8.5 scenario (see Appendix 3).

3.10 Köyceğiz Lake

3.10.1 Introduction

Köyceğiz Lake, an alluvial lake located in the Western Mediterranean Basin of Türkiye, is part of the Köyceğiz-Dalyan Lagoon system. It connects to a series of lakes via the Dalyan Channel and discharges into the Mediterranean Sea. Originally part of the sea, it was isolated from the Dalaman River, which then flowed along a different course. Our analysis of Landsat data from 1985 onward showed that the lake area fluctuated between 66 and 19 km^2 by 2021, averaging 31 km^2 in 2023 (Fig. 12). The lake is fed by three main streams: Namnam, Yuvarlak, and Kargıcak, contributing to its hydrological dynamics.

Köyceğiz Lake was designated as a Specially Protected Area in 1988 and is recognized for its unique ecological and limnological characteristics, shaped by spatiotemporal variations in chemical composition (Kazanci et al. 1992), salinity (Stumpp et al. 2014), and temperature from geothermal hot springs beneath the lake (Avşar et al. 2017). The surrounding sweetgum (*Liquidambar orientalis*) forests and the loggerhead turtle (*Caretta caretta*) nesting site at İztuzu Beach highlight the need for protection (Geven et al. 2016).

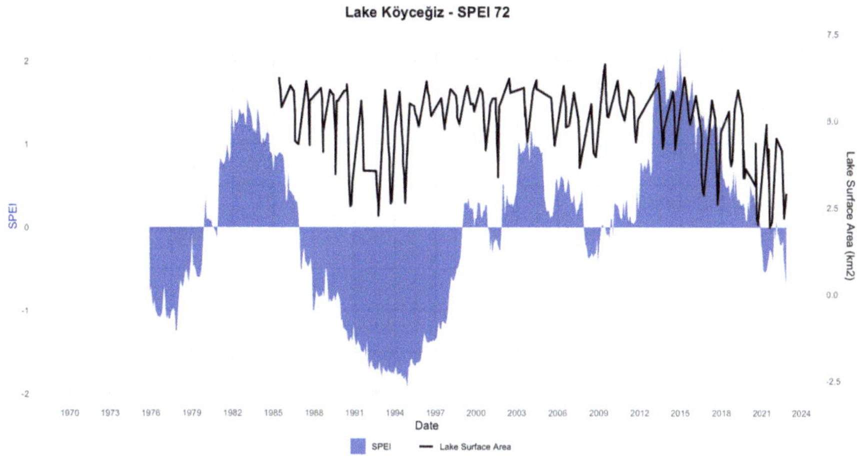

Fig. 12 SPEI and lake surface area for Köyceğiz Lake

3.10.2 Pressures

Tourism and aquaculture activities defined in the 2004 assessment are ongoing (KBAP 2025). Invasive species introductions are now also recognized as a threat, see Appendix 2.

Eutrophication of the lake, influenced by the Yuvarlakçay and Namnam streams, was recognized by 2006 (Taşeli 2009). The Yuvarlakçay stream, which had low to moderate organic pollution in 1992 (Kazanci and Dügel 2000), became a serious source of nitrite-nitrogen, total phosphate, and coliforms in 2006, similar to the Namnam stream (nitrate-nitrogen and ammonia–nitrogen) (Taşeli 2009). A study from 2011–2012 classified the lake water as being of poor quality regarding ammonia and orthophosphate levels (Özdemir et al. 2019). By 2018, the lake was already eutrophic, and with a slight (4%) increase in TP loads and a significant 65% reduction of the freshwater inflow, it was projected to become hypertrophic between 2018 and 2035 (Atasoy Aytış et al. 2023). Additionally, there is heavy metal pollution in the lagoon system, stretching from Lake Köyceğiz in the north to İztuzu Beach in the south (Arslan and Avşar 2020).

Agricultural land in the sub-basin is limited, but urbanization and tourism are the main problems for the lake, as evidenced by the high percentage of artificial areas (Table 12). The sub-basin water areas seem stable according to land cover analysis; however, the LSA analysis revealed a significant decrease (Fig. 12).

3.10.3 Fish Fauna

Köyceğiz Lake supports a diverse fish fauna, with 23 native, endemic, and invasive species (Yerli 1989; Balık et al. 2005; Yılmaz et al. 2006; Genç and Yılmaz 2023;

Table 12 Land cover changes in the Köyceğiz Lake sub-basin over the past three decades

Köyceğiz Lake	Corine		% Change	Sentinel-2		% Change
	1990	2018	1990–2018	2018	2023	2018–2023
Artificial areas	0.69	0.86	26	4.99	6.01	20
Agricultural areas	15.61	16.05	3	10.43	9.62	− 8
Forest/seminatural areas	76.47	76.01	− 1	78.21	78.05	0
Water	7.24	7.08	− 2	6.35	6.30	− 1

Çiçek et al. 2023). The native species include *A. anguilla* (CR), *A. fasciatus* (LC), *C. auratus* (NT), *Chelon labrosus* (Risso 1827) (NT), *C. ramada* (NT), *Chelon saliens* (Risso 1810) (NT), *C. carpio* (LC), *D. labrax* (NT), *M. cephalus* (LC), *Oedalechilus labeo* (Cuvier 1829) (LC), *Oxynoemacheilus* sp., *S. fluviatilis* (LC), and *S. glanis* (LC). The endemic species include *Barbus xanthos* Güçlü, Kalayci, Küçük and Turan 2020 (NE), *Capoeta aydinensis* Turan, Küçük, Kaya, Güçlü and Bektaş 2017 (NE), *Cobitis dorademiri* Erk'akan, Özdemir and Özeren 2017 (NE), *Knipowitschia byblisia* Ahnelt 2011 (LC), *Knipowitschia caunosi* Ahnelt 2011 (LC), *Ladigesocypris irideus* (Ladiges 1960) (NT), and *Squalius fellowesii* (Günther 1868) (LC) (IUCN 2024), and the invasive species include *A. boyeri*, *Coptodon zillii* (Gervais 1848), and *G. holbrooki*.

Endemic species, such as *B. xanthos*, *C. aydinensis*, and *L. irideus*, hold particular ecological and conservation value due to their restricted distribution (Freyhof et al. 2020; Çiçek et al. 2023). They face significant threats from habitat degradation, water abstraction, and pollution, which is exacerbated by invasive species like *A. boyeri*, *C. zillii*, and *G. holbrooki* (Freyhof et al. 2020; Genç and Yılmaz 2023).

Due to its connection to the Dalyan Lagoon, Köyceğiz Lake seasonally hosts marine species. However, these marine species are not included in this study's species list as the focus is on the lake's resident fish fauna.

3.10.4 Future

Köyceğiz Lake's surface area has fluctuated significantly, with a tendency to decrease (Fig. 12). Since 2020, the sub-basin appears to have entered a period of aridity, which is expected to worsen with the projected future changes in precipitation and temperature. Under the SSP2-4.5 scenario, assuming moderate socioeconomic development and gradual climate mitigation efforts, precipitation is projected to decline between 2021 and 2040 and then rise between 2081 and 2100. In the pessimistic scenario, a smaller decrease is expected between 2021 and 2040, followed by a significant decrease between 2081 and 2100. Temperature fluctuations within the sub-basin are projected to be more pronounced under both scenarios, across all evaluated timeframes (see Appendix 3).

Despite being designated a "Specially Protected Area" (MoEUCC 2025), one of the highest protection statuses, efforts to mitigate pollution in the lake have been unsuccessful.

3.11 Overall Assessment

Trend analyses indicate alarming changes in the ten KBAs. Nearly all the lakes exhibit a decreasing trend in lake surface area (LSA) despite an increasing Standardized Precipitation-Evapotranspiration Index (SPEI) (Table 13). Since increasing SPEI indicates wetness, LSA would normally be expected to increase. However, the observed LSA decline suggests that factors beyond precipitation, such as groundwater extraction and land use changes, drive the water loss in these ecosystems (Çolak et al. 2022).

All lake surface area change trends are negative. The decrease was lowest in Bafa Lake and steepest in Burdur Lake. For İznik, Salda and Burdur lakes (SMK Tau LSA $\cong -0.4$), vigilance and preventive measures can help maintain health and functionality. All lakes experienced a below average wet period, as also reflected by SMK Tau SPEI values. However, some lakes (e.g., İznik, Bafa, Acı, Burdur, Salda, Eğirdir, Köyceğiz) shows signs of possible entry to a dry period.

Precipitation and temperature changes projections suggest worsening conditions for the KBA lakes. Near-future precipitation trends (2021–2040) show an overall, albeit insignificant, decline. On the other hand, far-future (2081–2100) projections indicate even steeper reductions (Table 14; Fig. 13). A few lakes (İznik, Uluabat, and Manyas) show slight increases under the SSP2-4.5 scenario.

Conversely, temperature projections indicate a strong warming trend across all lakes under both scenarios (Table 15; Fig. 14). Projections of temperature change in the pessimistic SSP5-8.5 scenario suggest potential threats to lakes. This temperature

Table 13 Trends in LSA and SPEI

Lake	SMK Tau SPEI	SMK Tau LSA
Bafa Lake	0.361	− 0.138
Işıklı Lake	0.629	− 0.301
Köyceğiz Lake	0.207	− 0.303
Lake Acıgöl	0.657	− 0.334
Manyas Lake	0.527	− 0.342
Uluabat Lake	0.47	− 0.352
Eğirdir Lake	0.52	− 0.372
İznik Lake	0.435	− 0.382
Salda Lake	0.666	− 0.447
Burdur Lake	0.448	− 0.489

Table 14 Projected precipitation changes (mm) under different climate scenarios

Lake	Historical	SSP2-4.5				SSP5-8.5			
		2021–2040		2081–2100		2021–2040		2081–2100	
İznik Lake	603.0	580.4	– 3.7%	617.7	2.4%	598.7	– 0.7%	567.1	– 5.9%
Bafa Lake	723.1	709.4	– 1.9%	705.5	– 2.4%	711.8	– 1.6%	622.2	– 14.0%
Acıgöl Lake	535.7	510.6	– 4.7%	521.2	– 2.7%	522.5	– 2.5%	468.0	– 12.6%
Burdur Lake	525.8	502.2	– 4.5%	513.2	– 2.4%	515.1	– 2.0%	464.0	– 11.8%
Eğirdir Lake	558.7	535.6	– 4.1%	549.5	– 1.7%	548.8	– 1.8%	500.8	– 10.4%
Salda Lake	561.4	534.3	– 4.8%	547.0	– 2.6%	550.8	– 1.9%	490.9	– 12.6%
Uluabat Lake	632.0	616.1	– 2.5%	665.1	5.2%	629.5	– 0.4%	593.4	– 6.1%
Manyas Lake	688.5	679.6	– 1.3%	738.7	7.3%	686.3	– 0.3%	651.3	– 5.4%
Işıklı Lake	560.3	536.0	– 4.3%	549.1	– 2.0%	548.4	– 2.1%	495.5	– 11.6%
Köyceğiz Lake	963.4	921.3	– 4.4%	954.6	– 0.9%	951.5	– 1.2%	835.7	– 13.3%

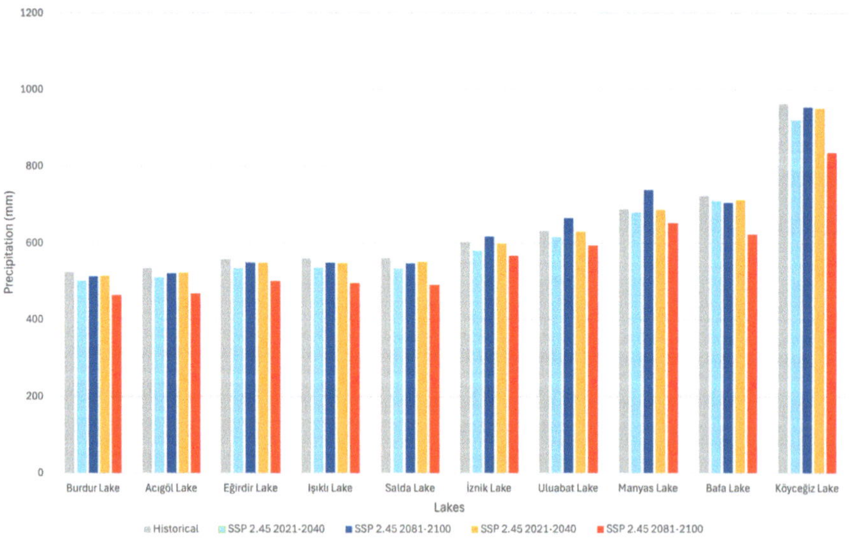

Fig. 13 Precipitation changes in the ten lakes based on SSP2-4.5 and SSP5-8.5 for the periods 2021–2040 and 2081–2100

Table 15 Projected temperature changes (°C) under different climate scenarios

Lake	Historical	SSP2-4.5				SSP5-8.5			
		2021–2040		2081–2100		2021–2040		2081–2100	
İznik Lake	13.1	15.4	17.1%	17.6	34.2%	15.8	20.4%	20.5	55.9%
Bafa Lake	17.5	19.6	11.8%	21.8	24.1%	20.0	14.0%	24.4	39.1%
Acıgöl Lake	11.8	14.1	19.2%	16.4	38.8%	14.5	22.3%	19.5	65.1%
Burdur Lake	11.3	13.6	19.8%	15.9	40.3%	14.0	23.1%	19.0	67.3%
Eğirdir Lake	10.5	12.9	22.6%	15.3	45.4%	13.2	26.1%	18.3	74.7%
Salda Lake	10.5	12.6	20.9%	14.9	42.6%	13.0	24.4%	18.0	71.7%
Uluabat Lake	13.5	15.7	16.6%	18.0	33.2%	16.2	19.9%	20.9	54.8%
Manyas Lake	13.8	16.0	16.5%	18.2	32.4%	16.5	19.5%	21.1	52.9%
Işıklı Lake	10.7	13.1	21.8%	15.4	43.8%	13.4	25.3%	18.6	73.1%
Köyceğiz Lake	16.1	18.2	12.8%	20.4	26.2%	18.6	15.0%	23.2	43.7%

increase is expected to intensify evaporation rates and further stress these freshwater ecosystems.

3.12 Discussion

The selected ten KBAs examined in this study are critical ecosystems that support a significant proportion of Türkiye's freshwater fish fauna. However, they face multiple anthropogenic and climate-induced pressures, endangering their ecological integrity. Across the ten lakes a total of 71 native and endemic fish species (see Appendix 1, Tables 16 and 17), nearly 17% of Türkiye's freshwater fish fauna, have been recorded, highlighting their conservation importance.

This chapter updates the threats previously identified in KBA assessments. While key pressures such as water abstraction, habitat modification, and pollution were already recognized (KBAP 2025), our findings highlight additional and emerging threats, including invasive species proliferation, groundwater depletion, and increasing salinization due to climate change. The revised threat assessments, based on the IUCN threats schema, reveal significant gaps in previous evaluations, emphasizing the need for a more comprehensive and dynamic approach to KBA protection. Updated threats for each KBA are given in Appendix 2.

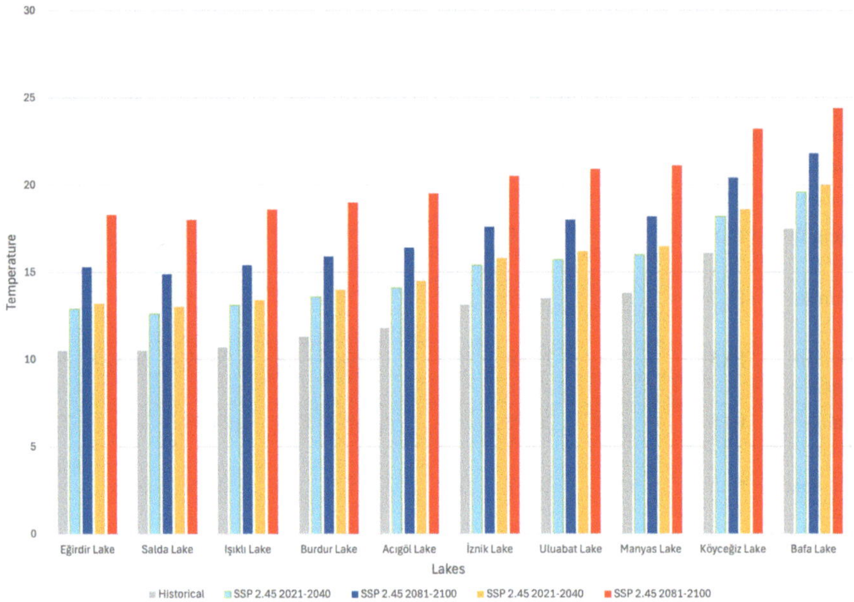

Fig. 14 Temperature changes in the ten lakes based on SSP2-4.5 and SSP5-8.5 for the periods 2021–2040 and 2081–2100

Many lake ecosystems of the Mediterranean Region are under threat due to climate changes impacts coupled with anthropogenic effects. One devastating example of this is Lake Marmara (Gölmarmara) situated in a closed basin, formed by alluvial deposits. The lake was turned into a reservoir for irrigation purposes by the State Hydraulic Works with barriers and flow control in both directions (Murathan et al. 2023; Gunacti et al. 2023) and declared a "Wetland of National Importance" in 2017; however, soon after this, in 2021, it dried up completely.

Given our findings, Türkiye's Mediterranean KBAs require urgent and targeted conservation interventions. Addressing water management challenges, mitigating invasive species impacts, and implementing habitat restoration efforts are crucial to ensure long-term survival of endemic and native fish species. Integration of climate projections into conservation planning is equally vital to safeguard these unique freshwater ecosystems from accelerating climate change impacts.

In 2021, the first National Water Forum was held with public institutions and organizations, non-governmental organizations, metropolitan municipalities, and water and sanitation authorities and stakeholders. Although academic participation was limited (9%), eleven working groups identified short, medium, and long-term water-related strategies. In this study, it was concluded that to identify water-related strategies, a comprehensive analysis must be conducted and as with any analysis, data must be interpreted with care. During the land cover analysis, we found that the CLC2018 data has some major inconsistencies for the selected region. Some of the rural areas that should be classified as artificial are classified differently, mainly as agricultural

Table 16 Species list for lakes

Species	İznik Lake	Bafa Lake	Lake Acıgöl	Burdur Lake	Eğirdir Lake	Salda Lake	Uluabat Lake	Manyas Lake	Işıklı Lake	Köyceğiz Lake
Alburnoides manyasensis								1		
Alburnus carinatus							1	1		
Alburnus nicaeensis	1									
Alosa maeotica							1	1		
Anatolichthys anatoliae									1	
Anatolichthys iconii					1					
Anatolichthys saldae						1				
Anatolichthys sureyanus				1						
Anatolichthys transgrediens			1							
Anguilla anguilla		1					1	1		1
Aphanius fasciatus		1								1
Barbus niluferensis							1	1		
Barbus oligolepis	1						1	1		
Barbus xanthos										1
Blicca bjoerkna							1	1		
Capoeta aydinensis										1
Capoeta pestai					1					
Capoeta tinca	1									
Chelon auratus		1								1
Chelon labrosus										1

(continued)

Table 16 (continued)

Species	İznik Lake	Bafa Lake	Lake Acıgöl	Burdur Lake	Eğirdir Lake	Salda Lake	Uluabat Lake	Manyas Lake	Işıklı Lake	Köyceğiz Lake
Chelon ramada		1								1
Chelon saliens										1
Chondrostoma meandrense		1							1	
Clupeonella muhlisi							1			
Cobitis dorademiri										1
Cobitis phrygica						1				
Cobitis puncticulata							1	1		
Cobitis taenia	1									
Cobitis turcica					1					
Cyprinus carpio	1	1			1	1	1	1	1	1
Dicentrarchus labrax		1								1
Egirdira nigra					1					
Esox lucius							1	1	1	
Garra klatti					1				1	
Gasterosteus aculeatus	1									
Gobio maeandricus									1	
Gobius niger		1								
Knipowitschia byblisia										1
Knipowitschia caucasica	1						1	1		
Knipowitschia caunosi										1

(continued)

Table 16 (continued)

Species	İznik Lake	Bafa Lake	Lake Acıgöl	Burdur Lake	Eğirdir Lake	Salda Lake	Uluabat Lake	Manyas Lake	Işıklı Lake	Köyceğiz Lake
Ladigesocypris irideus										1
Luciobarbus kottelati		1								
Mugil cephalus		1					1			1
Neogobius fluviatilis								1		
Oedalechilus labeo										1
Oxynoemacheilus anatolicus						1				
Oxynoemacheilus angorae	1									
Oxynoemacheilus germencicus		1								
Oxynoemacheilus marmaraensis							1	1		
Oxynoemacheilus mediterraneus					1					
Oxynoemacheilus mesudae									1	
Oxynoemacheilus sp.										1
Petroleuciscus borysthenicus							1	1		
Proterorhinus semilunaris	1						1	1		

(continued)

Table 16 (continued)

Species	İznik Lake	Bafa Lake	Lake Acıgöl	Burdur Lake	Eğirdir Lake	Salda Lake	Uluabat Lake	Manyas Lake	Işıklı Lake	Köyceğiz Lake
Pseudophoxinus burduricus						1				
Pseudophoxinus handlirschi					1					
Rhodeus amarus							1	1		
Rutilus frisii	1									
Rutilus rutilus	1						1	1		
Salaria pavo		1								
Salariopsis fluviatilis	1	1					1			1
Scardinius erythrophthalmus							1	1		
Seminemacheilus ispartensis					1					
Silurus glanis	1	1					1	1		1
Squalius carinus									1	
Squalius cii	1						1	1		
Squalius fellowesii										1
Syngnathus abaster		1					1			
Tinca tinca	1						1		1	
Vimba mirabilis		1								
Vimba vimba	1				1		1	1		

Table 17 Presence of invasive and introduced species

Species	İznik Lake	Bafa Lake	Lake Acıgöl	Eğirdir Lake	Salda Lake	Uluabat Lake	Manyas Lake	Işıklı Lake	Köyceğiz Lake
Atherina boyeri	1	1		1	1	1			1
Carassius gibelio	1	1		1		1	1	1	
Coptodon zillii									1
Gambusia holbrooki	1	1	1	1	1	1	1	1	1
*Knipowitschia caucasica**		1		1				1	
Lepomis gibbosus	1	1				1	1		
Oncorhynchus mykiss								1	
Pseudorasbora parva				1		1	1		
Sander lucioperca				1				1	
*Tinca tinca**								1	

*Introduced species—non-native for the lake

areas, but these are already listed as artificial in CLC1990. A similar problem has been observed and documented for Spain due to a change in map production methodology after 2012 (García Álvarez and Camacho Olmedo 2023).

To minimize the error in estimating the land cover change, we merged the artificial areas of 1990 with the 2018 map and predicted the percentage change based on this. According to our analysis, the original CLC2018 underestimated artificial areas by, on average, 23%, with a range of 8–67%. We are not sure of the cause of this classification change, but researchers need to be cautious when using CLC2018 for Türkiye. Water level quantification also seems to exhibit some inconsistencies. perhaps reflecting the fact that water level is a snapshot and thus does not represent the entire year. To better interpret the water surface area changes, monthly/weekly data are needed rather than yearly values.

Time-series of satellite images used for lake surface area detection might have some inconsistencies. Due to cloudy seasons, there are gaps in time intervals. Furthermore, water detection indices may have misclassified water pixels due to natural artifacts (Feyisa et al. 2014). Therefore, corrected Level-2 images and minimized cloud covers are needed.

3.12.1 Fish Fauna

The KBA lakes in Türkiye, which are critical habitats for a diverse endemic and native fish species, face increasing threats from climate change and human activities. Climate-induced factors like reduced rainfall, rising temperatures, and prolonged droughts have exacerbated habitat degradation, particularly in hydrologically fragile sub-basins such as Burdur, Salda, and Acıgöl (Davraz et al. 2019a; Çolak et al. 2022). In these lakes, shrinking water levels and increasing salinity have significantly stressed endemic species like *A. transgrediens* and *A. saldae*, leading to declining populations in their already restricted habitats (Güçlü and Küçük 2012; Freyhof 2014e).

Anthropogenic influences, including water abstraction, dam construction, and agricultural runoff, have compounded these challenges. For example, the unregulated water abstraction for irrigation in Eğirdir, Burdur and Salda lakes has led to the desiccation of springs critical for the survival of endemic fish such as *P. burduricus* and *O. anatolicus*. Similar pressures in Işıklı Lake have fragmented habitats and altered water quality, threatening endemic species like *C. meandrense* and *G. maeandricus* (Freyhof 2014c; Çiçek et al. 2023; IUCN 2024).

Invasive species pose a significant threat to the KBA lakes in Türkiye, where they outcompete native and endemic fauna, alter food web dynamics, and degrade habitat quality (Tarkan et al. 2015; Mangit et al. 2018; Giannetto and İnnal 2021; Çiçek et al. 2023). The presence of invasive species across the ten studied lakes (see Appendix 1, Table 17) reveals the extent of biological invasions and their ecological impacts.

Among the most widespread invasive species, *C. gibelio* is present in seven of the ten lakes, where it has established dominance and significantly disrupted native fish populations. *C. gibelio* introduction has been associated with severe ecological shifts, including the extinction of *A. nicaeensis* in İznik Lake (Freyhof 2014a) and significant declines in native fish populations, disrupting the ecological balance of aquatic ecosystems (Gaygusuz et al. 2007; Yalçın-Özdilek et al. 2019). A long-term study of Ömerli Dam Lake demonstrated that increasing *C. gibelio* densities over six years led to a significant decline in native fish populations, emphasizing its negative impact on local fish communities (Tarkan et al. 2012). Similarly, *G. holbrooki*, is recorded in eight of the ten lakes, highlighting its ability to rapidly colonize diverse freshwater environments. *G. holbrooki* is considered one of the world's 100 most invasive species due to its broad dietary preferences, adaptability to various habitats, and high reproductive potential, allowing it to rapidly establish dense populations (Lowe et al. 2000). This species exerts a significant pressure on aquatic ecosystems through predation, competition, and aggressive behavior, leading to declines in fish populations (Pyke 2008). In Acıgöl, *G. holbrooki* presence has been associated with increasing pressures on the critically endangered *A. transgrediens* (Güçlü and Küçük 2012; IUCN 2024), further threatening its survival in this fragile ecosystem. Another highly impactful invasive species, *A. boyeri*, recorded in six lakes, has been linked to the displacement of native species and changes in food web dynamics, particularly in Eğirdir, İznik, and Uluabat Lakes. Eğirdir Lake, in particular, has suffered major biodiversity losses due to invasive species. The introduction of *S. lucioperca* has led to the extinction of *P. handlirschi*, a species unable to withstand the predation pressure from this top predator (Yerli et al. 2013; Küçük 2012).

The biodiversity impact of invasive species is particularly pronounced in lakes with high invasive species richness, such as Eğirdir, İznik, Uluabat, and Bafa, where multiple invasive species co-occur. The effects of these introductions are further compounded by habitat degradation, pollution, and hydrological changes, making conservation efforts increasingly urgent. *C. gibelio*, *A. boyeri*, and *G. holbrooki* have demonstrated strong adaptability, allowing them to dominate aquatic ecosystems and marginalize native species, in some cases leading to extinction. Ribeiro and Leunda (2012) highlight that non-native freshwater fish are among the leading drivers of biodiversity loss in Mediterranean freshwater ecosystems, affecting native species through competition, predation, habitat modification, and hybridization. Their impact is particularly severe in water bodies with high endemic fish diversity, where introduced species often cause irreversible ecological shifts. Given the widespread presence of invasive species in these KBA lakes, sustainable fisheries management and invasive species control must be prioritized. Measures such as regular monitoring, targeted removal programs, and habitat restoration are essential to mitigate their impact and preserve native fish populations. Without immediate intervention, the loss of native and endemic species will continue, further jeopardizing the ecological integrity of these critical freshwater ecosystems.

4 Conclusion

Water management has always been intricately linked to economic development, with growing populations increasing the demand for natural resources. In the context of Türkiye's KBA lakes, managing these resources ethically and sustainably has become more critical than ever, especially under the escalating impacts of climate change. While climate change undeniably influences water scarcity, viewing it as the sole driver oversimplifies the complex interplay of factors, including agricultural water abstraction, dam construction, pollution, and invasive species introductions, which contribute substantially to the degradation of lake ecosystems.

The fish fauna of Türkiye's KBA lakes, which includes numerous endemic species, is particularly vulnerable to these pressures. Endemic species face severe threats from habitat loss, water abstraction, and increasing salinity. Some endemic species are now extinct, while others, are listed as Critically Endangered (CR) and Endangered (EN) by the IUCN. The introduction of invasive species has further exacerbated these challenges by outcompeting native species and disrupting the ecological balance.

Addressing these multifaceted challenges requires a holistic and integrated approach. Effective conservation strategies should include regulating water abstraction, adopting sustainable agricultural practices, and implementing habitat restoration programs. Additionally, targeted measures to mitigate the impacts of invasive species and adapt to climatic shifts are essential to preserve the ichthyofaunal assemblages of these lakes. Protecting endemic species, which are essential indicators of ecological health and biodiversity, should be a priority in these efforts. The ecological health of Türkiye's KBA lakes is fundamentally intertwined with the survival of their endemic and native fish species, many of which are under significant threat. Only through such coordinated efforts can we ensure the resilience and long-term sustainability of these critical biodiversity hotspots, safeguarding them, while maintaining their ecological and economic importance.

Acknowledgements We are grateful to Anne Mette Poulsen for the linguistic improvements in the chapter. M. K. and E. J. were supported by TÜBİTAK program BIDEB2232, project number 118C250. FM was supported by TÜBİTAK, project number 122Y303.

Appendix 1: Fish Species

See Tables 16 and 17.

Appendix 2: Pressures

KBA ID—namc	Level 1	Level 2	Level 3	Status
777—İznik Lake	Natural system modifications	Dams and water management/use	Abstraction of surface water (commercial use)	Ongoing
777—İznik Lake	Natural system modifications	Dams and water management/use	Dams (size unknown)	Ongoing
777—İznik Lake	Pollution	Agricultural and forestry effluents	Herbicides and pesticides	Ongoing
777—İznik Lake	Pollution	Industrial and military effluents	Heavy metal pollution	Recent
777—İznik Lake	Natural system modifications	Dams and water management/use	Abstraction of surface water (agricultural use) (Leeves)	Recent
778—İznik Lake	Invasive and other problematic species, genes and diseases	Invasive non-native/alien species/diseases	Named species	Recent
759—Bafa Lake	Agriculture and aquaculture	Marine and freshwater aquaculture	Scale unknown/ unrecorded	Ongoing
759—Bafa Lake	Natural system modifications	Other ecosystem modifications		Ongoing

(continued)

(continued)

KBA ID—name	Level 1	Level 2	Level 3	Status
759—Bafa Lake	Residential and commercial development	Tourism and recreation areas		Ongoing
759—Bafa Lake	Pollution	Industrial and military effluents	Heavy Metal Pollution	Recent
759—Bafa Lake	Pollution	Agricultural and forestry effluents	Nutrient loads	Recent
759—Bafa Lake	Natural system modifications	Dams and water management/use	Abstraction of surface water (agricultural use) (Salinization)	Recent
759—Bafa Lake	Natural system modifications	Dams and water management/use	Abstraction of surface water (agricultural use) (Flood control on rivers)	Recent
759—Bafa Lake	Invasive and other problematic species, genes and diseases	Invasive non-native/alien species/diseases	Named species	Recent
763—Acıgöl Lake	Biological resource use	Hunting and collecting terrestrial animals	Motivation Unknown/ Unrecorded	Unknown
763—Acıgöl Lake	Energy production and mining	Mining and quarrying		Ongoing
763—Acıgöl Lake	Natural system modifications	Dams and water management/use	Abstraction of surface water (unknown use)	Ongoing
763—Acıgöl Lake	Pollution	Industrial and military effluents	Heavy Metal Pollution	Recent
763—Acıgöl Lake	Pollution	Agricultural and forestry effluents	Nutrient loads	Recent
763—Acıgöl Lake	Natural system modifications	Dams and water management/use	Abstraction of surface water (commercial use)(Sodium production)	Recent
763—Acıgöl Lake	Invasive and other problematic species, genes and diseases	Invasive non-native/alien species/diseases	Named species	Recent
766—Burdur Lake	Natural system modifications	Dams and water management/use	Abstraction of surface water (agricultural use)	Ongoing

(continued)

(continued)

KBA ID—name	Level 1	Level 2	Level 3	Status
766—Burdur Lake	Pollution	Agricultural and forestry effluents	Nutrient loads	Recent
766—Burdur Lake	Natural system modifications	Dams and water management/use	Abstraction of Ground Water (agricultural use)	Recent
766—Burdur Lake	Natural system modifications	Dams and water management/use	Abstraction of surface water (agricultural use) (Salinization)	Recent
767—Eğirdir Lake	Invasive and other problematic species, genes and diseases	Invasive non-native/alien species/diseases	Named species	Ongoing
767—Eğirdir Lake	Residential and commercial development	Tourism and recreation areas		Ongoing
767—Eğirdir Lake	Natural system modifications	Dams and water management/use	Abstraction of surface water (agricultural use)	Recent
767—Eğirdir Lake	Pollution	Agricultural and forestry effluents	Nutrient loads	Recent
767—Eğirdir Lake	Natural system modifications	Dams and water management/use	Abstraction of surface water (agricultural use) (Channelization)	Recent
767—Eğirdir Lake	Pollution	Industrial and military effluents	Heavy Metal Pollution	Recent
767—Eğirdir Lake	Pollution	Agricultural and forestry effluents	Herbicides and pesticides	Ongoing
767—Eğirdir Lake	Invasive and other problematic species, genes and diseases	Invasive non-native/alien species/diseases	Named species	Recent
770—Salda Lake	Natural system modifications	Dams and water management/use	Abstraction of ground water (agricultural use)	Ongoing
770—Salda Lake	Residential and commercial development	Tourism and recreation areas		Recent
770—Salda Lake	Pollution	Industrial and military effluents	Heavy Metal Pollution	Recent
770—Salda Lake	Natural system modifications	Dams and water management/use	Abstraction of Surface Water (agricultural use) (Artificial lakes)	Recent

(continued)

(continued)

KBA ID—name	Level 1	Level 2	Level 3	Status
770—Salda Lake	Invasive and other problematic species, genes and diseases	Invasive non-native/alien species/diseases	Named species	Recent
782—Uluabat Lake	Transportation and service corridors	Roads and railroads		Ongoing
782—Uluabat Lake	Pollution	Industrial and military effluents	Heavy Metal Pollution	Recent
782—Uluabat Lake	Pollution	Agricultural and forestry effluents	Nutrient loads	Recent
782—Uluabat Lake	Invasive and other problematic species, genes and diseases	Invasive non-native/alien species/diseases	Named species	Recent
783—Manyas Lake	Pollution	Agricultural and forestry effluents	Herbicides and pesticides	Ongoing
783—Manyas Lake	Pollution	Agricultural and forestry effluents	Nutrient loads	Recent
783—Manyas Lake	Invasive and other problematic species, genes and diseases	Invasive non-native/alien species/diseases	Named species	Recent
790—Işıklı Lake	Biological resource use	Hunting and collecting terrestrial animals	Persecution/ control	Unknown
790—Işıklı Lake	Natural system modifications	Dams and water management/use		Recent
790—Işıklı Lake	Invasive and other problematic species, genes and diseases	Invasive non-native/alien species/diseases	Named species	Recent
801—Köyceğiz L.	Agriculture and aquaculture	Annual and perennial non-timber crops	Agro-industry farming	Ongoing
801—Köyceğiz L.	Biological resource use	Hunting and collecting terrestrial animals	Intentional use (species being assessed is the target)	Unknown
801—Köyceğiz L.	Residential and commercial development	Tourism and recreation areas		Ongoing
801—Köyceğiz L.	Transportation and service corridors	Shipping lanes		Ongoing

(continued)

(continued)

KBA ID—name	Level 1	Level 2	Level 3	Status
801—Köyceğiz L.	Pollution	Industrial and military effluents	Heavy Metal Pollution	Recent
801—Köyceğiz L.	Pollution	Agricultural and forestry effluents	Nutrient loads	Recent
801—Köyceğiz L.	Invasive and other problematic species, genes and diseases	Invasive non-native/alien species/diseases	Named species	Recent

Appendix 3: Maps for Precipitation and Temperature Projections

See Figs. 15, 16, 17, 18, 19, 20, 21, 22, 23 and 24.

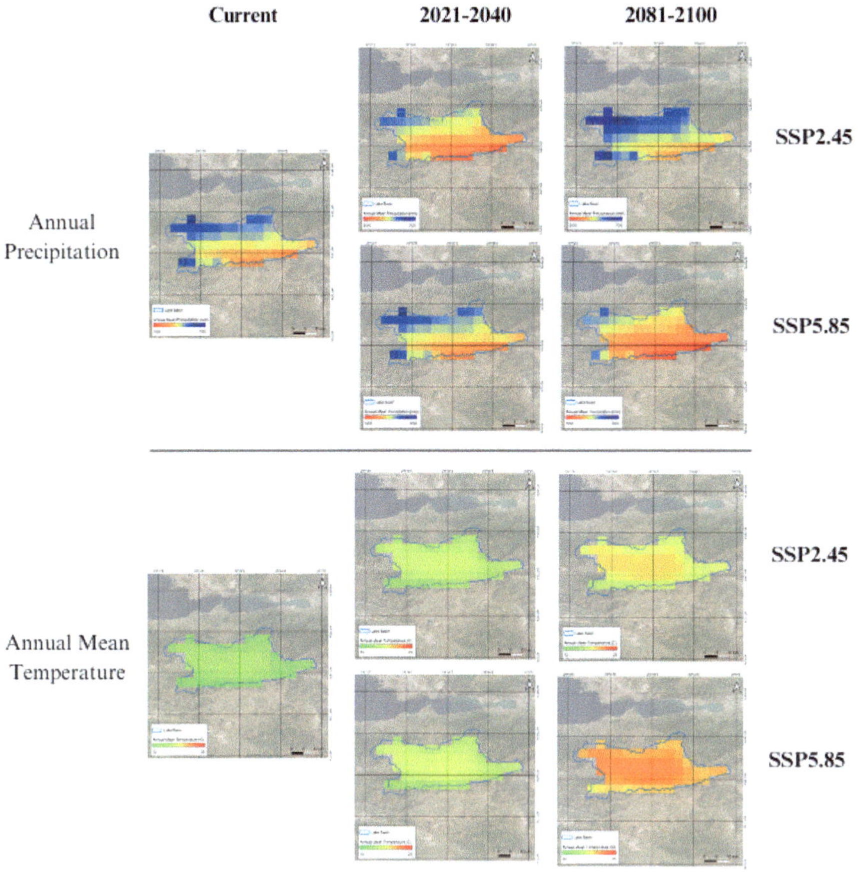

Fig. 15 Precipitation and temperature projections for İznik Lake based on SSP2.45 and SSP5.85 for the periods 2021–2040 and 2081–2100

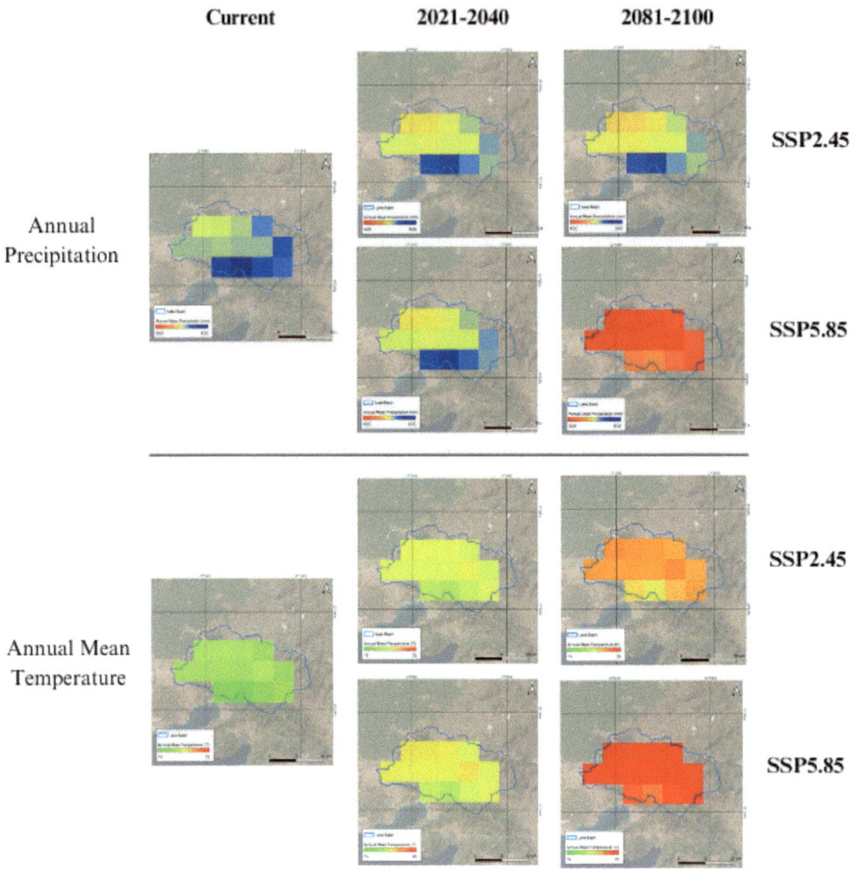

Fig. 16 Precipitation and temperature projections for Bafa Lake based on SSP2.45 and SSP5.85 for the periods 2021–2040 and 2081–2100

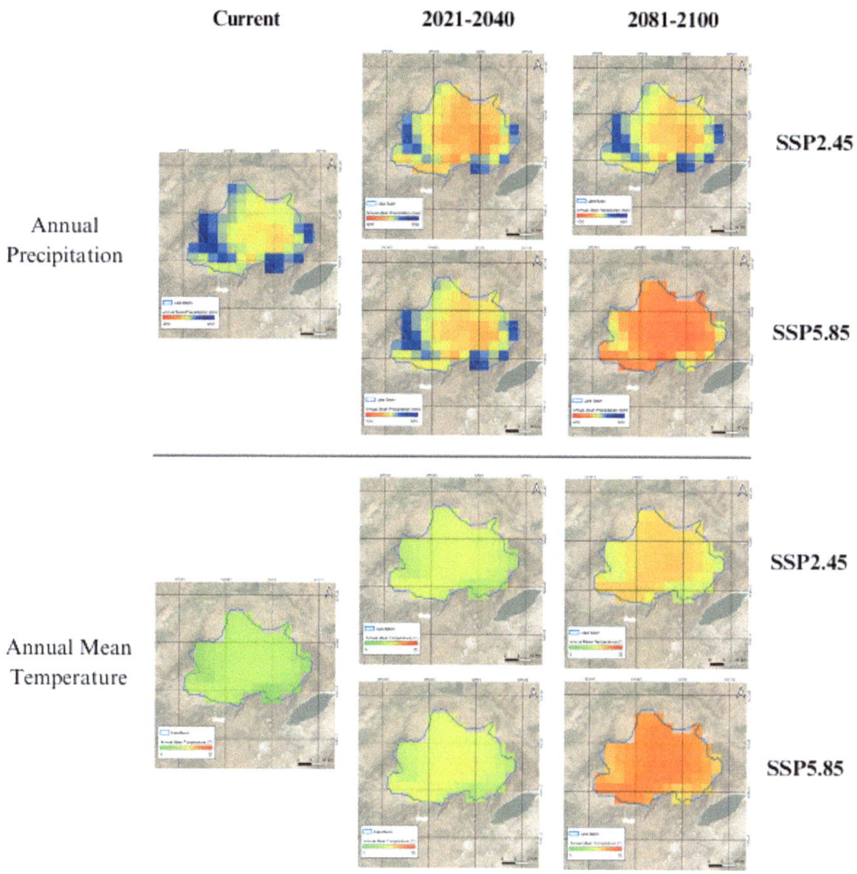

Fig. 17 Precipitation and temperature projections for Lake Acıgöl based on SSP2.45 and SSP5.85 for the periods 2021–2040 and 2081–2100

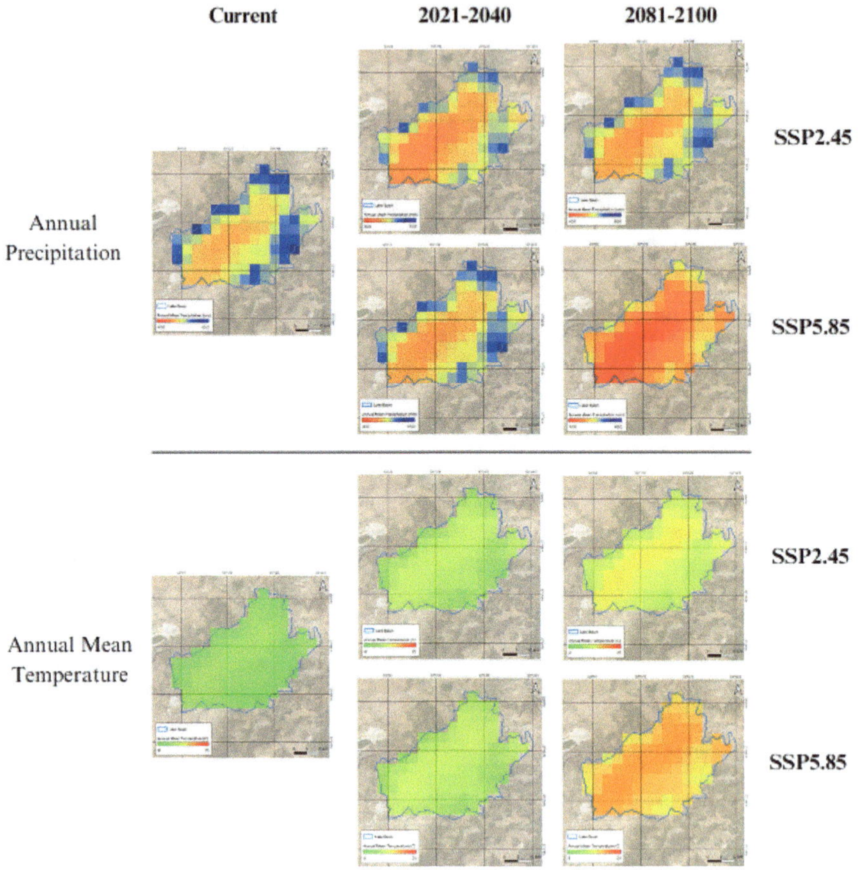

Fig. 18 Precipitation and temperature projections for Burdur Lake based on SSP2.45 and SSP5.85 for the periods 2021–2040 and 2081–2100

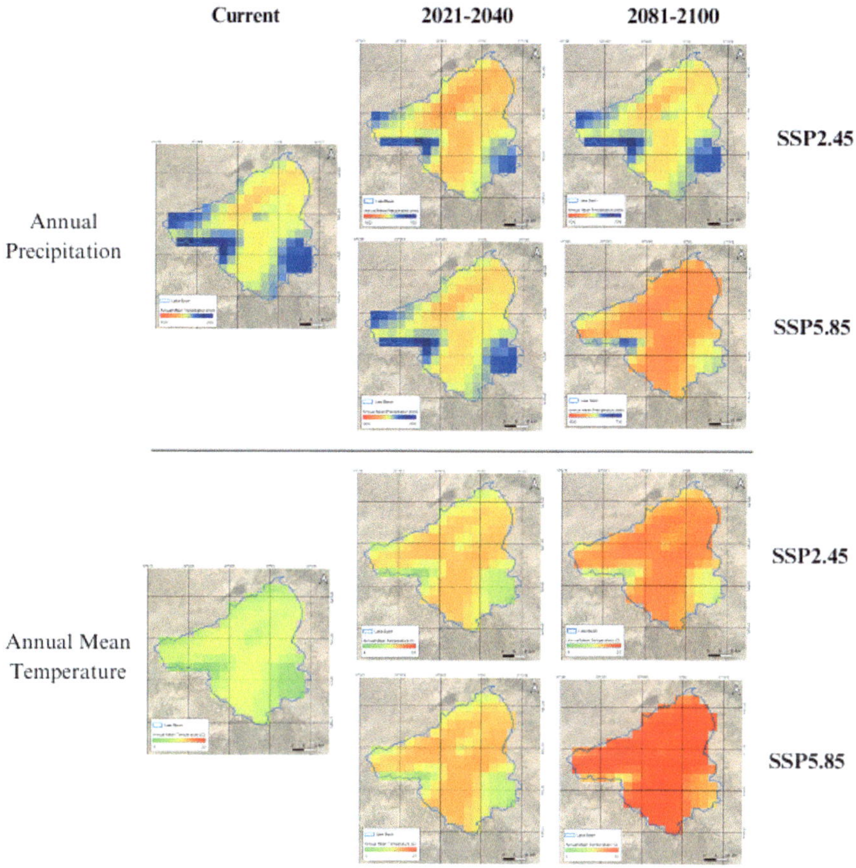

Fig. 19 Precipitation and temperature projections for Eğirdir Lake based on SSP2.45 and SSP5.85 for the periods 2021–2040 and 2081–2100

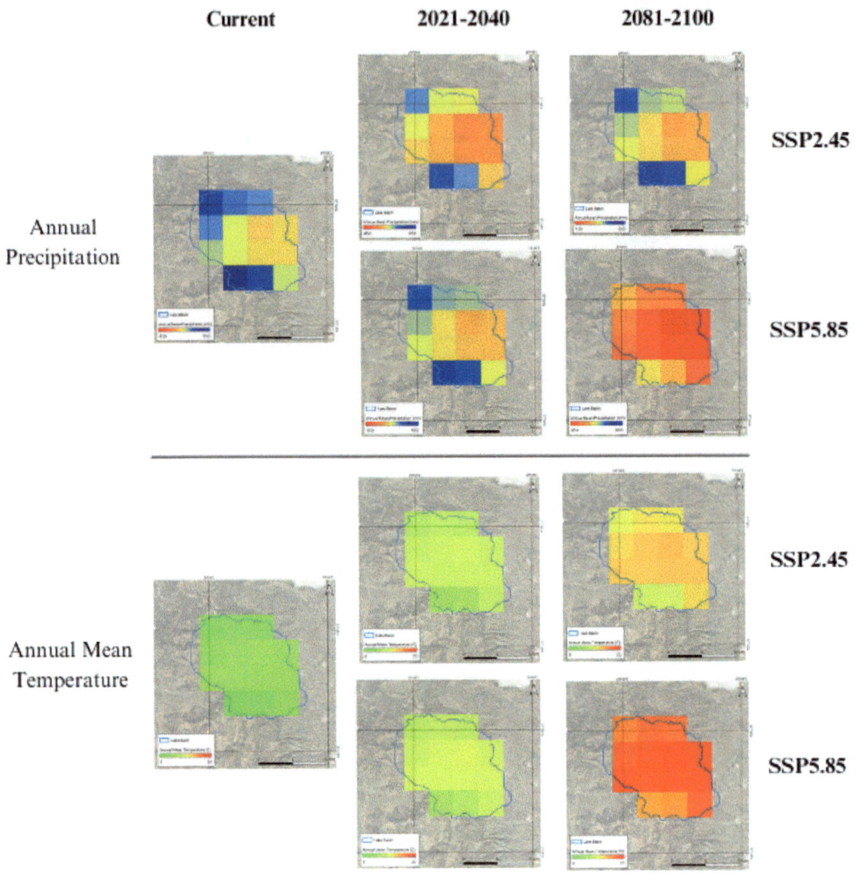

Fig. 20 Precipitation and temperature projections for Salda Lake based on SSP2.45 and SSP5.85 for the periods 2021–2040 and 2081–2100

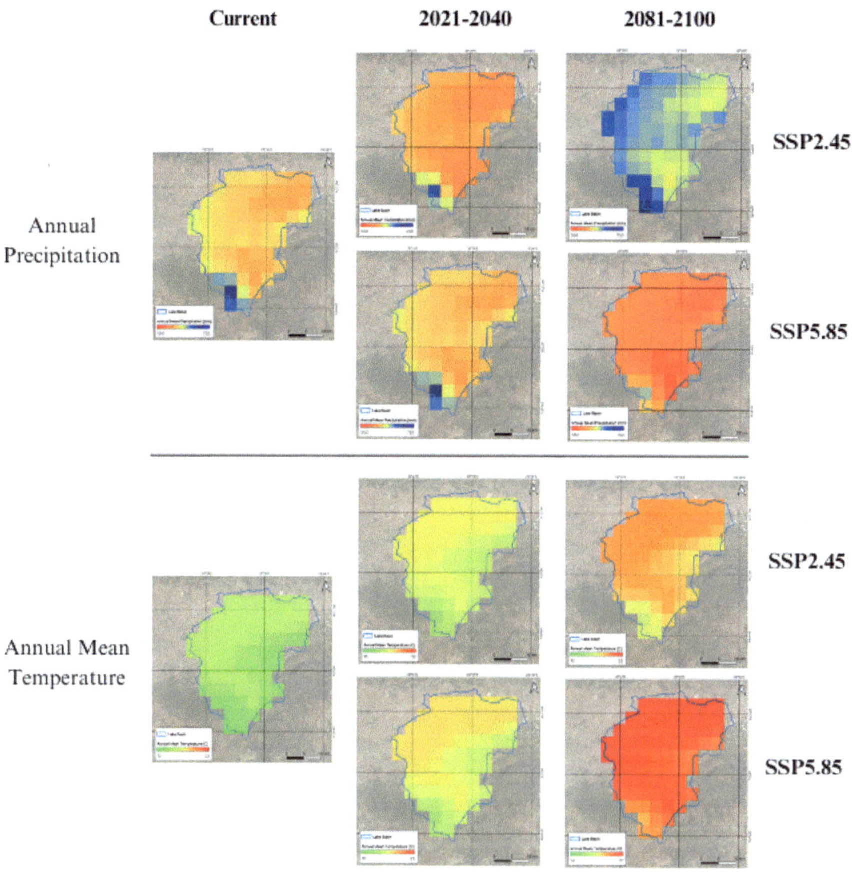

Fig. 21 Precipitation and temperature projections for Uluabat Lake based on SSP2.45 and SSP5.85 for the periods 2021–2040 and 2081–2100

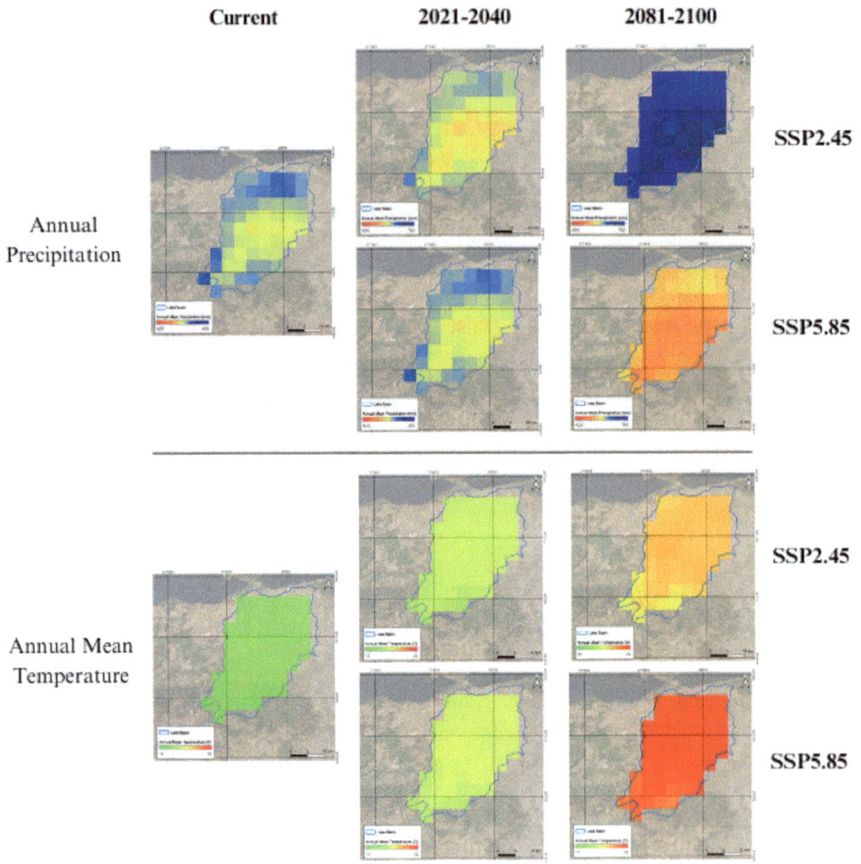

Fig. 22 Precipitation and temperature projections for Manyas Lake based on SSP2.45 and SSP5.85 for the periods 2021–2040 and 2081–2100

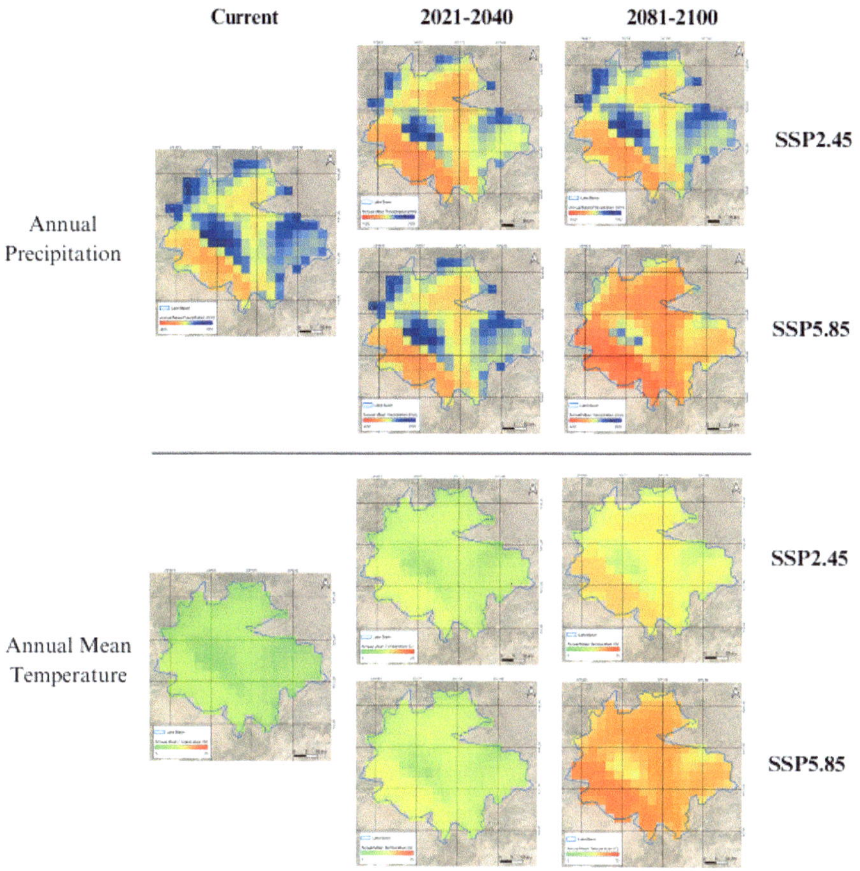

Fig. 23 Precipitation and temperature projections for Işıklı Lake based on SSP2.45 and SSP5.85 for the periods 2021–2040 and 2081–2100

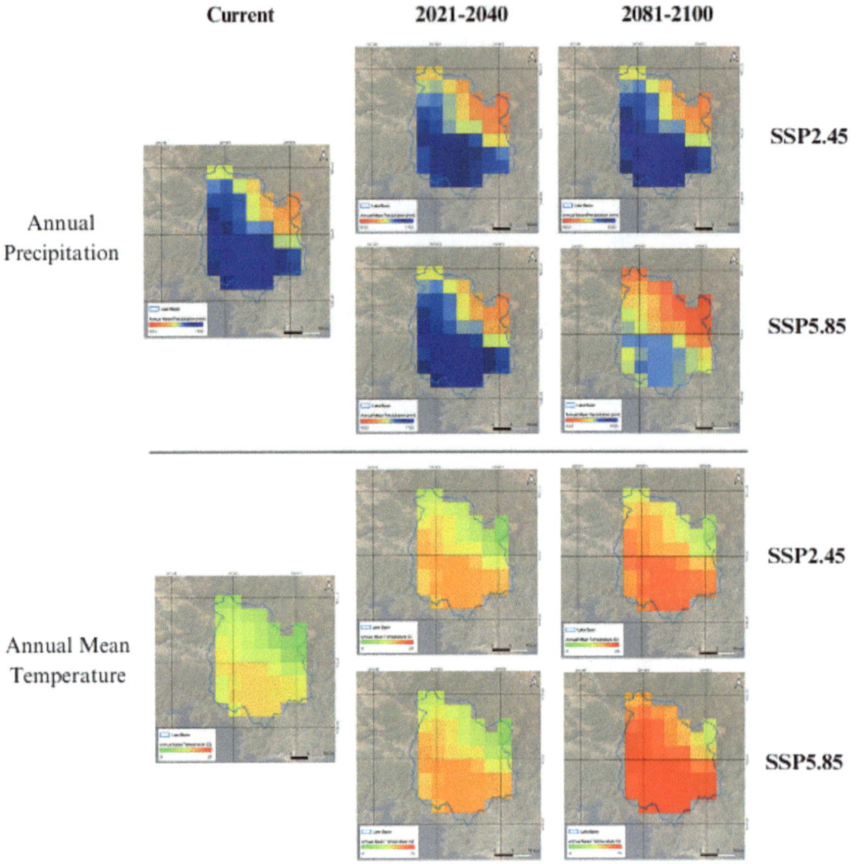

Fig. 24 Precipitation and temperature projections for Köyceğiz Lake based on SSP2.45 and SSP5.85 for the periods 2021–2040 and 2081–2100

References

Akbaba G, Boyacı YÖ (2015) The Seasonal Change of Macrobenthic Fauna in Işıklı Lake (Denizli). Süleyman Demirel Üniv Eğirdir Su Ürün Fak Derg 11(2):8–19 (in Turkish)

Akkoyunlu A (2003) Evaluation of eutrophication process in Lake İznik. Fresen Environ Bull 12:801–807

Aksever F, Büyükşahin S (2017) Assessment of variations in water quality using statistical techniques: a case study of Işıklı Lake, Çivril/Denizli, Turkey. Arab J Geosci 10(143). https://doi.org/10.1007/s12517-017-2877-4

Akşıray F (1982) Environmental effects on restricted distribution of pearl mullet, *Chalcalburnus tarichi* (Pall., 1881) produced in Lake Burdur. Atatürk Üniversitesi Fen Fakültesi Dergisi, Biyoloji Kongresi Tebliğleri, vol 1(1), pp 20–28

Algül F, Beyhan M (2020) Concentrations and sources of heavy metals in shallow sediments in Lake Bafa, Turkey. Sci Rep 10(1):11782. https://doi.org/10.1038/s41598-020-68833-2

Altınsaçlı S, Griffiths HI (2001) Ostracoda (Crustacea) of Lake Uluabat (Apolyont Gölü) (Bursa Province, Turkey). Limnologica 31(2):109–117

Andrews MB, Ridley JK, Wood RA, Andrews T, Blockley EW, Booth B, Sutton RT et al (2020) Historical simulations with HadGEM3-GC3.1 for CMIP6. J Adv Model Earth Syst 12(6):e2019MS001995

Arı Y (2019) Fisheries in Lake Manyas (Balıkesir): culture, ecology and sustainability. Aegean Geogr J 28(2):179–198 (in Turkish)

Arslan Ş, Avşar Ö (2020) Assessment of heavy metal pollution in Köyceğiz-Dalyan coastal lagoon watershed (Muğla) SW Turkey. Arab J Geosci 13(15):719. https://doi.org/10.1007/s12517-020-05690-3

Atalay I (2003) Effects of the tectonic movements on the karstification in Anatolia, Turkey. Acta Carsolog 32(2). https://doi.org/10.3986/ac.v32i2.348

Atasoy Aytış E, Soyer E, Ertürk A (2023) Assessing eutrophication vulnerability in Köyceğiz Lake: climate change and basin developments impact trophic status. Aquat Res 6(3):201–226. https://doi.org/10.3153/AR23020

Ataol M (2010) The water level changings in Burdur Lake. Coğrafi Bilimler Dergisi 8(1):77–92 (in Turkish)

Avşar Ö, Avşar U, Arslan Ş, Kurtuluş B, Niedermann S, Güleç N (2017) Subaqueous hot springs in Köyceğiz Lake, Dalyan Channel and Fethiye-Göcek Bay (SW Turkey): locations, chemistry and origins. J Volcanol Geoth Res 345:81–97. https://doi.org/10.1016/j.jvolgeores.2017.07.016

Ayaz S, Aktaş Ö, Atasoy Aytış E, Dağli S, Akca L (2012) Point and diffuse sources of pollution and surface water quality in Marmara Basin of Turkey. In: 1st century watershed technology: improving water quality and the environment conference proceedings, 27 May–1 June 2012, Bari, Italy 12–13665. https://doi.org/10.13031/2013.41407

Balaban C (2010) The Fish Fauna of manyas bird lake and certain biological features of the species. Master's thesis. Balıkesir University, Institute of Science, Department of Biology, 157p (in Turkish)

Balık S, Ustaoğlu MR, Sarı HM, İlhan A, Topkara ET (2005) The fish fauna of Yuvarlakçay (Köyceğiz, Muğla). Ege J Fish Aquat Sci 22(1–2):221–223

Berber S, Şaşı H, Topkara E, Cengiz Ö (2011) The Fish Fauna of Apolyont Lake (Bursa). Aquat Sci Eng 26(1):27–55 (in Turkish)

Beyhan M, Kaçıkoç M (2014) Evaluation of water quality from the perspective of eutrophication in Lake Eğirdir, Turkey. Water Air Soil Pollut 225(7):1994. https://doi.org/10.1007/s11270-014-1994-x

Blondel J, Aronson J, Bodiou J-Y, Boeuf G (2010) The Mediterranean region: biological diversity in space and time, 2nd edn. Oxford University Press

Bohn VY, Piccolo MC (2018) Standardized precipitation evapotranspiration index (SPEI) as a tool to determine the hydrological dynamic of plain regions (Argentina). Geosciences 37(3), Article 3. https://doi.org/10.5016/geociencias.v37i3.12236

Brückner H, Herda A, Kerschner M, Müllenhoff M, Stock F (2017) Life cycle of estuarine islands—from the formation to the landlocking of former islands in the environs of Miletos and Ephesos in western Asia Minor (Turkey). J Archaeol Sci Rep 12:876–894. https://doi.org/10.1016/j.jasrep.2016.11.024

Bulut C (2023) Heavy metal accumulation in aquatic macrophytes and plankton, water-sediment in Lake Iznik. Turk J Fish Aquat Sci 24(2). https://doi.org/10.4194/TRJFAS24469

Bulut C, Kubilay A (2018) The determination with Trophic State indices of water quality in Egirdir Lake. SDU Fac Arts Sci J Soc Sci 14(4):324–338. https://doi.org/10.22392/-egirdir.415073 (in Turkish)

Bulut E, Aksoy A (2008) Impact of fertilizer usage on phosphorus loads to Lake Uluabat. Desalination 226(1–3):289–297. https://doi.org/10.1016/j.desal.2007.02.112

Çaldırak H, Kurtulus B, Canoglu M, Tunca E (2017) Assessment of heavy metal contamination and accumulation patterns in the coastal and deep sediments of Lake Salda, Turkey. Fresen Environ Bull 25:8047–8061

Ceylan S, Bulut İ (2019) Tourism pressure, protection and sustainability at Salda Lake which is special environmental protection area. Turk Geogr Rev 73:79–89. https://doi.org/10.17211/tcd.637091 (in Turkish)

Çelik MA, Gülersoy AE (2013) Investigation of effects of land use activities around Işıklı Lake (Denizli, West Anatolia) on the Lake. SDU Fac Arts Sci J Soc Sci 2013(29):191–200

Çiçek E, Sungur S, Fricke R, Secer B (2023) Freshwater lampreys and fishes of Türkiye; an annotated checklist, 2023. Turk J Zool 47(6):324–468

Çınar Ş, Küçükkara R, Balık İ, Çubuk H, Ceylan M, Erol KG, Bulut C et al (2013) Detection of Fish Fauna, species composition and distribution of commercial fishing according to species in Lake Uluabat. J Fish Sci 7(4):309–316. https://doi.org/10.3153/jfscom.2013034 (in Turkish)

Çolak MA, Öztaş B, Özgencil İK, Soyluer M, Korkmaz M, Ramírez-García A, Akyürek Z et al (2022) Increased water abstraction and climate change have substantial effect on morphometry, salinity, and biotic communities in lakes: examples from the semi-arid Burdur Basin (Turkey). Water 14(8). https://doi.org/10.3390/w14081241

Davraz A, Aksever F (2023) Evaluation of Salda Lake hydrology parameters. AKU J Sci Eng 23(4):1029–1044. https://doi.org/10.35414/akufemubid.1233689 (in Turkish)

Davraz A, Sener E, Sener S (2019a) Evaluation of climate and human effects on the hydrology and water quality of Burdur Lake, Turkey. J Afr Earth Sci 158:103569. https://doi.org/10.1016/j.jafrearsci.-2019.103569

Davraz A, Varol S, Sener E, Sener S, Aksever F, Kırkan B, Tokgözlü A (2019b) Assessment of water quality and hydrogeochemical processes of Salda alkaline lake (Burdur, Turkey). Environ Monit Assess 191(11):701. https://doi.org/10.1007/s10661-019-7889-y

Demir H (2015) Investigation of Fish Fauna and bio-ecological characteristic economical species in Bafa Lake. Muğla Sıtkı Koçman University, Graduate School of Natural and Applied Sciences Department of Fishery Sciences. 146p (in Turkish)

Di Dario F (2018) *Clupeonella muhlisi* (errata version published in 2019). In: The IUCN red list of threatened species, p e.T98471539A143839033. https://doi.org/10.2305/IUCN.UK.2018-2.RLTS.T98471539A143839033.en

Dorak Z, Koker L, Saglam O, Akcaalan R, Albay M (2017) Determination of zooplankton community structure, biomass and trophic state of a shallow turbid lake. Fresenius Environ Bull 26(1-A):834–845

Eken G, Isfendiyaroğlu S, Bozdoğan M (2006) Türkiye'nin önemli doğa alanları. Doğa Derneği, Ankara, Turkey

Elmaci A, Ozengin N, Teksoy A, Topac FO, Baskaya HS (2009) Evaluation of trophic state of Lake Uluabat, Turkey. J Environ Biol 30(5):757–760

Erdoğan S (2011) A chemical reaction to a physical impact: Lake Bafa wetland ecosystem (Turkey) case. Ankara Üniv Çevrebilim Derg 3(1):001–008. https://doi.org/10.1501/Csaum_0000000040

Erkakan F, Atalay-Ekmekçi FG, Nalbant TT (1999) A review of the genus *Cobitis* in Turkey (Pisces: Ostariophysi: Cobitidae). Hydrobiologia 403:13–26

Erkmen B, Yerli SV, Erk'akan F, Kolankaya D (2013) Persistent organochlorine pesticide residues in water and sediment samples from Lake Manyas Turkey. J Environ Biol 34:171–176

European Environment Agency (2020a) CORINE land cover 1990 (raster 100 m), Europe, 6-yearly—version 2020_20u1, May 2020. Copernicus Land Monitoring Service. https://land.copernicus.eu/pan-european/corine-land-cover/clc1990

European Environment Agency (2020b) CORINE land cover 2018 (raster 100 m), Europe, 6-yearly—version 2020_20u1, May 2020. Copernicus Land Monitoring Service. https://land.copernicus.eu/pan-european/corine-land-cover/clc2018

Eyring V, Bony S, Meehl GA, Senior CA, Stevens B, Stouffer RJ, Taylor KE (2016) Overview of the coupled model intercomparison project phase 6 (CMIP6) experimental design and organization. Geosci Model Dev 9(5):1937–1958. https://doi.org/10.5194/gmd-9-1937-2016

Feyisa GL, Meilby H, Fensholt R, Proud SR (2014) Automated water extraction index: a new technique for surface water mapping using Landsat imagery. Remote Sens Environ 140:23–35

Fick SE, Hijmans RJ (2017) WorldClim 2: new 1-km spatial resolution climate surfaces for global land areas. Int J Climatol 37(12):4302–4315. https://doi.org/10.1002/joc.5086

Freyhof J (2014a) *Alburnus nicaeensis*. In: The IUCN red list of threatened species, p e.T19018670A19222798. https://doi.org/10.2305/IUCN.-UK.2014-1.RLTS.T19018670A19222798.en

Freyhof J (2014b) *Pseudophoxinus handlirschi*. In: The IUCN red list of threatened species, p e.T40747A19007061. https://doi.org/10.2305/IUCN-.UK.2014-1.RLTS.T40747A19007061.en

Freyhof J (2014c) *Chondrostoma meandrense*. In: The IUCN red list of threatened species, p e.T61191A19009010. https://doi.org/10.2305/-IUCN.UK.2014-1.RLTS.T61191A19009010.en

Freyhof J (2014d) *Vimba mirabilis*. In: The IUCN red list of threatened species, p e.T19452809A19848824. https://doi.org/10.2305/IUCN.UK.2014-1.RLTS.T19452809A198488

Freyhof J (2014e) *Aphanius transgrediens*. In: The IUCN red list of threatened species, p e.T1850A19006201. https://doi.org/10.2305/IUCN.-UK.2014-1.RLTS.T1850A19006201.en

Freyhof J (2014f) *Pseudophoxinus egridiri*. In: The IUCN red list of threatened species, p e.T39281A19006956. https://doi.org/10.2305/IUCN.-UK.2014-1.RLTS.T39281A19006956.en

Freyhof J (2014g) *Seminemacheilus ispartensis*. In: The IUCN red list of threatened species, p e.T19451077A19848348. https://doi.org/10.2305/-IUCN.UK.2014-1.RLTS.T19451077A19848348.en

Freyhof J (2014h) *Cobitis phrygica*. In: The IUCN red list of threatened species, p e.T19085501A19222988. https://doi.org/10.2305/IUCN.UK.2014-1.RLTS.T19085501A19222988.en

Freyhof J (2014i) *Alburnus carinatus*. In: The IUCN red list of threatened species, p e.T19018407A19222763. https://doi.org/10.2305/IUCN.-UK.2014-1.RLTS.T19018407A19222763.en

Freyhof J (2014j) *Pseudophoxinus burduricus*. In: The IUCN red list of threatened species, p e.T54126285A54128761. https://doi.org/10.2305/-IUCN.UK.2014-1.RLTS.T54126285A54128761.en

Freyhof J (2014k) *Barbus niluferensis*. The IUCN red list of threatened species, p e.T19021245A19222823. https://doi.org/10.2305/IUCN.-UK.2014-1.RLTS.T19021245A19222823.en

Freyhof, J. (2014l). *Aphanius anatoliae*. In: The IUCN red list of threatened species, p e.T1843A19005979. https://doi.org/10.2305/IUCN.UK.2014-1.RLTS.T1843A19005979.en

Freyhof J (2014m) *Gobio maeandricus*. In: The IUCN red list of threatened species, p e.T19088403A19223103. https://doi.org/10.2305/IUCN.-UK.2014-1.RLTS.T19088403A19223103.en

Freyhof J (2014n) *Squalius carinus*. In: The IUCN red list of threatened species, p e.T19451245A19849535. https://doi.org/10.2305/IUCN.UK.2014-1.RLTS.T19451245A19849535.en

Freyhof J, Bergner L, Ford M (2020) Threatened freshwater fishes of the Mediterranean Basin Biodiversity hotspot: distribution, extinction risk and the impact of hydropower. EuroNatur and RiverWatch, vol i–viii, pp 1–348

Fricke R, Bilecenoglu M, Sari HM (2007) Annotated checklist of fish and lamprey species (Gnathostomata and Petromyzontomorphi) of Turkey, including a Red List of threatened and declining species. Stuttg Beitr Z Natur Ser A (Biol) 706:1–169

Fricko O, Lutz W, Popp A, Cuaresma JC, Kc S, Leimbach M, Jiang L, Kram T, Rao S, Emmerling J, Tavoni M (2017) The shared socioeconomic pathways and their energy, land use, and greenhouse gas emissions implications: an overview. Glob Environ Change 42:153–168. https://doi.org/10.1016/j.gloenvcha.2016.05.009

García Álvarez D, Camacho Olmedo MT (2023) Analysing the inconsistencies of CORINE status layers (CLC) and layers of changes (CHA) (1990–2018) for a Spanish case study. Ann GIS 29(3):369–386. https://doi.org/10.1080/19475683.2023.2166583

Gaygusuz Ö, Tarkan AS, Gaygusuz ÇG (2007) Changes in the fish community of the Ömerli Reservoir (Turkey) following the introduction of non-native gibel carp *Carassius gibelio* (Bloch, 1782) and other human impacts. Aquat Invas 2(2):117–120

Genç TO, Yılmaz F (2023) Distribution and taxonomic status of freshwater fishes of Muğla (South-West Türkiye): updated checklist. Mugla J Sci Technol 9(1):71–79

Geven F, Ozdeniz E, Kurt L, Bolukbasi A, Ozbey BG, Ozcan AU, Turan U (2016) Habitat classification and evaluation of the Köyceğiz-Dalyan special protected area (Muğla/Turkey). Rend Lincei 27(3):509–519. https://doi.org/10.1007/s12210-016-0510-1

Giannetto D, Innal D (2021) Status of endemic freshwater fish fauna inhabiting major lakes of Turkey under the threats of climate change and anthropogenic disturbances: a review. Water 13(11):1534. https://doi.org/10.3390/w13111534

Giorgi F (2006) Climate change hot-spots. Geophys Res Lett 33(8). https://doi.org/10.1029/2006GL025734

Girgin S, Kazancı N, Dügel M (2004) On the limnology of deep and saline Lake Burdur in Turkey. Acta Hydrochim Hydrobiol 32(3):189–200. https://doi.org/10.1002/aheh.200400529

Gorelick N, Hancher M, Dixon M, Ilyushchenko S, Thau D, Moore R (2017) Google earth engine: planetary-scale geospatial analysis for everyone. Remote Sens Environ 202:18–27

Güçlü SS, Küçük F (2012) Two threatened endemic fish species of the world: *Aphanius splendens* and *Aphanius transgrediens* (Cyprinodontidae), from Turkey. Biolog Diver Conserv 5(3):44–47

Gülcü-Gür B (2017) The investigation of heavy metal levels in water and sediment from Işikli Lake (Turkey) in relation to seasons and physico-chemical parameters. J Aquac Eng Fish Res 3(2):87–96. https://doi.org/10.3153/JAEFR17012

Gülle İ, Küçük F, İnnal D, Güçlü SS (2016) Fish of Burdur province: biodiversity inventory, population and Habitat Status. Mehmet Akif Ersoy Univ J Sci Inst 7(Suppl. 1):232–239

Gunacti MC, Gul GO, Cetinkaya CP, Gul A, Barbaros F (2023) Evaluating impact of land use and land cover change under climate change on the Lake Marmara system. Water Resour Manage 37(6):2643–2656. https://doi.org/10.1007/s11269-022-03317-8

Gündoğan İ, Mordoğan H, Helvacı C (1995) Türkiye'deki Acı Güllerden Sodyum Sülfat Üretimi. Paper presented at the Endüstriyel Hammaddeler Sempozyumu, İzmir, Türkiye, 21–22 Apr 1995 (in Turkish)

Hampel FR (1974) The influence curve and its role in robust estimation. J Am Stat Assoc 69(346):383–393. https://doi.org/10.2307/2285666

Hanedar A, Tanik A (2023) Presenting water quality characteristics of Lake Salda, Turkey. Eur J Eng Appl Sci 6(1):41–49. https://doi.org/10.55581/ejeas.1250016

Hersbach H, Bell B, Berrisford P, Hirahara S, Horányi A, Muñoz-Sabater J, Thépaut J-N et al (2020) The ERA5 global reanalysis. Q J R Meteorol Soc 146(730):1999–2049

Hirsch RM, Slack JR, Smith RA (1982) Techniques of trend analysis for monthly water quality data. Water Resour Res 18(1):107–121

İleri S, Karaer F, Kâtip A, Onur S (2014) Sığ göllerde su kalitesi değerlendirmesi, Uluabat Gölü örneği. Uludağ Üniv Mühend Fak Derg 19(1):47–57 (in Turkish)

IUCN (2016) A global standard for the identification of key biodiversity areas, version 1.0. IUCN, Gland, Switzerland

IUCN (2024) The IUCN red list of threatened species (version 2024–2). Retrieved 2 Jan 2025, from https://www.iucnredlist.org

Karagüzel R, Taşdelen S, Akyol E (1995) An analysis of the level fluctuations of Egirdir Lake SW Turkey. Paper presented at the international symposium on water resources management. İzmir, Turkey, pp 701–710

Karafistan A, Arik-Çolakoğlu F (2005) Physical, chemical and microbiological water quality of the Manyas Lake, Turkey. Mitig Adapt Strat Glob Change 10:127–143

Karaman M, Budakoglu M, Tasdelen S (2017) Nitrate contamination in the groundwater of the Lake Acıgöl Basin, SW Turkey. Paper presented at the 19th EGU general assembly, Vienna, Austria, 23–28 Apr 2017, p 17257

Karaman M (2021) Comparison of thresholding methods for shoreline extraction from Sentinel-2 and Landsat-8 imagery: extreme Lake Salda, track of Mars on Earth. J Environ Manage 298:113481

Karami F, Balci N, Guven B (2019) A modeling approach for calcium carbonate precipitation in a hypersaline environment: a case study from a shallow, alkaline lake. Ecol Complex 39:100774. https://doi.org/10.1016/-j.ecocom.2019.100774

Karra K, Kontgis C, Statman-Weil Z, Mazzariello JC, Mathis M, Brumby SP (2021) Global land use/land cover with Sentinel 2 and deep learning. Paper presented at the 2021 IEEE international geoscience and remote sensing symposium IGARSS, pp 4704–4707

Kasparek M (1988) Bafa see: nature and history in the Turkish Aegean. Heidelberg

Katip A, Ileri S, Karaer F, Onur S (2015) Determination of the trophic state of Lake Uluabat (Bursa-Turkey). Ekoloji 24(97):24–35. https://doi.org/10.5053/ekoloji.2015.07

Katip A, Karaer F, Başkaya HS, Ileri S, Sarmaşik S (2012) Fraction distribution and risk assessment of heavy metals and trace elements in sediments of Lake Uluabat. Environ Monit Assess 184(9):5399–5413. https://doi.org/10.1007/s10661-011-2348-4

Kazanci N, Dügel M (2000) An evaluation of the water quality of Yuvarlakçay Stream, in the Köycegiz-Dalyan protected area, SW Turkey. Turk J Zool 24(1):69–80

Kazanci N, Girgin S, Dügel M (2004) On the limnology of Salda Lake, a large and deep soda lake in southwestern Turkey: future management proposals. Aquat Conserv Mar Freshwat Ecosyst 14(2):151–162

Kazanci N, Plasa R-H, Neubert E, İzbirak A (1992) On the limnology of Lake Köycegiz (SW Anatolia). Zool Middle East 6(1):109–126. https://doi.org/10.1080/09397140.1992.10637619

KBAP (Key Biodiversity Areas Partnership) (2025) World database of key biodiversity areas. Retrieved from https://keybiodiversityareas.org/

Kesici E, Kesici C (2006) Eğirdir Gölü (Isparta)'nın doğal yapısına yapılan müdahalelerin gölün ekolojik yapısına etkileri. Ege J Fish Aquat Sci 23(1/1):99–103 (in Turkish)

Kesmez CÖ, Dalkiran N (2023) Evaluation of Lake Uluabat (Bursa) surface water quality using multivariate statistical methods. Acta Aquat Turc 19(4):341–357. https://doi.org/10.22392/actaquatr.1267921 (in Turkish)

Kılıç M, Kılıç S, Pamukoğlu Y, Kırkan B, Erkaymaz T (2023) Element, pesticide and ion pollutants determination and method validation in Eğirdir Lake/Turkey. J Eng Sci Des 11(3):1154–1166. https://doi.org/10.21923/jesd.1185333

Koçak F, Aydın-Önen S, Açık Ş, Küçüksezgin F (2017) Seasonal and spatial changes in water and sediment quality variables in Bafa Lake. Environ Earth Sci 76(17):617. https://doi.org/10.1007/s12665-017-6950-9

Köker L (2022) Health risk assessment of heavy metal concentrations in selected fish species from İznik Lake Basin, Turkey. Environ Monit Assess 194(5):372. https://doi.org/10.1007/s10661-022-10046-3

Korkmaz M (2017) The Investigation of geographic variations of *Barbus* spp. (Pisces: Cyprinidae) distributed in Turkey. Ph.D. thesis. Hacettepe University, Department of Biology, 107p (in Turkish)

Korkmaz M, Mangıt F, Yerli S (2017) Main growth parameters of a threatened species Chondrostoma holmwoodii (Boulenger, 1896) from Tahtalı Reservoir, İzmir, Turkey. J Limnol Freshw Fish Res 3

Korkmaz M, Mangıt F, Dumlupınar İ, Çolak MA, Akpınar MB, Koru M, Pacheco JP, Ramírez-García A, Yılmaz G, Amorim CA, Özgencil İK, İnnal D, Yerli SV, Özkan K, Akyürek Z, Beklioğlu M, Jeppesen E (2023) Effects of climate change on the habitat suitability and distribution of endemic freshwater fish species in semi-arid central anatolian ecoregion in Türkiye. Water 15(8), Article 8. https://doi.org/10.3390/w15081619

Kosswig C, Geldiay R (1952) Fish of Lake Eğirdir. Balık ve Balıkçılık, İst Üniv Fen Fak Hidrobiyoloji Arş Enst Yayınları 3(1):3–14

Küçük F (2012) Extinct endemic fishes of Turkey: *Alburnus akili* (Gövce) and *Pseudophoxinus handlirschi* (Kavinne) (Pisces: Cyprinidae). Turk J Fish Aquat Sci 12(2):345–347. https://doi.org/10.4194/1303-2712-v12_2_21

Küçük F, Gülle İ, Güçlü SS, Çiftçi Y, Erdoğan Ö (2013) A new *Pseudophoxinus* (Teleostei, Cyprinidae) species from southwestern Anatolia, with remarks on the distribution of the genus in western Anatolia. ZooKeys 320:29–41

Küçük F, Sari HM, Demir O, Gülle İ (2009) Review of the ichthyofaunal changes in Lake Eğirdir between 1915 and 2007. Turk J Zool 33(3):277–286

Kurtul I, Kaya C, Kaykaç H, Ilhan A, Düzbastilar FO, Tosunoğlu Z, Haubrock PJ et al (2024) How fish populations in Lake Bafa (Western Anatolia) respond to ecological shifts. Aquat Conserv Mar Freshwat Ecosyst 34(5):e4154. https://doi.org/10.1002/aqc.4154

Lehner B, Grill G (2013) Global river hydrography and network routing: Baseline data and new approaches to study the world's large river systems. Hydrol Process 27(15):2171–2186

Leroy S, Kazancı N, İleri Ö, Kibar M, Emre O, McGee E, Griffiths HI (2002) Abrupt environmental changes within a late Holocene lacustrine sequence south of the Marmara Sea (Lake Manyas, N-W Turkey): possible links with seismic events. Mar Geol 190(1):531–552. https://doi.org/10.1016/-S0025-3227(02)00361-4

Lionello P, Scarascia L (2018) The relation between climate change in the Mediterranean region and global warming. Reg Environ Change 18(5):1481–1493. https://doi.org/10.1007/s10113-018-1290-1

Lowe S, Browne M, Boudjelas S, De Poorter M (2000) 100 of the world's worst invasive alien species: a selection from the global invasive species database, vol 12. Invasive Species Specialist Group, Auckland, p 12

Mangit F, Korkmaz M, Yerli SV (2018) Morphological variation of pumpkinseed (*Lepomis gibbosus*) with emphasis on allometry. Turk J Zool 42(1):53–61

Masson-Delmotte VP, Zhai P, Pirani SL, Connors C, Péan S, Berger N, Scheel Monteiro PM et al (2021) IPCC, 2021: summary for policymakers. In: Climate change 2021: the physical science basis. Contribution of working group I to the sixth assessment report of the intergovernmental panel on climate change

Médail F, Quézel P (1999) Biodiversity hotspots in the Mediterranean Basin: setting global conservation priorities. Conserv Biol 13(6):1510–1513

Mercan D, Arslan N (2023) Use of biomonitoring tools to detect water quality-dependent ecosystem (macroinvertebrate) responses in lentic systems: the examples of Lakes Iznik and Manyas, Türkiye. Oceanol Hydrobiol Stud 52(1):79–89. https://doi.org/10.26881/oahs-2023.1.06

MoAF (Ministry of Agriculture and Forestry) (2025) Nature tourism web application. Retrieved from https://ekotaban.tarimorman.gov.tr/

MoEUCC (Ministry of Environment, Urbanization and Climate Change) (2017) Bursa Province 2017 Environmental Status Report. Retrieved from https://webdosya.csb.gov.tr/db/ced/icerikler/-2017_bursa_cdr_son-20180913120048.pdf

Murathan A, Varlı D, Göktaş MA, Kuzucu A (2023) Relationship between groundwater and surface water in Marmara Lake. Paper presented at the HİDRO'2023 symposium, KTÜ, Trabzon, 25–27 Mayıs 2023 (in Turkish)

Mutlu H, Kadir S, Akbulut A (1999) Mineralogy and water chemistry of the Lake Acıgöl (Denizli), Turkey. Carbonat Evapor 14:191–199

Oruçoğlu K, Beyhan M (2019) Assessment of heavy metal pollution in the Lakes of Lakes District—Turkey. BilgeSCI 3(1):10–20. https://doi.org/10.30516/bilgesci.449984 (in Turkish)

Özbayram EG, Koker L, Akçaalan R, Aydın F, Ertürk A, Ince O, Albay M (2021) Contrasting the water quality and bacterial community patterns in shallow and deep lakes: Manyas vs. Iznik. Environ Manag 67(3):506–512. https://doi.org/10.1007/s00267-020-01357-7

Özdemir N, Demirak A, Gadirova E, Koç C, Temizer İA (2019) Investigation of some physico-chemical properties of Köyceğiz, Bafa, and Limni Lakes (Muğla-Turkey). Eur J Sci Technol 17:338–345. https://doi.org/10.31590/ejosat.587665

Özseven A, Akkurt I, Günoğlu K (2020) Determination of some dosimetric parameters in Eğirdir Lake, Isparta, Turkey. Int J Environ Sci Technol 17:1503–1510

Özuluğ M, Altun Ö, Meriç N (2005) On the fish fauna of Lake İznik (Turkey). Turk J Zool 29(4):371–375

Pyke GH (2008) Plague minnow or mosquito fish? A review of the biology and impacts of introduced *Gambusia* species. Annu Rev Ecol Evol Syst 39(1):171–191

Ribeiro F, Leunda PM (2012) Non-native fish impacts on Mediterranean freshwater ecosystems: current knowledge and research needs. Fish Manage Ecol 19(2):142–156

Saç G, Gaygusuz Ö, Ertürk A (2020) Invasive fish threat in Manyas Lake (Turkey). Biolog Divers Conserv 13(2):115–120

Sarı HM, Balık S, Bilecenoğlu M, Türe G (1999) Recent changes in the fish fauna of Lake Bafa, Aegean region of Turkey. Zool Middle East 18:67–76. https://doi.org/10.1080/09397140.1999.10637783

Sarı HM, Ilhan A, Balık S (2008) Some biological characteristics of *Hemigrammocapoeta kemali* (Hanko, 1924) in Işıklı Lake (Denizli-Turkey). J Appl Ichthyol 24(1):88–90

Sarmaşık SS (2012) Hydrodynamic modelling of Lake Uluabat. Master's thesis. Uludağ Üniversitesi, Turkey (in Turkish)

Şener Ş, Şener E, Bulut C (2023) Appraisal of heavy metal contents, spatial–temporal variation, toxic metal pollution, and health risk in water and sediment of Uluabat Lake (Ramsar Site, Turkey). Environ Sci Pollut Res 30(54):115246–115265. https://doi.org/10.1007/s11356-023-30490-z

Stumpp C, Ekdal A, Gönenc IE, Maloszewski P (2014) Hydrological dynamics of water sources in a Mediterranean lagoon. Hydrol Earth Syst Sci 18(12):4825–4837. https://doi.org/10.5194/hess-18-4825-2014

Sukatar A, Ertaş A, Akgül R, Tüney Kızılkaya İ (2021) Assessment of the ecological and trophic status of Lake Bafa (Turkey) based on phytoplankton. Ege J Fish Aquat Sci 38(2):135–146. https://doi.org/10.12714/egejfas.38.2.01

Şener E (2016) Determination of seasonal changes in Burdur Lake surface water temperature using Landsat 8 Satellite images. J Eng Sci des 4(2):67–73

Tarkan AS, Gaygusuz Ö, Gürsoy Gaygusuz Ç, Sac G, Copp GH (2012) Circumstantial evidence of gibel carp, *Carassius gibelio*, reproductive competition exerted on native fish species in a mesotrophic reservoir. Fish Manage Ecol 19(2):167–177

Tarkan AS, Marr SM, Ekmekçi FG (2015) Non-native and translocated freshwater fish. FiSHMED Fish Mediter Environ 3:1–28

Taşeli BK (2009) Response of lake water quality to wastewater inputs from land-based fish farm located on Yuvarlakçay Creek in Köyceğiz-Dalyan specially protected area, Turkey. Environ Monitor Assess 157(1–4):557–574. https://doi.org/10.1007/s10661-008-0555-4

Temiz F, Durduran SS (2016) Monitoring coastline change using remote sensing and GIS technology: a case study of Acıgöl Lake, Turkey. IOP Conf Ser Earth Environ Sci 44:042033. https://doi.org/10.1088/1755-1315/44/4/042033

Tuygun GT, Salgut S, Elçi A (2023) Long-term spatial-temporal monitoring of eutrophication in Lake Burdur using remote sensing data. Water Sci Technol 87(9):2184–2194

Uca Avci ZD, Karaman M, Ozelkan E, Kumral M, Budakoglu M (2014) OBIA based hierarchical image classification for industrial lake water. Sci Total Environ 487:565–573. https://doi.org/10.1016/j.scitotenv.2014.04.048

Uysal R, Yağcı MA, Yeğen V, Alp A, Yağcı A (2008) Işıklı Gölü'ndeki (Çivril-Denizli) Turna (*Esox lucius* L., 1758) Populasyonunun Büyüme Özellikleri. E.U. J Fish Aquat Sci 25(4):259–265 (in Turkish)

Üzücek, N. (2005). The first inflatable rubber dam in Turkey for protecting and restoring the aquatic environment. In: Proceedings of the international conference on water resources and environmental research, pp 2183–2187

Varol S, Davraz A, Şener Ş, Şener E, Aksever F, Kırkan B, Tokgözlü A (2021) Assessment of groundwater quality and usability of Salda Lake Basin (Burdur/Turkey) and health risk related to arsenic pollution. J Environ Health Sci Eng 19(1):681–706. https://doi.org/10.1007/s40201-021-00638-5

Vicente-Serrano SM, Beguería S, López-Moreno JI (2010) A multiscalar drought index sensitive to global warming: the standardized precipitation evapotranspiration index. J Clim 23(7):1696–1718. https://doi.org/10.1175/2009JCLI2909.1

Weissova B, Polla S (2019) The hinterland of Nikaia/Nicaea/Iznik: analyzing the hellenistic, roman, and late antique Bithynian landscape through remote sensing and GIS techniques. eTopoi J Anci Stud 8:21–49. https://doi.org/10.17169/refubium-25618

Wildekamp RH, Küçük F, Ünlüsayın M, Neer VW (1999) The genus Aphanius Nordo, 1827 (Pisces: Cyprinidontidae) and its species in Turkey with remarks on systematics. J Turk Zool 23:23–44

Xu H (2006) Modification of normalised difference water index (NDWI) to enhance open water features in remotely sensed imagery. Int J Remote Sens 27(14):3025–3033. https://doi.org/10.1080/01431160600589179

Yağcı MA, Alp A, Akın Ş, Yağcı A, Yeğen V, Uysal R, Cesur M et al (2021) The effects and interactions of three invasive fish species introduced to the aquatic ecosystem of a Turkish lake (Eğirdir Lake). Aquat Ecol 55(4):1273–1283

Yağcı MA, Alp A, Yağcı A, Uysal R, Bilgin F, Cesur M, Yegen V (2022) Length-weight relationships of ten fish species from Lake Eğirdir, Turkey. J Anat Environ Anim Sci 7(2):105–109

Yağcı MA, Yeğen R, Uysal R, Yağcı A, Cesur M, Bostan H, Çetinkaya S (2008) Fish fauna and fisheries of Lake İznik (Bursa-Turkey). Rev Hydrobiol 2:159–168 (in Turkish)

Yalçın Özdilek Ş, Partal N, Jones RI (2019) An invasive species, *Carassius gibelio*, alters the native fish community through trophic niche competition. Aquat Sci 81(2):29. https://doi.org/10.1007/s00027-019-0623-6

Ye X, Li X, Xu C-Y, Zhang Q (2016) Similarity, difference and correlation of meteorological and hydrological drought indices in a humid climate region—the Poyang Lake catchment in China. Hydrol Res 47(6):1211–1223. https://doi.org/10.2166/nh.2016.214

Yerli SV (1989) Investigations on economic fish populations in Köyceğiz Lagoon system. Ph.D. thesis. Hacettepe University Institute of Science and Technology, 267p (in Turkish)

Yerli SV, Alp A, Yeğen V, Uysal R, Apaydın Yağcı M, Balık İ (2013) Evaluation of the ecological and economical results of the introduced alien fish species in Lake Eğirdir, Turkey. Turk J Fish Aquat Sci 13(5):795–809. https://doi.org/10.4194/1303-2712-v13_5_03

Yılmaz F, Barlas M, Yorulmaz B, Özdemir N (2006) Muğla iç su balıkları üzerine taksonomik bir çalışma. Ege J Fish Aquat Sci 23(1–2):27–30

Yılmaz G, Çolak MA, Özgencil İK, Metin M, Korkmaz M, Ertuğrul S, Jeppesen E et al (2021) Decadal changes in size, salinity, waterbirds, and fish in lakes of the Konya Closed Basin, Turkey, associated with climate change and increasing water abstraction for agriculture. Inland Waters 11(4):538–555

Yoğurtçuoğlu B, Ekmekçi FG (2014) Conservation action for the Acıgöl Toothcarp, *Aphanius transgrediens*. Saving Freshwater Fishes and Habitats, Newsletter of the IUCN SSC/WI Freshwater Fish Specialist Group, vol 4, p 30

Yozukmaz A, Yabanlı M (2023) Heavy metal contamination and potential ecological risk assessment in sediments of Lake Bafa (Turkey). Sustainability 15(13):9969. https://doi.org/10.3390/su15139969

Lakes of Albania—Ecology, Biodiversity and Conservation Approaches

Lulëzim Shuka, Enkeleda Berberi, Maria Špoljar, and Spase Shumka

Abstract This chapter provides an analysis of status and trends in ecology, freshwater biodiversity and conservation within natural lakes of Albania using a combination of: (i) quantitative data; (ii) qualitative data from a literature review; and (iii) evidence from the peer-reviewed scientific literature. This contribution considers biodiversity of the Albanian standing freshwaters in the broadest sense, focusing specifically on Prespa, Ohrid and Shkodra Lake ecosystems. During the writing of the chapter, it became apparent that any conclusions drawn from the data would be necessarily limited due to a general lack of quantitative, nationally consistent, observational data for aquatic species and habitats of the large Balkan Lakes. For this reason, proxies, for example habitat trends, have been employed to examine likely trends in biodiversity that is highly affected by anthropogenic activities and climate changes. These trends are also explored further within the context of the ecoregion considerations. With this approach, we attempt to explore key trends in lake and other aquatic ecosystems, with coverage of the following areas: (i) native, rare and endemic species at risk; (ii) water oscillations; (iii) habitat loss and fragmentation; (iv) pollution, nutrients, and eutrophication; and (v) future climate implications. Finally, given to that fact large lakes in Albania are part of the protected area system, chapter addresses conservation needs and necessity for building proper resilience.

Keywords Biodiversity · Conservation · Endemic species · Alien and invasive · Resilience

L. Shuka
Faculty of Natural Sciences, University of Tirana, Tirana, Albania
e-mail: lulezim.shuka@fshn.edu.al

E. Berberi · S. Shumka (✉)
Faculty of Biotechnology and Food, Agricultural University of Tirana, Tirana, Albania
e-mail: sshumka@ubt.edu.al

E. Berberi
e-mail: eberberi@ubt.edu.al

M. Špoljar
Department of Biology, Faculty of Science, University of Zagreb, Zagreb, Croatia
e-mail: maria.spoljar@biol.pmf.hr

K. Stefanidis et al. (eds.), *Lakes of the Eastern Mediterranean and Balkans*,
https://doi.org/10.1007/978-3-032-02168-7_3

1 Introduction

At the global scale, biodiversity is in serious decline (Sayer et al. 2025), with freshwater ecosystems being particularly affected by different human interventions (Albert et al. 2021). After Sayer et al. (2025) on the basis of monitored natural inland wetlands (including peatlands, marshes, swamps, lakes, rivers and pools, among others), 35% of wetland area was lost between 1970 and 2015, at a rate three times faster than that of forests (Gardner and Finlayson 2018). Of the remaining wetland habitats, 65% are under moderate-to-high levels of threat and 37% of rivers over 1000 km are no longer free-flowing over their full length. Declines are continuing, generally out of sight and out of mind, despite the importance of the freshwater realm. Freshwaters support over 10% of all known species, including approximately one-third of vertebrates and one-half of fishes, while only covering less than 1% of the surface of the Earth (Strayer and Dudgeon 2010). In addition to supporting the culture, commerce, and lifestyles of billions of people worldwide, this diversity of freshwater species offers vital ecological services (such as flood control, nutrient cycling, and climate change mitigation) and can be utilized as bioindicators of wetland condition.

1.1 Biotic Components of Albanian Lakes, Objectives of the Chapter

The objective of this chapter is to provide an analysis of status and trends of freshwater bodies of Albania, their biodiversity within Albania's lakes using a combination of: (i) quantitative data; (ii) qualitative data from a literature review; (iii) evidence from the peer-reviewed scientific literature and (iv) authors experiences and provided scientific inputs within more than two decades. This chapter considers physical settings, particularities and biodiversity in the broadest sense, focusing specifically on natural lake ecosystems. Being located at the southwestern part of the Balkan Peninsula and influenced by Mediterranean and continental climate the large Balkan Lakes of Prespa, Ohrid and Skadar/Shkodra are at the crossroad of accelerated climate implications, so the other objective of this contribution is to provide key findings and opinions on conservation needs, adaptation and resilience.

All large lacustrine ecosystems of Albania are either crucial part of the protected areas system or stand at their buffer areas (Shuka et al. 2009a, b; Skarbøvik et al. 2010; Shumka et al. 2023a). It is essential to incorporate ecological sustainability within the context of a national ecological network of protected areas and ecosystem services (e.g., regulating, supporting, and cultural services), particularly in light of current developments. This is further complicated by the catchment continuum; communities are longitudinally linked but often fall under different legislative authorities and even across national borders (der Sleen and Albert 2021). The lack of effective planning and sector integration exacerbates pressures on the sustainability of fisheries linked to factors including inappropriate land use practices, water pollution, habitat modification and non-native species introductions, but also the overexploitation of fish resources.

This approachaims to explore key trends in natural lake ecosystems, with coverage of the following areas: (i) lakes challenges from different receptors and perspectives; (ii) habitats and biodiversity at risk; (iii) fluctuations in water level and hydrological regimes altered due to different reasons, including inappropriate water use practices; (iv) habitat loss and fragmentation; (v) nutrients enrichment, eutrophication and the need to prevent further deterioration of human resource use, and (vi) future climates and literature projections regarding Albanian lakes as a part of entire Mediterranean basin.

1.2 Linking Lake and River Systems with Map of Biogeographic Regions of Europe

At the global scale the freshwaters cover less than one per cent of the earth's surface, while yet they are home to approximately six per cent of all known animal species (Strayer and Dudgeon 2010; Markovic et al. 2017). The IUCN Red List includes 128,918 species of which 35,765 (28%) are threatened with extinction. Regarding to the freshwater species (n = 25,007), more than 29% are currently at risk of extinction (IUCN 2013), primarily due to a combination of factors including pollution, unsustainable land use, overexploitation of freshwater resources, anthropogenic disruption of hydrological habitat connectivity and introduction of alien species (Shumka et al. 2008; Oikonomou et al. 2014; Mantyka-Pringle et al. 2014; IUCN 2021). These threats along with the growing demand for freshwater resources driven by rapid tourism and housing development, suggest that freshwater ecosystems and the biodiversity they support are, and will remain, among the most threatened globally (Palmer et al. 2008).

According to the map of global ecoregions (Abell et al. 2008; Shumka et al. 2023a), the Albanian rivers are part of the 420 Southeast Adriatic Drainage. As it is known, this map of freshwater ecoregions is based on the distributions and compositions of freshwater fish species, reflecting keymajor ecological and evolutionary patterns (Shumka et al. 2023a, b). Beside this, the complexity of the larger Drini and other nearby basins, adjacent water drainage makes the delineation a challenging issue (Shumka et al. 2023a).

The exceptional natural values characterized by Lakes of Prespa, Ohrid and Shkodra represent the basis for the protection of this unique ecosystem. So, Lakes of Prespa owns status of National Park (IUCN Management Category II) in Albania and Greece; In Montenegro Lake Skadar/Shkodra has been a National Park (IUCN Management Category II) since 1983 and the Albanian government (Council of Ministers' Decision, No. 684 dated 02.11.2005) proclaimed the Albanian part of Lake Shkodra a "Managed Natural Reserve" (IUCN Category IV). Lake Ohrid in Albania is proclaimed as a landscape protected area (IUCN Management Category

II) and in North Macedonia it is partly included in the National Park of Galičica (IUCN Management Category II). At the international level, since 1995, the Lakes Prespa and Shkodra have been included on the Ramsar List (as a wetland area of international significance), while Lake Ohrid area after extension of 2018 is entirely part of UNESCO List as a World Heritage site.

The distinctive ecological conditions of the basins, whose significance has been acknowledged across national borders, are the consequence of the abiotic and biotic elements of Albania's large Balkan Lakes and other bodies of water as well as their location beginning karst terrain in the outer part of the Dinaric Alps. These principles show how cultural and ecological values have coexisted throughout history. The blazing history of the huge lake region and its significance for the growth of local developments in all littoral countries are demonstrated by the numerous monastery complexes, churches, hermitages, fortresses, and other cultural monuments.

2 Habitat Richness, Rare Plant Species, Lake Shore Development

2.1 Habitat Richness of Albanian Lakes

Freshwater ecosystems are of great importance for the European Community due to the diversity of habitat types and biodiversity that they host. The habitat types of large inland lakes, man-made reservoirs or irrigation ponds as standing freshwater habitat types represent the natural or artificial resources with great economic, climatic, cultural, aesthetic, scientific and educational importance (Dudgeon et al. 2005; Schröter et al. 2005; Grzybowski and Glińska-Lewczuk 2019). These resources are also utilized for purposes such as drinking water, irrigation, aquaculture, and industrial use.

Being under the impact of the continuous unsustainable use of the services that these ecosystems offer as well as under rapid climate change, the freshwater habitats are listed among the most endangered ecosystems in the world. Increasing threats to European aquatic and wetland habitats as result of global warming, hydrological alterations, pollutants from intensive agriculture, spreadof invasive species and overexploitation of landscape and biological resources (Ortmann-Ajkai et al. 2018), have classified them as third most threatened habitats (46% of them area assessed as Critically Endangered, Endangered, Vulnerable) in the Red List of European Habitats (Janssen et al. 2016) among seven major habitat groups of Europe.

The largest lakes of Shkodra, Ohri and Prespaare the most important in terms of diversity of living organism's andrare plant species and macrophytes (Kashta 2007; Shuka et al. 2008). The high biological diversity of Albanian lakes and reservoirs is supportedby the presence of numerous types of aquatic habitats (Fig. 2a); even within one lake there are found diverse habitat types (Pavlides 1997; Shuka et al. 2008, 2021). Based on our observations and literature data (Kashta 2007, 2009;

Kashta et al. 2015; Xhulaj and Shuka 2007; Schneider et al. 2020; Fotiadis et al. 2020),13 habitat types are recorded in the freshwater ecosystems of Albania.

The biggest tectonic-karstic lakes of Shkodra, Ohrid, Prespes, and karstic lakes of Belshi are home to the greatest variety of habitat types in Albania's freshwater ecosystems. With the exception of Lake Dushku, which is comparable to Albania's major lakes and tectonic-karstic lakes in terms of the variety of habitat types and their floristic composition, the number is lower in lakes of glacial origin (Shumka et al. 2024b). The major species, the lakes in which they are located, and the habitat types of Community concern (Council Directive 92/43/EEC) in Albanian lakes are listed. Nonetheless, the majority of Albania's glacial lakes is oligotrophic and has clear, pure waters (Shumka et al. 2024b).

3130. Oligotrophic to Mesotrophic Standing Waters with Vegetation of the *Littorelleteauniflorae* and/or *Isoeto-Nanojuncete*

Description: The habitat includes standing oligotrophic to mesotrophic freshwaters of lakes and ponds with very few nutrients due to the poor substrate (Shuka et al. 2005; Shumka et al. 2024b) and their bank vegetation belonging to the *Littorelletaliauniflorae* order (Fig. 1c, d, g).

Floristic composition: The dominant species of this habitat types are *Centaurium pulchellum, Juncus compressus, Eleocharis acicularis* and *Sparganium erectum.* Other species: *Cyperus fuscus* and *C. michelianus, Eleocharis quinqueflora, E. palustris, Sparganium emersum, S. ramosum, Schoenoplectus tabernaemontani,* etc.

Distribution: This habitat type occurs in the glacial lakes of ultramafic substrate such as lakes in Lura, Balgjaj, Martaneshi, Shebenik and Valamara Mountains. It covers also small surface inthe Ohrid, Lukova, Funari andJezerca Lakes.

3140. Hard Oligo-Mesotrophic Waters with Benthic Vegetation of *Chara* spp.

Description: The habitat is characterized by clear waters with high transparency and often blue, and poor in nutrients (Xhulaj and Shuka 2007; Kashta et al. 2015). The bottom of the habitat types is inhabited by green algae of the group *Charophuta*, dominated from *Chara* and *Nitella* species.

Floristic composition: The floristic structureof the habitat types is inhabited by*Chara globularis, C. fragifera, C. ohridana, Nitellasyncarpa, N. flexilis, N. hyaline*and *Nitellopsis obtusa.* Other species: *Charaaspera, C. kokeillii, C. vulgaris, Nitellaconfervacea, N. mucronate, N. opaca, N, tenuissima*and *Tolypella glomerata.*

Distribution: This habitat type was observed in the Ohrid and Shkodra Lakes, Lake of Dushku and Lakes of Belshi, Blue Eye, Gjergjevicaand Gjançireservoirs.

3150. Natural Eutrophic Lakes with *Magnopotamion* or *Hydrocharition*-Type Vegetation

Description: This habitat type is dominated mainly by free-floating macrophytes of classes *Lemnetea* and *Potamogetonetea* (Shuka et al. 2008; Kashta et al. 2015;

Fig. 1 **a** Lake of Ohrid, **b** Lake of Shkodra, **c** Great Prespa Lake, **d** Lake of Lukova, Pogradec, **e** Lake of Lulashi, Jezercë, **f** Lake of Merhoja, **g** Black Lake of Lura, **h** Grama Lake and **j** Bistrica Reservoir. *Author of photos* D. Shuka a, b, c, j and L. Shuka d, e, f, h

Fotiadis et al. 2020). It was found in the eutrophic waters of the lakes where high nutrient concentrations from agricultural runoff or wastewater discharges are present.

Floristic composition: The dominant species of the habitat types are *Hydrocharismorsus-ranae, Myriophyllum spicatum* and *M. verticillatum, Nymphaea alba* and *Nuphar lutea, Potamogeton natans* and *P. perfoliatus, Trapa natans, Lemna spp., Utricularia* spp. Other species: *Baldellia ranunculoides, Elodea canadensis, Lemnagibba, L. minor, L. trisulca, Spirodela polyrrhiza, Nymphoides peltate, Utricularia australis, U. vulgaris, U. minor, Azollafiliculoides, P. amphibia, P. coloratus, P. crispus, P. gramineus, P. lucens, P. nodosus, P. pectinatus, P. praelongus, P. pusillus, Sagittaria sagittifolia, Salvinia natans* and very rare *Hippuris vulgariss* (Fig. 2c, d, f, g).

Distribution: This habitat type was observed in the Shkodra, Ohrid and Prespa Lakes as well as in the glacial Lake of Lily in Valamara, in two Lakes ofMartaneshi area, in two Lakes of Balgjaj-Kacni and two Lakes in Lura (Shumka et al. 2024b). The habitat is found also in karstic lakes of Belshi (Belshi, Dega, Ibza, Merhoja, Log and Red Lakes) and Blackswamp ofRoskoveci.

3160. Natural Dystrophic Lakes and Ponds

Description: The dystrophic Natural lakes and ponds are charactersed by brown tinted water due to peat and humic acids (Bunce et al. 2013), generally placed on peaty soils or marshes with evolutionary trend towards bogs. Vegetation cover of habitat belong to the order *Utricularietalia* and *Potamogetonetalia* in case of glacial lakes.

Floristic composition: Habitat is dominated from *Utricularia vulgaris* and *Nuphar lutea, Carex pseudocyperus* and *C. rostrata.* Other species: *Utricularia australis, Nuphar alba, Carex davalliana, C. echinate, C. vesicaria, Myriophyllum spicatum, M. verticillatum* and *Potamogeton natans.*

Distribution: This habitat has limited distribution surface within Lesser Prespa Lake and it was foundin the glacial lakes of Dushku, Lake of flowers in Lura, Dragani Lake and Black swamp of Roskoveci and Kashta. It was also found in some ponds, closed to the karstic lakes of BelshiPlateau, such as Grekani and Seferani lakes.

3260. Water Courses of Plain to Montane Levels with *Ranunculion fluitantis* and *Callitricho-Batrachion* Vegetation

Description: The habitat type is composed from species of the genus *Ranunculus* sect. *Batrachium* and species of the genus *Callitriche* that occupies the waters on the spring lakes or the water courses in the alpine lakes. The habitat occurs also in the dams or reservoirs with running waters.

Ranunculus species of the sec. *Batrachium* in the lakes are often companied with *Nymphaeetum albae, Ranunculion aquatilis, Potamogeton perfoliatus* and *Myriophyllum spicatum-Potamogeton perfoliatus* plant communities.

Floristic composition: Habitat is dominated from *Ranunculus trichophyllus* and *Callitriche palustris* or *Nymphaea alba, Myriophyllum spicatum* and *Potamogeton*

Fig. 2 **a** Lesser Prespa Lake, a mosaic of Habitat types, **b** Habitat types: *Salix alba* and *Populus alba* galleriesand *Reedbeds* in Lake of Shkodra, **c** Natural eutrophic lakes with *Magnopotamion* or *Hydrocharition*-type vegetation with dominance of *Nymphoides peltate* in Great Prespa Lake and **d** with dominance of *Nymphaea alba* and *Nuphar lutea* in Lesser Prespa Lake, **e** Sedges community habitat type in Goat Lake, Balgjaj, **f** *Trapa natans* in Shkodra Lake, **g** *Salvinia natans*‘ carpets’ in Black Swamp of Roskoveci. *Author of photos* D. Shuka a, b, c, d and L. Shuka e, f, g

perfoliatus. Other species: *R. aquatilis, Callitriche cophocarpa, C. stagnalis, Ceratophyllum demersum*or *Potamogeton natans,* etc.

Distribution: The habitat type with *R. trichophyllus* and *C. palustris* were observed in shallow waters near the spring of Syri i Sheganit in Shkodra Lake, Dragani Lake and also in the semi-standing waters of dams and reservoirs with cold and fresh waters such as in Blue Eye and Viroi Lake in the south Albania. *R. trichophyllus* was the only aquatic plant speciesinthe Lake of Lady, Korabi Mt, that ocupay more than 30% of this glacial lake (Fig. 2d).

3270. Rivers with Muddy Banks with *Chenopodion rubric* P.P., and *Bidention* P.P. Vegetation

Description: According to Fotiadis et al. (2020) and Rigóand Barina (2021) this habitat type occursin the muddy banks of Great Prespa Lake and several reservoirs, where *Cyperus fuscus* and *Chenopodium rubrum* forms monotypic communities. This habitat types were created by water withdraw due to drought, its use for irrigation or unknown karst processes. The condition of this habitat type depends on fluctuations in the water level.

Floristic composition: The muddy banks of the lakes, reservoirs and ponds are mainly dominated by species such as *Chenopodium rubrum, Cyperus fuscus* and *Bidens frondosa.* Other species: *Bidens tripartita, Polygonum lapathifolium* and *Xanthium* spp.

Distribution: The habitat covers very limited surfaces in Great Prespa and Shkodra Lakes. It was recorded also in several lakes of Belshi Plateau (Dega, Karjan, Omlliand Marinza) as well as inKashtaand Black swamps, and Reservoirs of Murrizi, Kurjan and Katjel.

6420. Mediterranean Tall Humid Herb Grasslands of The *Molinio-Holoschoenion*

Description: Habitat type 6420 in Albanian freshwater ecosystems cover very small surfaces. The habitatincludes vegetation of humid grasslands of tall grasses and rushes that are classified under the alliance *Molinio-Holoschoenion* (Bunce et al. 2003; Fotiadis et al.2020). This habitat types take place closed to reedbeds.

Floristic composition: The dominant species of the habitat are *Scirpus holoschoenus, Cyperus longus, Juncus bufonius, Oenanthe aquatica, Rorippa amphibia, Sparganium erectum*etc. Other species: *Cardamine raphanifolia, Carexpseudocyperus, Cyperus longus, C. michelianus, C. rotundus, Inula viscosa, Juncus acutus, Oenanthe pimpinelloides, O. tenuifolia, Prunella vulgaris, Rorippa amphibia, R. silvestris, Schoenoplectus lacustris, Schoenus nigricans.*

Distribution: Habitat was reported from the Prespa Lakes, Shkodra Lake, and Lakes of Belshi and Black swamp of Roskoveci (Shuka et al. 2008; Rakaj 2009; Rakaj and Kashta 2010; Desfayes 2004, 2020).

91E0.* Alluvial Forests with *Alnus glutinosa* and *Fraxinus excelsior* (*Alno-Padion, Alnionincanae, Salicion alba*)

Description: The habitat type includes a wide range of different forest and riverine scrub communities in floodplains, along the banks of large rivers where they grow on the lowest parts of the alluvial plane. Generally, this habitat types developed in the contact zone between the rivers and lakes. Alluvial forests share a relatively high level of oxygen-rich groundwater that exhibits periodic fluctuations and regular flooding.

Floristic composition: The dominant species of the habitat are *Alnus glutinosa, Fraxinus angustifolia, Salix alba, Populus alba, P. nigra, Ulmus minor,* etc. Other species: *Agrostis stolonifera* agg., *Bidens tripartita, Equisetum palustre, Galiumpalustre, Lycopus europaeus, Lythrumsalicaria, L. tribracteatum, Mentha aquatica, M. pulegium, Polygonum hydropiper, Rorippa sylvestris, Ranunculus repens, Solanum dulcamara, Urtica dioica,* etc.

Distribution: Habitat covers very small surfaces in the Lakes of Shkodra, Ohrid and Ulza.

92A0. *Salix alba* and *Populus albagalleries*

Description: These forests are extensive along the shores of the lakes, as well as along rivers and canals, mainly in tectonic and karstic lake. They are dominated by *Salix alba.* Despite of this, the forests with *Salix alba* and *Populus alba* are very fragmented in the downstream section of the rivers or brooks.

Floristic composition: The dominant species of the habitat are *Salix alba* and/or *Populus alba, S. purpurea, S. triandra, S. viminalis, Quercus robur, Fraxinus excelsior,* etc. Other species: *Agrostis stolonifera* agg.,*Brachypodium sylvaticum, Arrhenatherum elatius, Cyperus fuscus, Equisetum arvense, Rumex* spp., *Ranunculus ficaria, Paspalum paspalodes, Arum italicum, Arundo donax, Dactylis glomerata, Clematis vitalba, Solanum dulcamara, Urtica dioica,* etc.

Distribution: Habitat covers very small surfaces in the Shkodra (Fig. 2b) and Ohrid Lakes and some lakes of Belshi Plateau, Lake of Viroi and Dushkuas well as Bistrica and Gjançireservoirs.

C3.21 and C3.23/D5.11 and D5.13—Reedbeds

Description: Habitats composed by reedbeds includes plant communities of *Phragmitetum* and *Typhetum angustifoliae* with absolute dominance of *Phragmites australis* and *Typha angustifolia.* The habitat often represents a mix of plant community between these species in the lake areas where the waters are standing and shallow (C3.21 and C3.23). According to Fotiadis et al. (2020), habitats with monotypic community species (D5.11 and D5.13) can be found also in the transition zones between shallow waters and muddy areas of the Prespalakes in case of water fluctuation and also in several reservoirs.

Floristic composition: This habitat type is dominated by *Phragmites australis* and *Typha angustifolia*. Other species: *Phragmites australis, Typha angustifolia, Typha latifolia, Carex pseudocyperus, Carex rostrata,* etc.

Distribution: Habitat covers large surfaces in the Prespes Lakes and in lakesof Shkodra, Ohrid and Dushku (Fig. 2a, b). The habitat was observed also in Grekani, Karjani, Sevilza and Seferaniand Long lakes of Belshi municipality. Habitats with reedbeds have been observed also in Kashta and Black swamps and Bistrica reservoir.

C3.2. Flowering Rush Communities

Description: The habitat with flowering rush communities represents small stands with *Butomus umbellatus* bank communities. In the transition zone of the lake waters and their banks there are found also with other companied species of the muddy bank.

Floristic composition: Habitat type is *Butomus umbellatus* and *B.umbellatuss* ubsp. *scutarensis*. Other species: *Alisma plantago-aquatica, Catabrosa aquatica, Lythrum salicaria, Myosotis scorpioides, Polygonum amphibium, Rorippa amphibia, R. sylvestris, Veronica anagallis-aquatica.*

Distribution: The habitat was observed in Great Prespa Lake (Fotiadis et al. 2020), Shkodra Lake and in Black swamp of Roskoveci and several lakes in Balgjaj, Lura and Martaneshi Mts range (Kashta 2007, 2009; Kashta et al. 2015; Shumka et al. 2024b).

D.512 *Schoenoplectetum lacustris*

Description: This habitat type is composed from plant communities dominated by *Schoenoplectus lacustris*. Habitat type occurs near the reedbeds and usually in shallow waters around littoral zone according to Pavlides (1997).

Floristic composition: *Schoenoplectus lacustris* subsp. *glaucus* and *S. lacustris* subsp. *lacustris* are dominant species of the association with otherspecies such as *Schoenoplectus tabernaemontani, Phragmites australis, Callitriche lenisulca, Carex pseudocyperus, Lythrum salicaria, Myosotis scorpioides, Polygonum lapathifolium, Solanum dulcamara, Sparganium erectum* and *Alisma plantago-aquatica*.

Distribution: The habitat was recorded in the large lakes of Great Prespa (Fotiadis et al. 2020) and Shkodra. It was observed also in the Lakes of Merhoja, Sevilza, Marinza and Seferani, all in Belshi municipality as well as in Black swamp of Roskoveci (Desfayes 2004, 2020; Rakaj 2009; Rakaj and Kashta 2010).

D5.21 Sedges Community

Description: This habitat represents the dense shore formations of *Caricetum acutiformis, Caricetum ripariae* and *Eleocharitetum palustris* with large coverage of *Carex* spp. (Fotiadis et al. 2020).The dominant species often mixed with small communities of *Agrostis stolonifera, Juncus articulatus, Carexhirta,* etc.

Floristic composition: The lakeshore vegetation is dominated by *Carex acuta, C. distans, C. paniculate, C. pendula, C. remota, C. riparia, C. vesicaria, C. vulpina.*

Other species: *Agrostis stolonifera, Juncus articulatus, Carexhirta*and less *Lythrum-salicaria, Nasturtium officinale, Myosotis scorpioides.*

Distribution: The habitat has been reported from Prespa, Shkodra and less in Ohri-dandDushku lakes. It was found also in most of glacial lakes on ultramafic rocks such as Valamara, Shebeniku, Martaneshi, Balgjaj (Fig. 2e) and Lura mountains (Shumka et al. 2024b). Small stands of habitat occur also in lakes of Karjan, Sevilza, Marinza and Seferaniin Belshi municipality and Bistrica (Fig. 21, j) and Bulo reservoirs (Desfayes 2004, 2020).

2.2 Species Richness and Rare Plants, Lake Shore Development

Floristic composition of aquatic vegetation of the lakes and reservoirs in Albania, exluding glacial lakes, belong to c. 226 plant taxa, from which 27-species belong to *Charophyta* and *Pteridophyta–Polypodiopsida* groups, and others to the *Angiospermae* (Desfayes 2004, 2020; Rakaj 2009; Rakaj and Kashta 2010; Kashta et al. 2015; Shuka et al. 2021; Shumka et al. 2024b). Outstanding species richness was recorded in the tectonic and karstic lakes. The contributionof aquatic plants in lake vegetation is c. 175 taxa, with the dominance of 151 macrophyte taxa, most of which have been found in the lakes of Shkodra, Ohrid and Lesser Prespa (Shuka et al. 2008; Kashta et al. 2015). So far, the flora of Reservoirs and natural wetlands of Albania is represented from 158 species (Rigo and Barina 2021), more than 2/3 of which are also found in the tectonic-karstic and glacial lakes of the country. Table 1 presents the floristic richness of natural lakes and marshes, where more than four plant species have been recorded.

The freshwater habitat types of natural lakes, swamps and reservoirs are very important for the conservation of many rare plant species or invertebrates and bird species (Mancinelli et al. 2019). According to Kashta (2007) and Rakaj (2009), 19 taxa of aquatic and other hygrophilous macrophytes have been found in these habitats which are considered rare or threatened for the country's flora. Some of these species, such as *Baldellia ranunculoides, Carex melanostachya, Hydrocharismorsus-ranae, Hydrocotyle vulgaris, Ludwigia palustris, Marsilea quadrifolia, Nymphoides peltata, Sagittaria sagittifolia, Spirodellapolyrhiza* and *Trapa natans*, have been recorded only in 2 or 3 localities in Albania. Six species such as *Caldesia parnassifolia, Carex acuta, C. elata, Cyperus glaber, Eleocharis acicularis, Fimbristylis dichotoma, Lindernia procumbens* have been reported only from Shkodra Lake so far, while *Hippuris vulgaris* was recorded only in the Lesser PrespaLake. *Eleocharis carniolica* and *Verbena supina* have been reported only fromShpati area and Bulo reservoir, respectively. All the species recorded in the reservoirs and ponds do not have high conservation interest (Rigo and Barina 2021). Recently, in some freshwater habitats, including Great Prespa Lake was observed *Azollafiliculoides*, forming 'green carpets' and blocing the grouth of other aquatic plant species.

Table 1 Floristic richness of natural freshwater bodies in Albania

	Name of the Lake	Species richness		Name of the Lake	Species richness
1	Shkodra Lake	112	14	Lake of Merhoja	7
2	Ohrid Lake	80	15	Kashtaswamp	7
3	Lesser Prespa	77	16	Lake of flowers (Lurë)	7
4	Great Prespa Lake	71	17	Lake of Kacnia 1	6
5	Black swamp, Roskovec	38	18	Lake of Cows	6
6	Lake of Dushku	19	19	Lake of Ladies	6
7	Bistrica reservoir	17	20	Lake of flowers (Kacni)	5
8	Lake of Karjan	17	21	Goat Lake	5
9	Lake of Grekan	15	22	Lake of flowers (Allaman)	5
10	Seferani Lake	12	23	Lake of Bruçi	5
11	Omlli Lake	11	24	Kurti Lake	5
12	Lake of Karjan	10	25	Lake of Dragani	4
13	Sevilza Lake	9	26	Lake of lilies (Valamara)	4

The most important species that occur in natural lakes and swamps of Albania are strictly protected species of Community interest, listed in Annex II and IV of the Habitat Directive (Council Directive 92/43/EEC), such as *E. carniolica* (II), *L. procumbens* (IV) and *Marsilea quadrifolia* (II, IV). Six other species (*C. parnassifolia, C. acuta, S. natans, T. natans, T. minima* and *T. shuttleworthii*), listed in the appendix I of Bern Convention (1979), have a very narrow distribution range in the Lakes of Shkodra and Prespa, where they are represented by populations with a small number of individuals.

Almost extinct species of the freshwater habitats in Albania can be considered *M. quadrifolia* and *Lyciumeuropaeum*, since they have not been found in the recorded localities for many years.

The fluctuation of the water level and persistent drought have increased the surface of mud vegetation in Great Prespa and most of glacial and karstic lakes and reservoirs, changing and modifying the lake shoreline. The spread of mud plants along lake/reservoir banks can develop ephemeral vegetation in the lake shoreline that can be a refuge for the invasive plants and those are connected to wet habitats.

2.3 Trends in Freshwater Ecosystems, Dam, Man-Made Lakes and Conversion of Floodplain Areas

According to Rigó and Barina (2021) the standing and running freshwater ecosystems in Albania are represented by 247 natural lakes, from which 103 have been glacial lakes, and about 800 artificial water reservoirs. Recent observations show that 19glacial lakes in Albania, due to the global warming and the low level of snow cover in the winter (Shuka et al. 2017; Špoljar et al. 2022), are dried or sustain water only during the winter and spring time. So, currently in Albania there are 84 glacial lakes with a surface more than 0.5 ha (Shumka et al. 2024b).The Albanian lakes and man-made reservoirs differ according to their formation origin, altitudinal elevation, and the type of their water supply and most importantly,, climate and the human activities.

The great human impact on water flow modification and climate change has also affected the Albanian artificial water bodies. So far, there are 630 reservoirs in the country, which is 21% fewer than before 1995. Of these, 65% of the existing dams are in a technically dangerous condition, and 524 of them have a depth ranging from 15 to 30 m (Anonymous 2003). The primary cause of the decline in the volume of man-made water bodies is the demolition and neglect of their dams.

The implementation of new management strategies (Muir et al. 2012) in terms of the conservation of freshwater habitat types (Piczak et al. 2023) is important to reduce biodiversity loss (Lynch et al. 2023), stabilizing ecological function and guarantee ecosystem services for a long time (Ge et al. 2018; Grzybowski and Glińska-Lewczuk 2019).

2.4 Examples of Changing Land Use

Significant changes have been reported during 30-years in land use and Albanian freshwater ecosystems, including the man-made water reservoirs and ponds and their watershed (Shuka et al. 2005, 2017; Zakkak et al. 2018). Land use changes and urbanization have posed serious threats to the lake shorelines of Shkodra and Ohrid, negatively impacting their habitats and conservation status. The overexploitation of the natural resources, e.g. forest cutting and overgrazing or using of the waters for irrigation and electricity production, has drastically modified the lakeshore and water volume of the lakes in Valamara, Martaneshi, Kacnia and Lura areas and almost all the karstic lakes of Belshi municipality (Zakkak et al. 2018; Shumka et al. 2024b).The high cover vegetation of tall floating and free-floating species in Lesser Prespa, Dushku and Shkodra Lakes, as well as in the lakes of Belshi Plateau and Roskoveci swamp is a reflection of the high level of nutrients discharged into these water bodies. Many of the changes and modifications that already have been occured in some of the lakes and reservoirs are irreversible e.g., Lesser Prespa Lake, Black swamp of Roskovec and some lakes in Martaneshi area. Mismanagement of forests and deforestation after 1992, overgrazing and the fires of the last 15-years have drastically changed the lake landscape and reduced the plant cover in the catchment of glacial lakes, increasing soil erosion and filling the lakes with alluvium. The drastic decrease of precipitation in the form of snow has been accompanied by the reduction of the flow of spring water into the lakes and the increase of the intensity of water evaporation, due to global warming (Malatesta et al. 2023; Shumka et al. 2024b; Wagner et al. 2019), is the main factor in reducing the volume of water in lakes and promoting the spread of macrophytes. This phenomenon is quite evident in the lakes of Prespa, the small lakes of Dumre Plateau and and all lakes of the Korabi, Lura, Balgjaj, Martaneshi and Valamara geoecosystems, part of the Drin-Bunë, Mati, Shkumbini and Semani basins.

3 Large Lakes Zooplankton Composition

As a mediator of energy exchange between phytoplankton and fish, zooplankton plays a crucial role in the pelagic food web of lakes. As such, it reacts quickly to trophic cascade alterations through either top-down or bottom-up control (Špoljar 2013). In case of large Balkan Lakes the role and contribution to ecosystem functioning needs further studies. For poorly understood aquatic ecosystems like the Albanian ones, this is even more crucial. Although zooplankton research of Balkan lakes started already within end of nineteenth century (Shumka 2021), our contemporary knowledge is far from complete.

According to various researches on Albanian inland water, 140 taxa of bdelloids and monogononts make up the Rotifera fauna (Shumka 2021). Trichocerca (15 species), Lecane (16 species), Brachionus (15 species), Keratella (7 species), Polyarthra (7 species), and Lepadella (6 species) are the most commonly recorded rotifers. At the regional level, it is important to note that Petkovic (1978) listed 205 rotifer taxa, only for Lake Scadar/Shkodra, according to the authors of Montenegro. Additionally, the extremely high number of species found in Lake Shkodra at the moment must take into account the most recent taxonomic revisions, citing a large number of synonyms and corrections (Segers et al. 2012; Jersabek and Leitner 2013). Regarding the occurrence the examined complex of rotifer species corresponds to the Palearctic biogeographical area (Segers et al. 2012).

18 zooplankton taxa were found in the pelagic zone of Lake Prespa as a result of various investigations (Guseska et al. 2012; Shumka 1994, 2000). Of them, 10 belonged to Rotifera, 5 to Cladocera, 2 to Copepoda, and 1 to Mollusca. The rotifers were dominated by following species: *Keratella quadrata* (Muller,1786), *Keratella cochlearis* (Gosse, 1851), *Trichocerca capucina* (Wierz. et Zach., 1893), *Polyarthra vulgaris* Carlin,1943, *Fillinia longiseta* (Ehrenberg, 1834), *Kellicottia longispina* (Kellicott, 1879), *Gastropus stylifer* Imhof,1891, *Asplanchna priodonta* Gosse, 1850, *Synchaeta pectinata* Ehrenberg, 1832, *Brachionus diversicornis F. homoceros* (Wierzejski, 1891); Cladocera: *Diaphanosoma brachyurum* (Liévin, 1848), *Daphnia cucullata* G.O. Sars, 1862, *Daphnia galeata* G.O. Sars, 1864, *Leptodora*

kindtii (Focke, 1844), *Bosmina longirostris* O.F. Muller, 1785; Copepoda: *Arctodiatomus steindachneri* Richard 1897, *Mesocyclops leuckarti* Claus, 1857; Mollusca: *Dreissena polymorpha* (Pallas, 1771) (larvae). Some of the observed species are cosmopolitans, and their dispersion region is broad. The only endemic species in the Western Balkans is Arctodiatomus steindachneri, which has a small range in Lake Ohrid, Prespa, and Janina (Kiefer and Fryer 1978). It was initially identified as Diaptomus steindachneri in the Janina Lake (Greece) by Richard (1897). According to Richard's description, Doflein (1921) provided evidence of *A. steindachneri* in the Lake Prespa. Karvounaris (1979) described it in Small Prespa.

Certain composition and uniqueness are reflected in the zooplankton of the ancient, tectonic, oligomictic, and oligotrophic Lake Ohrid (North Macedonia and Albania) (Shumka 2001; Tasevska et al. 2017). Regarding the environmental conditions, it exhibits distinct seasonal and regional dynamics. The lake's metalimnion was found to be the most productive area, with the highest levels of zooplankton abundance and phytoplankton biomass. 16 species of low-abundance pelagial zooplankton have been found, including two endemic copepods, *Cyclops ochridanus* (Kiefer 1932) and *Arctodiaptomus steindachneri* (Richard 1897). In the zooplankton assemblage, the Copepods achieved an impressive 60% share after Tasevska et al. (2017). The most prevalent rotifer, *Kellicottia longispina* (Kellicott, 1879), in the summer and copepod nauplii, *Eudiaptomu sgracilis* (Sars, 1862) and *Cyclops ochridanus* in the spring and autumn, make up the majority of microphagous zooplankton. Over the course of the investigation, these microphagous zooplankton populated aphotic hypolimnion because of their needs for the bacterio-detritus suspension. Copepodites and mature copepods in the metalimnion were commonly used as raptorials, and they were strongly and favorably impacted by temperature, dissolved oxygen, and, most importantly, phytoplankton biomass. There were relatively few, isolated, and irregular obligate predators among rotifers, Asplanchna priodonta (Gosse, 1850) and Ploesoma truncatum (Levander, 1894), and cladocerans, Leptodora kindtii (Focke, 1844).

Rotifers, copepods, and cladoceran microcrustaceans are the most prevalent taxa of freshwater metazooplankton in Lake Shkodra (Shumka et al. 2018). About 58% of all zooplankton species are rotifers, whereas just 29 and 54 species of Copepoda and Cladocera, respectively, are represented in this group. With distinct presences in various lake sections, the copepod and cladoceran species present in the lake are widespread throughout the surrounding area. There are observable differences in the abundance and composition of rotifers in Lake Skadar's plankton. The horizontal distribution of the zooplankton, which reflects the main water motions in Lake Shkodra, showed a definite resemblance across the several study periods. Due to the local winds, the water flows and motions from the north to the south are sufficiently substantial and only slightly altered. It is evident that the primary factors influencing the diversity and horizontal distribution of zooplankton in Lake Shkodra must be explained by the movement of water masses, the impact of river inflows, temperature variations, the low nutrient content, and variations in nutrient content, given the paucity of research and vegetation influences (in the heavily populated northern and southeast zones of the lake).

4 Lakes State and Trends in Freshwater Biodiversity with Focus to Fish of Special Interest

In recent decades, almost 20% of the world's freshwater fish have become endangered or threatened due to environmental and human-caused stresses; however, some estimates suggest that this number may be underestimated (Bräutigam 1999). The trends and difficulties of freshwater biodiversity in light of climate change and the growing number of non-native species are highlighted in this segment, which focuses on freshwater fish species.

Three large Lakes of the Balkans i.e. Lakes of Prespa, Lake Ohrid and Lake Shkodra/Skadar comprise the major freshwater ecosystems within Albanian territories interconnected through River Drin (Matzinger et al. 2006). The beginning of the water flow towards the Adriatic Sea is Lake Prespa, which is made up of two lakes (Micro and Macro Prespa) connected by a narrow channel with controlled water flow that crosses the alluvial isthmus separating them. Additionally, water moves from Lake Prespa to Lake Ohrid via subterranean karst caverns. The largest lake in South-East Europe by volume is Lake Ohrid, which is shared between Albania and North Macedonia. The Black Drin River runs north through North Macedonia as Lake Ohrid's sole surface outflow. One branch of the Drin River travels west through Albania, joining the Buna River about 1 km from its source at Lake Shkodër, closes to the Albanian city of Shkodra. The largest lake in South-East Europe by surface area is Lake Skadar/Shkodër, which is shared by Albania and Montenegro. The Morača River, which flows through Montenegro's city of Podgorica, is the biggest river that empties into the lake. Lake Shkodër is drained by the Buna River, which empties into the Adriatic Sea.

Since new alien species are establishing themselves in natural ecosystems and different marine species are frequently found in the lower and middle sections of water courses, particularly in transitional waters, it is difficult to determine the precise number of freshwater fish species (Shumka et al. 2024a). According to a review of the available data, 101 species of freshwater fish from 25 families have been identified (Shumka et al. 2023a). Of these, 25 are introduced species, 16 are amphidromous species that frequently enter freshwaters, and 76 are native freshwater fishes.

The Drini basin, which includes Lakes Ohrid and Skadar, has the greatest number of species in terms of species richness (Shumka et al. 2023a). The Vjosa basin comes in second. The Danubian basin has the fewest species, mostly at higher elevations, and is only weakly represented in Albania. From a biogeographic perspective, the majority of Albania was a single district, bounded to the north by the Drini basin and to the south by Vjosa. Several authors have called this district the Albanian, South Adriatic, or Southeastern Adriatic region (Bianco 1990; Economidis and Daoulas 2003; Geiger et al. 2014; Oiknomou et al. 2014; Zogaris et al. 2009; Oikonomou et al. 2014; Zogaris and Economou 2017).

4.1 Beyond of Intralacustric Speciation of Lake Ohrid

Stankovic's well-known work, "The Lake Ohrid and its living world," from 1961, goes into great detail about the ecological analyses and evaluations of the lake. The near shore or littoral zone community and the offshore or pelagic community in the deep waters in the center of Lake Ohrid are two distinct plant and animal communities. Different groupings of livings define each of these communities.

The unique ecosystem of Lake Ohrid is intimately linked to the development of its extraordinarily rich biodiversity. Like the majority of lakes, Lake Ohrid's ecosystem reflects the ecoclimatological environment and watershed (Wagner et al. 2017). The subaquatic entry of cool, pure, and oxygen-rich water into the lake enabled the existence and development of the lake's vast biodiversity. By providing oxygen, nutrients, and ions as well as by establishing clearly distinct yet consistent border conditions, the subaquatic springs help to build particular habitats for endemic species (Matzinger et al. 2006).

The fact that certain endemic taxa are only found around the springs suggests that the spring water may have had a significant role in the development of Lake Ohrid's distinct ecosystem. The subaquatic springs provide a sizable amount of the water budget, and the water through-flow is poor because of Lake Ohrid's lengthy hydraulic residence time (Wagner et al. 2023). Because of Lake Ohrid's oligotrophic condition and consequently limited algal productivity, it is crucial to dilute the lake waters with pure spring water. With a transparency of 14 m, the lake is still very oligotrophic, although its mean total phosphorus concentration has increased to 4.5 mg m^{-3} (Matzinger et al. 2006; Wagner et al. 2023).

Tectonic in origin, the Ohrid Lake is categorized as a mountainous lake with clear water (Wagner et al. 2023; Wilke et al. 2016, 2020). There are 17 native fish species in the lake, 10 of which are endemic (two of which are members of the Salmonidae family). Archaic, endemic, and relict forms that date back to the Tertiary period are highly diverse and abundant in Lake Ohrid's benthic ecology. Only found in Lake Ohrid, the rounded sponge *Ochridaspongia rotunda* is a well-known endemic and relict form that, when combined with three other sponges, represents an 80% enemismat rate. The class Gastropoda exhibits both high variety and a high percentage of endemism; 31 gastropod species (68.8%) are endemic, whereas Bivalvia has an 8% endemism rate (estimated to be 86% for the entire lake, where 50 known gastropod species are endemic). Oligochaeta (33%), Hirunidea (54%), Copepoda (13%), Ostracoda (63%), Amphipoda (60%), and Isopoda (75%), are other classes that exhibit a significant degree of endemism. Endemism is present also among the microalgae species. Of the 550 species of diatoms, about 90 are rare, tertiary relict, or endemic, with 146 species being classified as endemic (Wagner et al. 2023). The Ohrid Lake's important significance in providing a high number of wintering palaearctic water birds has earned it recognition as a Wetland of International Importance.

From a fisheries standpoint, Lake Ohrid is very important due to its size and the diversity of its fish fauna. The fish populations in this lake are of far superior quality and commercial worth than those in other lakes in the Balkan region; they even outperform the shallow, extremely productive lakes of the Aegean zone. Six non-native species, or a total of twenty-one taxa, make up the fish fauna, which includes 17 native species from four families: Salmonidae (2), Cyprinidae (12), Cobitidae (2), and Anguillidae (1). They are all non-migratory species. It is worth to note that after construction of electric hydro power plants along Drini River the migration route for the European eel (*Anguilla anguilla*) was affected, while eel stocking in this lake has been happen 20 years ago on North Macedonian side of the lake In terms of both commercial value and conservation relevance, the Ohrid trout (*Salmo letnica*) and the Ohrid belvica (*Salmo ohridanus*) are two of the most significant species of trout found in Lake Ohrid.

4.2 Fish of Conservation Importance-Do We Lose Some Iconic Species as Ohrid Trout?

One may argue that Lake Ohrid offers priceless archives of tectonic, climate, environmental, and evolutionary histories, building on the considerations in the previous section (Albrecht and Wilke 2008; Wagner et al. 2013, 2017). Lake Ohrid has recently gained international recognition as a place of significance for long-term reconstruction of the palaeoclimate. It is increasingly being used to test theories about biodiversity and the evolution of endemism in ancient lakes because of its extensive sediment record and strategic location between the Mediterranean and Western Europe (Reid et al. 2019).

Scientists from the International Astronomical Union named a lake on Titan, a moon orbiting Saturn, Ohrid in 2010! This is obviously no accident! Two features immediately stand out to visitors to Lake Ohrid, which is home to many endemic species and the occurrence of speciation. The lake itself, with its breathtaking sweeping vista of the water's surface encircled by towering mountains, comes first, followed by the lengthy history of the lake and settlements coexisting. Scientists have just reaffirmed this after discovering what may be one of the first sedentary populations in Europe. Around 8000 years ago, a settlement of stilt homes is thought to have existed on a portion of the lake's Albanian side (Kuta 2024).It is thought to be the oldest European lakeside hamlet ever found, according to archaeologists.

Following North Macedonia's side of Lake Ohrid, which was designated a World Heritage Site in 1979, Albania's side of the lake was also named a UNESCO World Heritage Site in 2019 as an extension mixed property. The site was expanded to include the cultural and historic area of Ohrid in 1980. Thus, this location of "outstanding universal value" has been recognized for conservation by UNESCO's esteemed list for forty-five years. The unique Ohrid trout represents both the threats and the possibility of survival in Lake Ohrid, which has experienced issues over the last 20 years, from overtourism to development.

Thus, contemporary development plans (often no plans, or swiftly updated, and not unified among littoral countries!) are pushing freshwater biodiversity into impoverishment, and the once-celebrated global scientific relevance value is gradually becoming obsolete. Climate change is making matters worse; simulations show that freshwater bodies in the Mediterranean, including Lake Ohrid, are among the most vulnerable ecosystems. The Ohrid trout (*Salmo letnica*) has a distinct demarcation based on statistical data of fish populations gathered from official sources.

Since the early 2000s, the Lake Ohrid trout population has been steadily declining. Due to overfishing and ongoing high fishing pressure, the Lake Ohrid trout and Lake Ohrid belvica are both in danger of experiencing an irreversible population reduction (Spirkovski et al. 2017).Even though both nations that share Lake Ohrid have worked to increase the trout population through a cooperative restocking program, the ongoing poaching of young specimens has prevented the effects from being seen. The carp population is in comparatively good condition, the eel population is kept healthy with stocking material, and the bleak population is growing. It is important to note that, in terms of captures and in previously identified habitats, the population of *Chondrostoma ohridanus* (Ohridnase) has likewise dramatically decreased over almost 2.5 decades. With the barbel population (Barbusrebeli), a comparable circumstance has been noted. Though their status can fluctuate based on changes in their habitats over time, the remaining endemic but commercially undervalued species are in generally healthy condition.

4.3 Freshwater Invaders in Expansion

Invasive and non-native fish species have posed a significant threat to the native fish fauna in recent decades. Shumka et al. (2008) published Albania's first list of non-native fish species. The arrival of Chinese carps—grass carp, silver carp, and bighead carp—coincided with the invasion of gibel carp. In 1959, fingerlings of silver carp and grass carp were brought to Albania from China, marking the first known introduction to the Balkan region. In 1968, bighead carp and possibly Wuchang bream (*Megalobrama amblycephala*, Cyprinidae) were brought in, marking the reintroduction of these fish (Shumka et al. 2008). Studies describing the general dangers connected to the introduction of alien fish species make up the majority of articles (Vilizzi, 2012). Although there is a continuing scientific debate regarding whether the introduction of alien species can only lead to an increase in biodiversity rather than causing major disturbances in ecosystems, it has been hypothesized that many introduced alien species have negative effects (Economidis et al. 2000).

The initial report on the existence of just two introduced species—*Gambusia holbrooki* and *Cyprinus carpio*—was given by Poljakov et al. (1958). According to Rakaj (1995), Albania is home to 20 non-native species. The importation of species for aquaculture, especially East Asian cyprinids and numerous salmonids, is responsible for the ten-fold growth in foreign species in Albania. The number of alien fish species has increased to 25 (Shumka et al. 2023a, b) after 10 more non-native species were discovered in later research (Shumka et al. 2020; Pietrock et al. 2022).

All of the freshly discovered surviving species, with the exception of *Lepomis gibbosus*, were previously listed in Rakaj (1995). However, it appears that two of the described species—Carassius carassius (Linnaeus, 1758) and *Gambusia affinis* (Baird and Girard, 1853)—were misidentified and are now regarded as representatives of others that were introduced. Additionally, the status of *Sabanejewia balcanica* (Karaman, 1922) and *Misgurnus fossilis* (Linnaeus, 1758), which were included in Rakaj (1995), is unclear. According to Šanda et al. (2008), the discovery of *S. balcanica* in Rakaj (1995) might have resulted from the incorrect identification of unusual *Cobitis ohridana* specimens. The same might be true for *M. fossilis*. However, *Misgurnus fossilis* differs from *Cobitis ohridana* in terms of morphology (different number of barbels around mouth). It is also probable that East Asian Misgurnus, rather than *M. fossilis*, was accidentally brought to Albania due to the introduction of several East Asian cyprinids (Table 2).

In Albania, eight introduced species have established themselves, whereas *Oncorhynchus mykiss* and *Hypophthalmichtys molitrix* are possibly the result of hatchery escapes and previous introductions that have not been shown to reproduce successfully. The majority of East Asian cyprinids, including *Hypophthalmichtys nobilis* (Richardson, 1845), *Ctenopharyngodon idella* (Valenciennes, 1844), *Megalobrama terminalis* (Richardson, 1846), *Parabramis pekinensis* (Basilewsky, 1855), and a few other species [*Thymallus thymallus* (Linnaeus, 1758), *Ameirus melas* (Rafinesque, 1820), and *Coregonus* sp.] reported in Rakaj (1995) have not been documented in recent years. The aforementioned taxa's introductions were most likely a failure.

Even if relatively few freshwater ecosystems have been preserved due to introduction, transfer, or stocking with alien species, the diversity of freshwater fishes in Albanian water bodies remains highly diverse and rich. Above all, we have a duty and responsibility to protect these special ecosystems for coming generations, which can only be accomplished by taking long-term conservation and management measures.

5 Freshwater Environments: Consequences for Biodiversity Conservation

The future of endangered and unusual species in freshwater habitats depends on the health and quality of the water catchment. Over the past 30 years, low vegetation cover has been noted in practically all lakes and reservoirs. It is also clear that invasive species and endangered native species have expanded more quickly due to the extended drought and high summer temperatures, moving up into cooler climates. Based on the impact of current threats such as climate change, pollution from use of the chemical fertilizers in agriculture and use of inland waters for irrigation and electricity production, no freshwater habitats can be considered out of these risks

Table 2 Alien fish species for multiple water basins in Albania

Species	English name	Drini Basin	Prespa	Mati	Ishmi	Erzeni	Shkumbini	Semani	Vjosa
Ameiurus nebulosus (Lesueur, 1819)	Brown bullhead	+							
Abramis brama (Linnaeus, 1758)	Freshwater bream						+	+	
Carassius auratus (Linnaeus, 1758)	Goldfish	+					+	+	
Carassius carassius (Linnaeus, 1758)	Crucian carp	+					+	+	
Carassius gibelio (Bloch, 1782)	Prussian carp	+	+		+	+	+	+	+
Carssius langsdorfi (Temminck and Schlegel, 1846)	Gin-buna carp	+						+	
Ctenopharyngodon idella (Valenciennes, 1844)	Grass carp	+	+				+	+	
Economidichthys pygmaeus (Holly, 1929)	Western Greece goby		+						
Gambusia holbrooki Girard, 1859	Mosquitofish	+	+	+	+	+	+	+	+
Hypophthalmichthys molitrix (Valenciennes 1844)	Silver carp	+	+					+	
Hypophthalmichthys nobilis (Richardson 1845)	Bighead carp	+						+	
Huso huso (Linnaeus, 1758)	Beluga	+							
Lepomis gibbosus (Linnaeus, 1758)	Pumpkinseed	+	+	+	+	+	+	+	+
Megalobrama terminalis (Richardson, 1846)	Black Amur bream	+						+	
Mylopharyngodon piceus (Richardson, 1846)	Black carp	+							
Oreochromis niloticus (Linnaeus, 1758)	Nile tilapia				+				

(continued)

Table 2 (continued)

Species	English name	Drini Basin	Prespa	Mati	Ishmi	Erzeni	Shkumbini	Semani	Vjosa
Oncorhynchus mykiss (Walbaum, 1792)	Rainbow trout	+		+		+	+		+
Parabramis pekinensis (Basilewsky, 1855)	White amur bream		+					+	
Perca fluviatilis (Linnaeus, 1758)	European perch	+							
Pseudorasbora parva (Temminck and Schlegel, 1846)	Topmouth gudgeon	+	+	+	+	+	+	+	+
Rhodeus amarus (Bloch, 1782)	Bitterling	+	+					+	
Sander lucioperca (Linnaeus,1758)	Pike-perch	+							
Silurusglanis (Linnaeus,1758)	Wels catfish		+						
Thymalus thymalus (Linnaeus,1758)	European grayling	+							
Tinca tinca (Linnaeus, 1758)	Tench	+	+					+	
		20	11	4	5	5	9	15	5

(Shuka et al. 2005; Trajanovska et al. 2014; Reid et al. 2019). According to Cantonati et al. (2020), among the aforementioned risks, climate change is the primary threat to Alpine, Arctic, and Boreal lakes, as well as to Albanian lakes, due to the country's Mediterranean climate.

According to Wagner et al. (2019) it is difficult to project the future of precipitation and climate, but current situation and the trends warna reduction in rainfell and higher temperatures (WWF 2022; Shumka et al. 2024a, b).The consequences of these risks could lead to the extinction of *M. quadrifolia* and *L. europaeum* and *H. vulgaris* that have not been observed so far and occur in very small population in one locality. The water regime of Prespes Lakes has great impact on the air humidity balance and conservation of threatenedendemic and nearendemic plants such as *Centaurea greaca, C. shumkana, C. soskae, C. triniifolia* subsp. *campylacme, Colchicum doerfleri* and *Heliospermum pusillum* subsp. *Chromodontum* (Malo and Shuka 2013; Wagenitz et al. 2018; Miszczak et al. 2022; Shuka et al. 2023) which occurs in the rocky slopes of the lakes.

The shoreline and catchment of the glacial lakes of Dashi, Sylbica, Grama and Lake of Flowers in Lura is the home-habitat of rare and endangered plants such as *Crocus bertiscensis, Galanthus elwesii, H. pusillum* subsp. *albanicum, H. oliverae* and *Ranunculus degenii* that makes them with high conservation interest (Shuka et al. 2009b; Miszczak et al. 2022; Raca et al. 2022). Climate change threatens these plants, while xerophilous plants, better adapted to the changing climate, are expected to colonize the habitats of the lakes and reservoirs.

References

Abell R, Thieme ML, Revenga C, Bryer M, Kottelat M, Bogutskaya N, Coad B, Mandrak N, Balderas SC, Bussing W et al (2008) Freshwater ecoregions of the world: a new map of biogeographic units for freshwater biodiversity conservation. BioScience 58(5):403–414. https://doi.org/10.1641/B580507

Albert JS, Destouni G, Duke-Sylvester SM, Magurran A, Oberdorf T, Reis R, Winemiller K, Ripple W (2021) Scientists' warns to humanity on the freshwater biodiversity crisis. Ambio 50:85–94. https://doi.org/10.1007/s13280-020-01318-8

Albrecht C, Wilke T (2008) Lake ohrid: biodiversity and evolution. Hydrobiologia 615:103–140. https://doi.org/10.1007/s010750-008-9558-y

Anonymous (2003) Vlerësimiirreziqevengafatkeqësitënë Shqipëri, Raportstudimi, UNDP, DFID, Tirana, 106p

Bianco PG (1990) Potential role of the paleohistory of the Mediterranean and Paratethys basins on the early dispersal of Euro-Mediterranean freshwater fishes. Ichthyolog Explor Freshw 1:167–184

Bräutigam A (1999) The freshwater crisis. World Conserv 30:4–5

Bunce RGH, Bogers MMB, Evans D, Jongman RHG (2013) Field identification of habitats directive Annex I habitats as a major European biodiversity indicator. Ecol Ind 33:105–110. https://doi.org/10.1016/j.ecolind.2020.10.004

Cantonati M, Guella G, Spitale D, Angeli N, Borsato A, Lencioni V, Filippi ML (2014) The contribution of lake benthic algae to the sediment record in a carbonate mountain lake influenced by marked natural water-level fluctuations. Freshw Sci 33(2):499–512

Desfayes M (2004) Additions to the vascular flora of Albania. Ann Bot Nuova Ser IV:155–168
Desfayes M (2020) Albania: aquatic and wetland species. Unpublished report. 19p
der Sleen V, Albert A (2021) Patterns in freshwater fish diversity. In: Reference module in earth systems and environmental sciences. https://doi.org/10.1016/B9778-0 12-819168-8.00056-6
Doflein F (1921) Mazedonien. Erlebnisse und Beobachtungeneines Naturforschersim Gefolge des Deutschen Heeres. Gustav Fischer Verlag, Jena, p 592
Dudgeon D, Arthington AH, Gressner MO, Kawabata ZI, Knowler DJ, Lévêque C, Naiman RJ, Prieur-Richard AH, Soto D, Stiassny ML, Sullivan CA (2005) Freshwater biodiversity: importance, threats, status and conservation challenges. Biol Rev 81:163–182. https://doi.org/10.1017/S1464793105006950
Economidis PS, Daoulas C (2003) *Barbusprespensis* Karaman, 1924. In: Banarescu PM, Bogutskaya NG (eds) The freshwater fishes of Europe, *Cyprinidae* 5/II, Part II, *Barbus*, pp 365–376. Wiebelsheim, Germany: Aula–Verlag
Fotiadis G, Sakellarakis FN, Mahmutaj E, Shyti I, Cano S, Dobblesteijn R (2020) Assessment and mapping of the Great Prespa Lake wetland habitat types in Albania. Final report, 38p
Gardner RC, Finlayson C (2018) Global Wetland Outlook: State of the World's Wetlands and their Services to People, Ramsar Convention Secretariat
Geiger MF, Herder F, Monaghan MT, Almada V, Barbieri R, Bariche M et al (2014) Spatial heterogeneity in the Mediterranean Biodiversity Hotspot affects barcoding accuracy of its freshwater fishes. Mol Ecol Resour 14:1210–1221. https://doi.org/10.1111/1755-0998.12257
Ge Y, Zhang K, Yang X (2018) Long-term succession of aquatic plants reconstructed from palynological records in a shallow freshwater lake. Sci Total Environ 643:312–323. https://doi.org/10.1016/j.scitotenv.2018.06.203
Grzybowski M, Glińska-Lewczuk K (2019) Principal threats to the conservation of freshwater habitats in the continental biogeographical region of Central Europe. Biodivers Conserv 28(14):4065–4097. https://doi.org/10.1007/s10531-019-01865-x
Guseska D, Tasevska O, Kostoski, G (2012) Zooplankton dynamic of Lake Prespa (Macedonia) Biologia 67/5: 939–944, Section Zoology. https://doi.org/10.2478/s11756-012-0092-z
IUCN (2013) IUCN Red List of Threatened Species. Version, 2013.2
IUCN (2021). IUCN Red List, 2017–2020 Report, 24p
Janssen JAM, Rodwell JS, Criado MG, Gubbay S, Haynes T et al (2016) European red list of habitats. Luxembourg: Publications Office of the European Union
Jersabek CD, Leitner MF (2013) The Rotifer world catalog. World Wide Web electronic publication. http://www.rotifera.hausdernatur.at/
Karvounaris D (1979) Contribution to the knowledge of the planktonic Cladocera and Copepoda in ten natural lakes of Macedonia (Greece). Thesis, University. Thesloniki, Greece, 158p (in Greek, English summary)
Kashta L (2007) Aquatic macrophytes in the Lake Shkodra-River Buna wetlands complex. Rivers and citizens. Cross Border Exp Environ Protect Sustain Dev 1:82–93
Kashta L (2009) Flora harofite (*Charophyta*) e rajonitte Shkodrës. Universitetii Shkodrës "Luigj Gurakuqi", Bul. Shk. Ser Shk Nat 59:97 106
Kashta L, Rakaj M, Zeneli V (2015) The diversity of aquatic macrophytes in the transboundary lakes of Shkodra, Ohrid and Prespa—Albanian part. BSHN (UT) 20:28–39
Kiefer F, Fryer G (1978) Das Zooplankton der Binnegewässer. 2. Teil: Freilebende Copepoda. Die Binnegewässer, Band 26, Teil 2, Schweizerbart, Stuttgart, p 380
Kuta S (2024) This 8,000-year-old village on stilts may be the oldest of its kind in Europe. Available on https://www.smithsonianmag.com/smart-news/is-this-the-oldest-village-in-europe-180982787/
Lynch AJ, Hyman AA, Cooke SJ, Capon SJ, Franklin PA, Jähnig SC et al (2023) Future-proofing the emergency recovery plan for freshwater biodiversity. Environ Rev. https://doi.org/10.1139/er-2022-0116
Malatesta L, De Sanctis M, Ammann E, Attorre F, Buffi F, Cambria VE, Shuka L, Fanelli G et al (2023) Bioregionalization of Albania: mismatch between the flora and the climate suggests that

our models of Southern European bioregions are in need of a revision. Folia Geobot 58(1):71–87. https://doi.org/10.1007/s12224-023-09432-7
Malo S, Shuka L (2013) Distribution of *Colchicum doerfleri* Halácsy, *Colchicum triphyllum* Kunze and *Colchicum bivonae* Guss., in Albania. Int J Ecosyst Ecol Sci 3(2):273–278
Mancinelli G, Mali S, Belmonte G (2019) Species richness and taxonomic distinctness of zooplankton in ponds and small lakes from Albania and North Macedonia: the role of bioclimatic factors. Water 11(11):2384. https://doi.org/10.3390/x11112384
Mantyka-Pringle CS, Martin TG, Moffatt DB, Linke S, Rhodes JR (2014) Understanding and predicting the combined effects of climate change and land-use change on freshwater macroinvertebrates and fish. J Appl Ecol 51:572–581. https://doi.org/10.1111/1365-2664.12236
Markovic D, Carrizo S, Carcher O, Walz A, David J (2017) Vulnerability of European freshwater catchments to climate change. Glob Change Biol 23:3567–3580
Matzinger A, Jordanoski M, Veljanoska-Sarafiloska E, Sturm M, Muller B, Wuest A (2006) Is Lake Prespa jeopardizing the ecosystem of ancient Lake Ohrid? Hydrobiologia 553:89–109. https://doi.org/10.1007/s1075000564279
Miszczak S, Shuka D, Shuka L, Migdalek G, Słomka A (2022) Low and high elevation *Heliosperma* species (*Caryophyllaceae*)—insight based on chromosome number, pollen characters and seed micromorphology. Phytotaxa 554(1):032–046. https://doi.org/10.11646/phytotaxa.554.1.2
Muir MC, Spray CJ, Rowan JS (2012) Climate change and standing freshwaters: informing adaptation strategies for conservation at multiple scales. Area 44(4):411–422. https://doi.org/10.1111/j.1475.2012.01130.x
Oiknomou A, Lperieur F, Leonardos ID (2014) Biogeography of freshwater fishes of the Balkan Peninsula, Hydrobiologia 738(1). https://doi.org/10.10507/s10750-014-1930-5
Ortmann-Ajkai A, Csicsek G, Hollós R, Magyaros V, Wágner L, Lóczy D (2018) Twenty-years' changes of wetland vegetation: effects of floodplain-level threats. Wetlands, 38, 591–604. https://doi.org/10.1007/s13157-018-1002-0
Palmer MA, Reidy Liermann CA, Nilsson C, Florke M, Alcamo J, Lake PS, Bond N (2008) Climate change and the world's river basins: anticipating management options. Front Ecol Environ 6:81–89
Pavlides G (1997) Aquatic and terrestrial vegetation of the Prepa area. Hydrobiologia 351:41–60
Petkovic S (1978) The contribution to the better knowledge of ecology and distribution of *Brachionus sessilis* Varga 1957 (Rotatoria). Poljoprivreda i Šumarstvo 24(2):17–37
Piczak ML, Perry D, Cooke SJ, Harrison I, Benitez S, Koning A et al (2023) Protecting and restoring habitats to benefit freshwater biodiversity. Environ Rev. https://doi.org/10.1139/er-2023-0034
Pietrock M, Ritterbusch D, Wolf-Christian Lewin, Shumka S, Spirkovski Z, Ilik-Boeva D,Brämick U, Peveling R (2022) The fish community of the ancient Prespa Lake (Southeast Europe): non-indigenous species take over. Fisheries & Aquatic Life. Arch Polish Fish 30:112–124. https://doi.org/10.2479/aopf-2022-0011
Poljakov GD, Filipi N, Basho K,Hysenaj A (1958) Pesquit e Shqiperise (Fishes of Albania). MihalDuri, Tirana
Raca I, Harpke D, Shuka L, Ranđelović V (2022) A new species of *Crocus* ser. *Verni* (*Iridaceae*) with 2n = 12 chromosomes from the Balkans. Plant Biosyst, 1–9. https://doi.org/10.1080/11263504.2020.1829735
Rakaj N (1995) Iktiofauna e Shqiperise (Ichthyofauna of Albania). Tirana: Shtëpia Botuese "Libri Universitar"
Rakaj M (2009) New aquatic macrophytes for the flora of Albania from the Lake Shkodra, Drini and Buna basins. J Int Environ Appl Sci 4(3):278–284
Rakaj M, Kashta L (2010) The threatened and rare plant species of the Lake Shkodra–Delta Buna hydrological system. In: BALWOIS 2010 conference on water observation and information system for decision support. Republic of Macedonia, Ohrid
Reid AJ, Carlson AK, Creed IF, Eliason EJ, Gell PA, Johnson PTJ, Kidd KA, MacCormack TJ, Olden JD, Ormerod SJ (2019) Emerging threats and persistent conservation challenges for freshwater biodiversity. Biol Rev 94:849–873. https://doi.org/10.1111/brv.12480

Richard J (1897) Entomostracés, recueillis par M. le Directeur Steindachner dans les lacs de Janina et de Scutari. Annalendes K. K. Naturhistorischen Hofmuseums 12(1):63–66. http://www.biodiversitylibrary.org/item/26870#page/81/mode/1up

Rigó A, Barina Z (2021) The floristic composition of irrigation ponds and water reservoirs in Albania after the long persistent drought of 2016–2017. In: 1st international electronic conference on biological diversity, ecology and evolution. MDPI, Basel, Switzerland. https://doi.org/10.3390/BDEE2021-09509

Sayer K, Fernando E, Jimenez R, Macfarlane N et al (2025) One-quarter of freshwater fauna threatened with extinction. Nature. https://doi.org/10.1038/s41586-024-08375-z

Schneider S, Trajanovska S, Biberdžić V, Marković A, Talevska M, Imeri A et al (2020) The Balkan macrophyte index (BMI) for assessment of eutrophication in lakes. Acta Zool Bulgarica 72(3):439–454

Schröter D, Cramer W, Leemans R, Prentice IC et al (2005) Ecosystem service supply and vulnerability to global change in Europe. Science 310(5752):1333–1337. https://doi.org/10.1126/science.1115233

Segers H, De Smet WH, Fischer C, Fontanelo D, Michaloudi E (2012) Towards a list of available names in zoology, partim phylum Rotifera. Zootaxa 3179:61–68. https://doi.org/10.11646/zootaxa.3179.1.3

Šanda R, Vukić J, Choleva L, Křížek J, Šedivá A, Shumka S, Wilson IF (2008) Distribution of loach fishes (Cobitidae, Nemacheilidae) in Albania, with genetic analysis of populations of *Cobitisohridana.* Folia Zool 57(1–2):42–50

Špoljar M (2013) Microaquatic communities as indicators of environmental changes in lake ecosystems. J Eng Res 1:29–42

Špoljar M, Shumka S, Tasevska O, Tomljanović T, Ostojić A, Balkić AG, Lajtner J, Pepa B, Dražina T, Ternjej I (2022) Small standing-water ecosystems in the transitional temperate climate of the Western Balkans. In: Pešić V, Milošević D, Miliša M (eds) Small water bodies of the Western Balkans, 1st edn. Springer, Cham, Switzerland, pp 21–51

Shuka L, Çullaj A, Shumka S, Miho A, Duka S, Bachofen R (2009a) Response of drinking-water reservoir ecosystems to anthropogenic impacts in Albania: trends of interrelationship. J Int Environ Appl Sci 4(4):478–486

Shuka L, Zekaj Z, Mullaj A (2009b) Biogeographical records of species of the genus *Crocus* L. in Albania. In: 5th Balkan botanical congress, Belgrade, Serbia. Book of abstracts. 7–11 Sept 2009, pp 52–53

Shuka L, Malollari I, Veizaj D (2005) Assessment of phosphates and nitrates effects in water's, quality of Ohrid and Prespa Lakes. J Environ Prot Ecol 6(4):761–770

Shuka L, Malltezi J, Mersinllari M, Vardhami I (2008) Dynamics of vegetation cover of Prespa Lakes and its watershed (Albanian side). In: Proceedings of the third international scientific conference on Balkan water observation and information system (BALWOIS), pp 27–31

Shuka L, Mullaj A, Hoda A, Kashta L, Miho A (2017) Overview on the flora and vegetation of the Albanian Alps—the degree of conservation and threats. Conference: In: 37th meeting of Eastern Alpine and dinaric society for vegetation ecology (EADSVE) Prizrcni, Kosovo, pp 13–16. https://doi.org/10.13140/RG.2.2.23064.70400

Shuka L, Shuka D, Diku A (2021) Bimët Endemikedheato me Përhapjetë Kufizuar, ParkuKombëtar "Prespa". Gent Grafik, Tiranë, 150p

Shuka D, Malo S, Hoda P (2023) Conservation status of the endemic and near endemic plant taxa of the Korça region, Albania. Thalassiasalentina 44(suppl):121–134

Shumka S (1994) Annual zooplankton dynamics of Lake Prespa. M.Sc. thesis. Institute of Biology, Faculty of Natural Sciencesand Mathematics, Skopje, University "St. Cyriland Methodius", 114p

Shumka S (2000) Zooplankton community as an indicator of Lake trophic state (Makro Prespa Lake). In: International symposium on sustainable deveopment of Prespa Region, Oteshevo, 23–25 June 2000, pp 23–28

Shumka S (2001) Feeding relations of Eudiaptomus gracilis (SARS) and influence on some parameters of their life cycle in Lake Ohrid. Verh Int Ver Limnol 27:3708–3711

Shumka S (2021) Checklist of rotifer species from Albania (phylum Rotifera). Opuscula Zool Budapest 52(1)

Shumka S, Paparisto A, Grazhdani S (2008) Identification of non-native freshwater fishes in Albania and assessment of their potential threats to the national biological freshwater diversity. In: BALWOIS conference, 21–31 May 2008, Ohrid, Republic of Macedonia. http://balwois.com/publication2008

Shumka S, Spoljar M, Taseska O (2018) The zooplankton of Shkodra/Scadar lake. In: Pešić V et al (eds) The Skadar/Shkodra Lake environment. Hdb Env Chem 80:239–254. https://doi.org/10.1007/698_2018_237

Shumka S, Shumka L, Trajce K, Ceci S (2020) First record of the Western Greece goby—*Economidichthys pygmaeus* (Holly, 1929), in Greater Prespa Lake (Albania). Ecol Monten 35:78–81. https://doi.org/10.37828/em.2020.35.6

Shumka S, Lalaj S, ŠandaR, Shumka L, Meulenbroek P (2023a) Recent data on the distribution of freshwater ichthyofauna in Albania. Croat J Fisher 81:33–44. https://doi.org/10.2478/cjf-2023-0004

Shumka S, Nagahama Y, Hoxha S, Asano K (2023b) Overfishing and recent risk for collapse of fishery in coastal Mediterranean lagoon ecosystem (Karavasta lagoon, southeastern Adriatic Sea). Fish Aquat Sci 26(4):294–303. https://doi.org/10.47853/FAS.2023.e25

Shumka S, Sanda R, Dhora Dh, Grapci-Kotori L (2024a) Fish resources of Inland waters and fisheries in Albania. In: Simic V, Simic S, Pesic V (eds) Management, sustainability, and conservation. Ecological sustainability of fish resources of Inland waters of the Western Balkans. Springer, pp 220–261. https://doi.org/10.1007/978-3-031-36926-1_7

Shumka S, Shumka L, Špoljar M, Shuka S (2024b) Evidence of climate change and the conservation needed to halt the further deterioration of small glacial lakes. Climate 12:124. https://doi.org/10.3390/cli12080124

Skarbøvik E, Shumka S, Mukaetov D, Udaya Sekhar U (2010) Harmonised monitoring of Lake Macro Prespa as a basis for integrated water resources management. Irrigat Drain Syst 24(3–4):223–238. https://doi.org/10.1007/s10795-010-9099-1

Spirkovski Z, Ilik-Boeva D, Talevski T, Trajcevski B, Palluqi A, Flloko A, Miraku T, Kapedani E, Ritterbusch D, Bramick U, Pietrock M, Peveling R (2017) Fish and fisheries Lake Ohrid—implementing the EU water framework directive in South-Eastern Europe. Deutsche Gesellschaft fur Internationale Zusammenarbeit, Bonn, Eschborn

Strayer DL, Dudgeon D (2010) Freshwater biodiversity conservation: recent progress and future challenges. J North Am Benthol Soc 29:344–358. https://doi.org/10.1899/08-171.1

Tasevska O, Špoljar M, Gušeska D, Kostoski G, Patcheva S, ElizabetaSarafiloska V (2017) Zooplankton in ancient and oligotrophic Lake Ohrid (Europe) in association with environmental variables. Croat J Fish 75:95–103. https://doi.org/10.1515/cjf-2017-0013

Trajanovska S, Talevska M, Schneider S (2014) Assessment of littoral eutrophication in Lake Ohrid by submerged macrophytes. Biologia 69(6):756–764. https://doi.org/10.2478/s11756-014-0365-9

Wagenitz G, Bergmeier E, Gregor T, Meierott L, Shuka L, Tan K (2018) A synopsis of the Centaurea soskae and triniifolia groups (Centaurea sect. Acrolophus) in the Prespa area and Northern Pindos. *Phytotaxa*, 348(2):77–89. https://doi.org/10.11646/phytotaxa.348.2

Wagner B et al (2017) The environmental and evolutionary history of Lake Ohrid (FYROM/Albania): interim results from the SCOPSCO deep drilling project. Biogeosciences 14:2033–2054

Wagner B et al (2019) Mediterranean winter rainfall in phase with African monsoon during past 1.36 million years. Nature 573:256–260

Wagner B, Tauber P, Francke A, Leicher N, Binnie SA, Cvetkoska A, Jovanovska E, Just J, Lacey JH, Levkov Z et al (2023) The geodynamic and limnological evolution of Balkan Lake Ohrid, possibly the oldest extant lake in Europe. Boreas 52:1–26. https://doi.org/10.1111/bor.12601

Wilke T, Wagner B, Albrecht C, Ariztegui D, Van Bocxlaer B, Delicado D, Francke A, Harzhauser M et al (2016) Scientific drilling projects in ancient lakes: integrating geological and biological histories. Global Planet Change 143:118–151

Wilke T et al (2020) Deep drilling reveals massive shifts in evolutionary dynamics after formation of ancient lake ecosystem. Sci Adv 6:eabb2943. https://doi.org/10.1126/sciadv.abb2943

WWF (2022) In: Almond REA, Grooten M, Bignoli J, Petersen T (eds) Living Planet Report 2022—building a nature positive society. WWF Gland Switzerland, Gland, Switzerland, p 115

Xhulaj M, Shuka L (2007) Important planta areas (IPAs) in biodiversity enabling activity. Ministry of Environment, Forestry and Water Administration, Tirana, pp 231–276

Zakkak S, Radović A, Panitsa M, Vassilev K, Shuka L, Kuttner M, Schindler S, Kati V (2018) Vegetation patterns along agricultural land abandonment in the Balkans. J Veget Sci 29(5):877–886. https://doi.org/10.1111/jvs.12670

Zogaris S, Economou AN, Dimopoulos P (2009) Ecoregions in the southern Balkans: should they be revised? Environ Manag 43:682–697. https://doi.org/10.1007/s00267-008-9243-y

Zogaris S, Economou AN (2017) The biogeographic characteristics of the River Basins of Greece. In: The handbook of environmental chemistry. https://doi.org/10.10007/698.2017.475

Lakes of Albania from Origin to Current Threats to Their Functional Ecosystems

Spase Shumka, Arjan Beqiraj, and Lulëzim Shuka

Abstract This chapter provides physical settings, particularities and other abiotic aspects in the broadest sense, focusing specifically on natural lake ecosystems. Being located at the southwestern part of the Balkan Peninsula and influenced by Mediterranean and continental climate the large Balkan Lakes of Prespa, Ohrid and Skadar/Shkodra are at the crossroad of accelerated climate implications. Lakes and their water ecosystems, both of natural or anthropogenic origin, dominate in the Albanian mosaic landscape and contributing to the high diversity of habitats and species as well as environmental heterogeneity. Albania has abundant water resources because to its climate, particularly its high rainfall and mountainous mountain relief. A significant volume of water from other nations drains through Albania since the topographical drainage system does not match the Albanian border with its neighbors. The entire area of wetlands is 970 km^2, or roughly 3% of the country's total land area. The majority of these aquatic environments are made up of lakes, coastal lagoons, and reservoirs. All of these include a high diversity of habitats and species, sensitive ecosystems that are still heavily impacted by human activity and are not well studied. Unquestionably, the three large transboundary lakes of Ohrid, Prespa, and Shkodra are the most significant aquatic complexes for Albania and the entire Balkan region. Their combined area, including the Drini watershed, is 19,352 km^2, of which 257 km^2 are in Greece, 3583 km^2 in North Macedonia, 4152 km^2 in Montenegro, 4360 km^2 in Kosovo, and 7000 km^2 are in Albania. From Prespa Lake to the Adriatic Sea's Buna delta, they form a cascade system.

Keywords Tectonic · Natural lakes · Climate change · Catchment · Hydro-ecological · Long-term data

---

S. Shumka (✉)
Faculty of Biotechnology and Food, Agricultural University of Tirana, Tirana, Albania
e-mail: sshumka@ubt.edu.al

A. Beqiraj
Faculty of Geology and Mining, Polytechnic University of Tirana, Tirana, Albania
e-mail: arjan.beqiraj@fgjm.edu.al

L. Shuka
Faculty of Natural Sciences, University of Tirana, Tirana, Albania
e-mail: lulezim.shuka@fshn.edu.al

K. Stefanidis et al. (eds.), *Lakes of the Eastern Mediterranean and Balkans*,
https://doi.org/10.1007/978-3-032-02168-7_4

1 Introduction

1.1 Lakes of Albania, Objectives of the Chapter

The objective of this chapter is to provide an analysis of status and trends of freshwater bodies of Albanian freshwater bodies using a combination of: (i) quantitative data; (ii) qualitative data from a literature review; (iii) evidence from the peer-reviewed scientific literature and (iv) authors experiences and provided scientific inputs within more than two decades. This chapter considers physical settings, particularities and biodiversity in the broadest sense, focusing specifically on natural lake ecosystems. Being located at the southwestern part of the Balkan Peninsula and influenced by Mediterranean and continental climate the large Balkan Lakes of Prespa, Ohrid and Skadar/Shkodra are at the crossroad of accelerated climate implications, so the other objective of this contribution is to provide key findings and opinions on conservation needs, adaptation and resilience.

There are several main lotic and lacustrine systems in Albania (Fig. 1). Listed from north to south these are: Ohrid-Drin-Skadar system (including River Buna) (A), rivers Mat (B), Ishëm (C), Erzen (D), Shkumbin (E), Seman (composed of two big inflows—Devoll and Osum) (F) and Vjosë (Aoos in Greek) (G), several short rivers flowing from the Acroceraune mountains to the southernmost part of the Adriatic Sea and the northernmost part of the Ionian Sea (H), the area around Butrint Lagoon (rivers Bistrica and Pavlla) (I). Most of the above mentioned lakes and rivers (A–G) belong to the Adriatic slope and the southernmost area (I) to the Ionian Sea slope (Shumka et al. 2024; Shuka and Tan 2024). Area H lies on the border of the Adriatic and Ionian Seas. Whereas there is no surface outflow of water from Prespa Lakes, there are underground connections with Lake Ohrid (Amataj et al. 2007). Only a very small area in the northernmost part of Albania, in the Albanian Alps (Bjeshkët e Namuna), belongs to the Danube basin (Dill 1993; Shumka et al. 2023). In Fig. 2 landscape features and current state view of the large transboundary lakes.

The discharge of all Albanian rivers is seasonally highly variable, being sometimes more than ten times less in summer than in winter. The beds of the main rivers are usually very wide, as a great amount of gravel and pebbles is deposited around the flow itself (Dill 1993), while the terrain is characterized by very diverse topographic textures, with flat lowlands in the West and mountains in the East.

The altitude increases gradually from West to East. Plains cover about 15% of the surface area, mainly in the West of the country, with hills of altitudes up to about 200 m a.s.l. In some places, the mountains are radially grouped as in the Albanian Alps or they form regularly oriented chains from the South-East to the North-West (Miho et al. 2013). In general, the mountains have steep slopes and flat crests, with steeper slopes in the Western region than in the Eastern part. Deep valleys often squeeze through narrow gorges forming canyons such as Bicaj, Seta, Osumi and Langarica ordeep gorges such as Drini, Shkumbini, Devolli or Kelcyra Gorge (Permeti), one of the biggest in Albania (Xhulaj and Shuka 2007). There the Vjosa River flows towards the Western Lowland. Karst formations are evident all over the country,

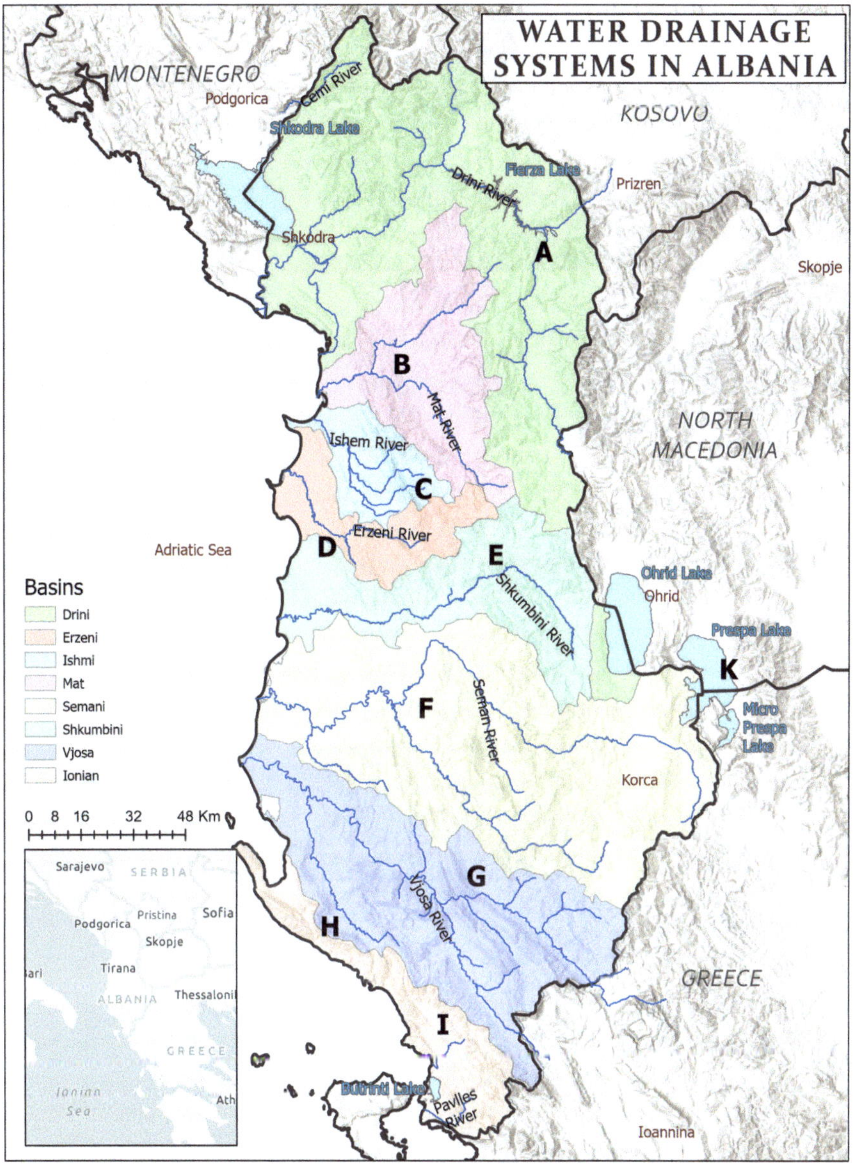

Fig. 1 Water basin systems of Albania. Ohrid-Drin-Skadar system (including River Buna) (A), Mat (B), Ishëm (C), Erzen (D), Shkumbin (E), Seman (F) and Vjosa (G), several short rivers flowing from the Cika mountain to the southernmost part of the Adriatic Sea and the northernmost part of the Ionian Sea (H), area around Butrint Lagoon (rivers Bistrica and Pavllo) (I)

Fig. 2 **a** Lake of Shkodra, north-western part, **b** Lake of Shkodra, the city and Albanian Alps in front, **c** Ohrid Lake catchment, **d** Ohrid Lake, Guri i Kuq-Çervenakë slopes, **e** catchment of Lesser Prespa Lake, **f** Great Prespa Lake, Zaveri swallow hole and **g** catchment of Great Prespa Lake. Author of photos: D. Shuka (**a**, **b**, **e**, **f**) and L. Shuka (**c**, **d**, **g**)

from the sea level up to the high altitudes. Karst topography is more pronounced at higher altitudes. Permanent ice is limited to mountain peaks higher than 1800 m a.s.l.

The climate is mainly of a Mediterranean subtropical character with high humidity; in the North and East it gradually varies to a moderate continental climate (Malatesta et al. 2023). Typically, the winter is relatively short, mild and wet, while the summer is long, hot and very dry. Precipitation is heavy, ranging from about 1300 mm per year in Saranda to 2000 mm per year in Shkodra, increasing from West to East. As a result of sudden strong rain events, brooks and torrents often develop with extreme erosion capacity. The number of sunny hours is high, so Tirana gets more than 330 sunny days per year yielding more than 2100 kWh m^{-2} $year^{-1}$ (Shuka et al. 2009; Miho et al. 2023).

Due to the climatic conditions, mainly to the heavy rainfall and the hilly mountain relief, Albania is rich in water resources. Since the topographical drainage system does not correspond to the Albanian border with neighboring countries, a considerable amount of water from other countries drains through Albania. From the whole watershed of 43,905 km^2, only 65% is within the national borders. Therefore, Albania is among the top countries in Europe related to the water quantity per capita; it amounts for, up to 13,000 m^3 per inhabitant per year.

Before 1960, wetlands covered more than 2300 km^2, equal to about 8% of the Albanian territory. Large-scale reclamation processes for agricultural purposes strongly reduced the total area of the wetlands since then to less than half (Miho et al. 2023; Kiri 2016; Shuka et al. 2009). Nevertheless, more than 1300 aquatic systems are scattered throughout the country: marine habitats, coastal lagoons, fluvial deltas, rivers, springs, lakes and ponds. Wetlands cover a total surface of 970 km^2, equal to about 3% of the whole national territory. The lakes, the coastal lagoons and the reservoirs represent the largest part of this aquatic habitats (Xhulaj and Shuka 2007). All of these areas support a high diversity of sensitive habitats and species, many of which are poorly known and continue to face significant human-induced impacts (Shumka et al. 2024). Among the freshwater bodies, glacial lakes represent some of the notable rinches within country, and currently threaten by climate changes (Shumka et al. 2024).

1.2 *Broader Implications for Aquatic Ecosystems Beyond the National Borders*

Lakes level regimes are driven by climate and basin controls, and vary considerably over space and time. Due to intensive land use changes before the 1990s, human interventions persisted for over four decades. So, the hydrological regimes that are influenced by factors including geology and topography (Poff et al. 2007), were forced by alterations, changes in water supply, nutrient enrichment, use for aquaculture purposes, aliens rapid invasion, etc. (Špoljar et al. 2022). The resulting hydrological regimes directly affect lake ecosystem characteristics in transboundary lakes (Catsadorakis and Malakou 1997), including the physical nature of lake habitats and river channels, sediment regimes, and prevailing water quality conditions which, in turn, drive the key aquatic ecosystem processes. The case of Lesser Prespa (shared between Greece and Albania) is an example of such variability that influenced the structure of aquatic habitats and the composition of ecological communities (Shuka et al. 2008, 2021), including plankton, plants, benthic macro-invertebrates and vertebrates, including fish, amphibians, reptiles, birds, and mammals.

However, the absence of thorough, reliable, and long-term data is a major obstacle to understanding and controlling the relationships between hydrological variability and the diversity and composition of aquatic ecosystems. Accurate estimates for both the present and the future are hampered by the fact that long-term biodiversity data are sometimes restricted to local scales.

From a practical and scientific perspective, the Prespa Lakes region and the Mali Thate–Galičica Mountain range are of tremendous interest. The area's significance for the people of three nearby nations—Albania, North Macedonia, and Greece—whose well-being is heavily reliant on the lake's resources is what gives it practical value. From a scientific perspective, the area is of significance because lakes Prespa and Ohrid create a complex yet shared hydrological system. The two lakes are separated by the high Mali Thate–Galičica mountain range, which is composed of karst rocks. Water from Lake Prespa flows underground through this massif into Lake Ohrid, which is located lower (Cvijić 1906; Anovski et al. 1991; Eftimi and Zoto 1997).

The field is currently seeing a very high level of research and practical attention. Due to a number of factors, Lake Prespa's water level dropped by about 9.50 m over the last six decades. The hydrology and hydrogeology of the lakes' catchment have never been thoroughly studied because of the intricate runoff mechanisms and the lack of scientific cooperation among specialists from the three transboundary nations (Popovska and Bonacci 2007).

2 From Origin to Current State and Trends in Hydrological Regimes Diversity of Lakes from Their Origin Point of View

According to Aliaj (2012), Prespa and Ohrid lakes are formed during the Pliocene Alpine orogeny about four to five million years ago. Great and Lesser Prespa lakes are divided by a dam. The average water level elevations for the Lake Lasser Prespa, Great Prespa and Ohrid are 851.6, 848.66 and 693.49 m a.s.l., respectively (Popovska and Bonacci 2007). The Prespa–Ohrid area, as part of the inner Alpine-folding region, was affected by extensional tectonics since Pliocene (Melo 2001) and is still under the stress of an active tectonic–seismic activity.

A progressive general uplifting of the Mali i Thate–Galičica Mts accompanied by a respective subsidence of both eastern and western depressions occurred during the Pliocene–Quaternary which brought to the formation of the big horst–graben system (Trocinoet al. 2010; Kiri 2021), composed of Prespa Lake graben to the east and Ohrid Lake and Korça field grabens to the west separated by the Mali i Thate–Galičica horst (Shuka et al. 2021; Parise et al. 2024).

Along the eastern and western borders of Galičica–Mali i Thate mountainous horst, there are notable regional tectonic faults that extend primarily south-north, causing a nearly 1500-m vertical shift against the water table of Ohrid Lake. On the western base of mountain slopes, Naum Springs, Tushemisht, and Zagorçan springs emerged as a result of other nearby active faults that produced a limestone relief with a relatively low elevation (Melo 2001).

The identical intensive faulting along the eastern side of Galičica–Mali i Thate Mts, conducted to the formation of the Prespa Lake graben, which is filled up by Pliocene deposits. The smaller grabens and horsts observed in the area of villages Gorica and Liqenas (Melo 2001) and the 30 m high vertical cliff of the Zaveri swallow hole are (Fig. 2f) consequences of the local tectonic fault activity.

Both field observations and recent earthquake data support the tectonic origin and ongoing tectonic activities of the Prespa and Ohrid basins (Hoffmann et al. 2010) which are considered as ancient lakes formed during the Pliocene, between 2–3 and 5 million years ago. Considering molecular clock analyses of DNA data derived from several ancient lake endemic species in combination with the findings for other endemic animal groups (Albrecht and Wilke 2008), the researchers proposed a Pliocene age of approximately 2–3 Ma for the Lake Ohrid, as well. On the other hand, there is no indication neither for desiccation phases of the Prespa Lake nor for lower lake levels during its long (ca. 48 ka) hydrological history (Wagner et al. 2010).

In-depth analyses of surface sediment samples and sediment successions are also used to examine contemporary processes and environmental changes across time in both lakes. Furthermore, a regional tephra-stratigraphic framework that includes tephra layers found in both lakes, as well as biogeochemical bulk analysis of specific metals, pigment concentrations, and grain size distributions, enable the clarification of contemporaneous changes in their trophic, chemical, and hydrologic states

Lake Ohrid is known for its high carbonate interglacial deposits, while Lake Prespa has rich organic matter interglacial sediments (Wagner et al. 2010). Consequently, during the interglacial eras, mesotrophic and oligotrophic conditions predominated in the deep Ohrid Lake and shallow Prespa Lake, respectively (Shuka et al. 2005; Xhulaj and Shuka 2007). Last but not least, although the biota records of lakes are the finest indicator of climatic and environmental changes, they are insufficiently detailed for Prespa and Ohrid lakes to reveal their age and place of origin.

2.1 Exploring from Hydro-Geology to Hydro-Ecological Trends

The catchment's geomorphological makeup, vegetation cover, and human activities all affect the freshwater ecosystems' water quality and future. Albania's standing freshwater habitats can be divided into three categories: (i) glacial lakes, (ii) lakes with tectonic-karstic origin, and (iii) karstic lakes, depending on their origin and the catchment's geomorphology.

In terms of the surface that occupy, lakes of tectonic-karst origin represent the largest number, more than 90, including 3-largest lakes, Shkodra, Ohrid and Prespa (Fig. 2), and also the lakes of the Dumre Plateau. According to Andreychouk et al. (2021) and Eftimi et al. (2022), the lakes of Dumre Plateau belong to the karst development related to the presence of a tectonically affected dome of Triassic evaporite rocks with gypsum dominance.

The catchment of Shkodra Lake is mainly built of limestone and dolomite rocks, while the catchment of Great and Lesser Prespa Lakes constitute from limestone formations (Fig. 2a, b, e–g), all with origin from the Mesozoic age. In contrast to the above, the Albanian part of the Ohrid Lake catchment presents a mosaic of geomorphological composition (Fig. 2c). Flysch sediments that often mixed with argillite and conglomerate or sandstone rocks occur in the periphery of watershed and at a distance from the lake shore. The western slopes of the catchment which descend directly in the lake are composed from ultramafic rocks which cover c. 24% of the surface and limestone formations in the upper parts (Geologic Map of Albania, Meçe and Aliaj 2000). The serpentinized harzburgites and dunites of the ultramafic rocks are the main source of chromium and lateritic nickel ore deposits (Dilek et al. 2008; Shuka and Hallaçi 2010). Currently, the deposits of chromium mines appear on the surface which, with the previous deposits of iron-nickel and coal minings (Fig. 2d), represents the main source of heavy metal pollution of the lake water.

The largest number of glacial lakes occurs in the northeast of Albanian mountain systems such as Valamara-Shebenik-Martanesh-Balgjaj-Allaman and Lura, whose catchment is composed by ultramafic rocks, originated from the Lower Triassic to Lower Jurassic age. According to Dilek et al. (2008), this formation includes tectonized harzburgite and dunite, accompanying with minor hierzolite, pyrexonite and gabro. The ultramafic formations are distinguished by a low cover vegetation compared with the high cover vegetation on granitic (Glacial lakes of Doberdol-Sylbica and Korabi Mts) and calcareous substrates (Stevanović et al. 2003; Shuka et al. 2008, 2017) such as glacial lakes in the Albanian Alps, Jezerca Mt (Fig. 3a–c), the catchment of Prespes Lake (Fig. 2e, g) and Bistrica Karstic lakes. The low cover vegetation on the ultramafic substrates and deforestations are the main reason of the increase of erosion phenomena, the filling of lakes with alluvial sediments and rapid increase of the macrophyte vegetation in the glacial lakes (Shumka et al. 2024).

Among the questions on the 'fate' of the hydro-ecological trends of Albanian lakes, the hydrology and geomorphology of Prespa and Pogradec catchment remains the most discussed.

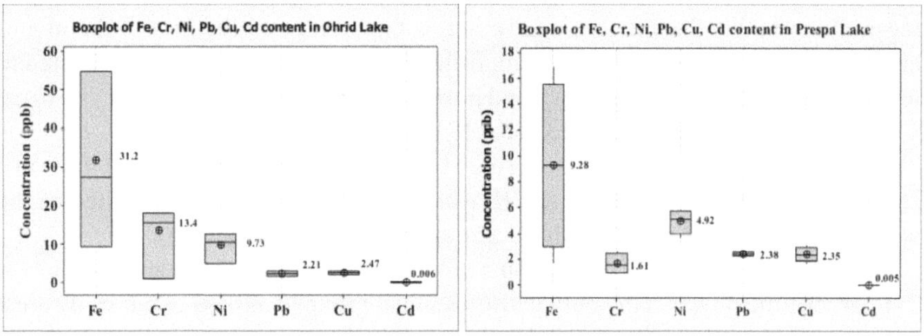

Fig. 3 Variation of metals concentration in Ohrid and Prespa Lakes

The Pliocene deposits, which include clay, sandstone, and conglomerate, fill the bottom of the Prespa and Ohrid lake graben structures, while the horst structure of the Mali Thate–Galičica Mountains is made of massive limestone from the Upper Triassic to Lower Jurassic (Eftimi et al. 2021; Kiri 2021; Parise et al. 2024).

Cvijić (1906) proposed the theory that the large karst springs of Tushemisht and St. Naum, which are situated on the southern edge of Lake Ohrid, are partially recharged by Prespa Lake due to the carbonate composition of the Mali Thate–Galičica Mountains and the higher elevation of the Prespa lake water table (about 155 m) in comparison to Ohrid lake. Although some unidentified karstic water volumes drain in the Ohrid Lake, the Tushemishit and St. Naum Spring discharge 2.5 m^3/s and 5.58 m^3/s, respectively (Amatajet al. 2007; Eftimi et al. 2014; Kiri 2016). The greatest proof of Lake Prespa's contribution to the recharge of Ohrid Lake, which was calculated to account for almost 25% of all inflows through subterranean karst canals, can be found at the Zaver swallow hole on the shore of Prespa Lake (van der Schriek and Giannakopoulos 2017).

The thick-bedded and massive limestones of the Mali Thate and Galičica Mts are highly karstified due to, among others, by the uplifting–down falling process (Ford and Williams 2007; Goldscheider and Drew 2007) which have produced complex and dense karst forms like Plateau surface on the top of mountains, blind valleys, caves and swallow holes. Apart from high difficulties of hydraulic parameters determination of karstic aquifers, several attempts are made to characterize them. Thus, Stamos et al. (2014) reported the following hydraulic parameters of limestone formations, estimated by several pumping tests of boreholes in the Greek territory: Transmissibility $T = 449.2\ m^2/day$, Hydraulic Conductivity $K = 22.5$ m/day, Storage Coefficient $S = 0.1–10\%$. Considering the highly heterogeneous nature of the karst aquifers, the above-reported values may be evaluated as apparent values (Eftimi et al. 2021), that, on the other hand, fit with the data provided for Dinaric karst by other researchers (Kresic and Stevanović 2010).

The high-elevation of the Mali Thate–Galičica karstic massif against its western drainage base, represented by Lake Ohrid shoreline, promoted high hydraulic gradients (Stevanović 2015) which, in turn, created conditions for the development of linear channels that discharge groundwater as big springs on the foot of the mountain slopes. Eftimi et al. (2021) distinguished three groups of springs: (a) Biljanini Springs, located near to Ohrid town and having a mean discharge of 1.5 m^3/s; (b) St. Naum–Tushemisht springs, located near to Pogradec town.

The St. Naum and Tushemisht springs have average discharge rate of 7.50 m^3/s and 2.5 m^3/s, respectively; (c) the Bilisht Valleygroup of springs, located in the western periphery of Mt. Ivan, whose average discharge is about 200 L/s.

Due to the dominance of the carbonate formations, mainly limestone, the HCO_3–Ca hydrochemical type is the only hydrochemical group identified in the study area. However, the karstic springs may be distinguished by comparing the main physical–chemical parameters, like Electrical Conductivity (EC), Total Dissolved Solids (TDS), temperature, etc. Eftimi et al. (2021) distinguish the above-mentioned spring groups based on the main physic-chemical parameters, as well.

Group 1 includes Biljanini and Galičica springs which have the less mineralized groundwater as shown by the EC (ca. 230 μS/cm) values. Their groundwater temperature is the lowest (8.1 °C in the small Galičica and 10.9 °C in Biljanini spring) compared to all the other studied karstic springs due to the higher elevation of their recharge areas. Group 2 includes St. Naum, Tushemisht and Zagorçan springs. These springs have higher (275–300 μS/cm) EC values, while the temperature ranges from 10.9 to 11.3 °C. Group 3 includes the springs of Devoll Valley, Big Golloborda, Progeri and Mançurisht. They have the highest (400–494 μS/cm) EC values. When compared spring waters with the Lake Prespa water which is characterized by the lowest (215 μS/cm) EC values, it is obvious that the respective increase of EC values in the karstic spring's groundwater is conditioned by their residence time within the karstic aquifer.

2.2 *Water Level, Oscillations and Other Phenomenon in Anthropocene*

According to monitoring data of the hydrological station at Stenje for the period 1951–2010, there was a decreasing but non-steady tendency of the Lake Prespa water level. The first sharp level decline started in 1986 reaching the maximum (app. 844 m a.s.l.) drop in 1995 (UNDP 2012) against maximum historical level of 852.91 m a.s.l. in 1963. A total decrease of 8.49 m in the Prespa Lake water level was reported by the Institute of Hydrometeorology of Tirana for the period 1963–2002 (Eftimi 2019). Popovska and Bonacci (2007) reported a water level decrease in Lake Prespa of 7.79 m for the period 1951–2000.

The above continuous water level decrease tendency of the Lake Prespa water was dedicated to three factors: (a) tectonic and seismic activities, (b) anthropogenic activities and (c) climate changes (Eftimi 2019; Popov and Anovski 2009; Matzinger 2006; Matzinger et al. 2006b).

On the contrary small water level fluctuations (ranging from 692.08 to 693.85 m a.s.l.) were reported for Lake Ohrid during the period from 2014 to 2019, indicating a relative stability. The water level fluctuations in Ohrid Lake do not correlate neither with that of Great Prespa lake nor with periodicity of precipitations in the region (Kiri 2021). A slight rise of the water level was observed during the period 2014–2019 in Lake Ohrid, with the highest water levels recorded in May and the lowest in November (Kiri 2021).

2.3 *What Is Going on with Great Prespa Lake?*

As mentioned above, a historical decline of water level in Prespa Lake is reported in different studies; even the rates of decrease are not always identical. A decreasing trend of 21.7 cm/yr is reported for the period 1963–2002 but a different (10.9 cm/yr) one was given for the period 1951–2000 (Popovska and Bonacci 2007). According to Eftimi (2019), the maximal (852.91 m a.s.l.) historical level was observed in 1963, while the minimal (844.12 m a.s.l.) in 2002. The recent and actual studies are mostly focused on the further clarification of the factors that contribute to this large water decline, which, in turn, has drawn the attention of the Albanian-Greece-Northern Macedonia governments and other world environmental organizations.

Following the dramatic water-level drop from 1987/88 to 1994/95, van der Schriek and Giannakopoulos analysis indicated that the sustained low lake levels below 847 m were caused by water abstraction, which accounted for approximately 72% of the decrease in lake surface evaporation, and amplified by climate-related inflow decreases, which accounted for approximately 28% of the decrease in lake surface evaporation (van der Schriek and Giannakopoulos 2017). Further on the same authors emphases that the fall itself was triggered by an extensive drought period. But the actual decrease was brought on by a protracted drought. Although estimates of the amount of water extracted for irrigation have been reported to be between 10 and 13 million m^3/year (Popov et al. 2009), which is equivalent to a drop of "only" 0.05 m of water level, overall studies of the water balance of the Prespa basin indicate that water extraction for irrigation, as well as for domestic or industrial use, cannot be considered the primary cause of the massive water loss of the lake. This could imply that the actual amount of water used for irrigation is likely significantly greater than what has been estimated. In actuality, it may be calculated that the water loss in the next years was more along the lines of at least 20 million m^3 annually when accounting for the irrigation projects created by the three bordering nations in the 1950s and 1960s. A study that examined air temperature from 1991 to 2020 (using data from the Pretor, North Macedonia) found that the mean annual temperature increased statistically significantly, but the maximal and minimal mean monthly temperatures did not change significantly. Additionally, during the analyzed period, no discernible changes were seen in the trend of Pretor's maximum monthly precipitation or yearly precipitation (Soria and Apostolova 2022). This has been enforced by other authors as well (Panagiotopoulos et al. 2014; Cvetkovska et al. 2014).

Among the factors controlling the Lake Prespa water level decline, the first one mentioned was the tectonic activity which, along with karstic solution processes, was thought to have widen the tectonic–karst pathways transmitting the water from Lake Prespa to Lake Ohrid. If this is the case, it might be expected that the discharge rate of large karst springs along the Lake Ohrid shoreline should be increased which has not been verified neither for St. Naum Spring (Stojov 2020) nor for the spring group of Tushemisht (Eftimi et al. 2021). Also geologically, this could not be acceptable because the tectonic–karstic processes need tens of thousands of years or more to happen (Chavkalovski 1997; Ford and Williams 2007; Parise et al. 2015).

Understanding the current and future of the lake's water level requires a thorough examination of Lake Prespa water balance. Unfortunately, due in large part to a lack of coordinated investigations by the three adjacent nations, the hydrology of the catchment has not been sufficiently studied, and accurate data for the current state are lacking (Trajkovski and Apostolova 2024; van der Schriek et al. 2023). Data from 25 years ago estimated that the lake's total inflow was approximately 16.9 m^3/s, or 534 million m^3 annually, of which 56% came from river runoff from multiple small streams, 35% from direct precipitation on the lake surface, and 9% from Micro Prespa. The remaining 46% of the water was lost through evaporation, 2% from irrigation, and 46% from outflow through the karst aquifer (Trajkovski and Apostolova 2024). The same author noted that these studies provided a kind of "status quo" water balance, when the lake's yearly water intake and outflow are comparable. Clearly, there has been a negative balance in recent decades. Therefore, "what is the cause of this effect and where does the water go?" is the key query. The annual water shortage has been estimated in a number of other studies. According to Popov et al. (2009), there was a deficit of about 32×10^6 m^3 annually over the 1990–2008 period since the estimated yearly inflow into Prespa Lake was 425.5×10^6 m^3 and the outflow was 458.3×10^6 m^3.

The anthropogenic impact on the level of Lake Prespa that represent the second factor, is mainly related to the increased use of its water for irrigation or other needs by the Albania, North Macedonia and Greece. In case of Albania, the water needed for irrigation of Korça plain was provided by the Lesser Prespa Lake. For this purpose, in 1976 the Devoll River in Albanian territory was diverted to flow to Lesser Prespa Lake where it discharged a water quantity of 30–70 million m^3/year during the period 1976–1996. The water quantity used for irrigation is relatively small comparing with the total volume of Great Prespa Lake and in addition, the water used for irrigation mostly was supplied by Devoll River which is part to another catchment. For this reason, Eftimi et al. (2021) concluded that the lake water used for irrigation can't be the reason for the decrease of Great Prespa Lake level. Another anthropogenic impact on the decline of Prespa Lake level could be the use of lake water for the massive irrigation of Resen plain in the territory of North Macedonia, or the use of Lesser Prespa water for irrigation in Greek territory.

The third factor affecting the Great Prespa Lake water level decline is considered to be climatic changes. The development of accurate climatic scenarios requires systematic monitoring of all the hydrologic-meteorological components which might be applied for the regional or local hydrological balances. In this context, the past hydrologic—meteorological monitoring data from all the three neighboring countries should be confronted and a common monitoring for the future should be implemented to better understand and manage the probable climate changes. As a result, many balance calculations of Prespa–Ohrid basin performed by different authors and institutions of three neighboring countries result in big discrepancies (Kolaneci 2003; Popovska and Bonacci 2007; Pano 1984). In particular, the lack of meteorological data on snowfall at high elevation stations and the absence of measured data on evaporation have been significant challenges. Comparing the data on precipitation with the data on the Lake Prespa water level decline, over a period of about 50 years, Selenica and Kolaneci (1997) and Milevski et al. (1997) also concluded that climate changes are the main factor of the lake water level decrease. The conclusions of Stojov's (2011, 2020) that the Prespa Lake suffers from continuous anthropogenic and climate changes, go to the same line with other opinions that stress the role of climate changes to the water level decline of the Great Prespa Lake.

To better understand the role climate changes to the hydrology of the Lake Prespa watershed and particularly, to the lake water level decline, Eftimi et al. (2021) presents interesting findings on the Zaver swallow hole which represents the "mouth" of an opened conduit that serves as pathway for drainage of Prespa lake water to shoreline springs of the Ohrid Lake. It's about a "road" connecting two sides of the Lake prolongation finishing to Zaver swallow hole which was seen only during the years 2002–2005 when the Lake arrived the lowest (844.5 m a.s.l.) known level and the "road" emerged on the water. According to authors, the "road", that is about 200 m long, 2.5–3.0 m wide and 0.5–3.0 m high, is constructed with local sandstone-conglomerate stones of different dimensions, from about 15–20 cm to big block more than 70–80 cm thrown without order. The authors believe that this "road" represents a dam constructed by local fishing people to keep lake level at higher elevation during climatic shortages similar to the actual ones (Fig. 2f). Leng et al. (2013) studied past climatic variability by analyzing the sediment isotope composition from Lake Prespa over the Last Glacial cycle. Their findings revealed that very low lake levels occurred around 1000 years ago, a period when the "road" of the Zaver swallow hole was likely constructed (Eftimi et al. 2021).

2.4 Trends in Lakes, Dam, Man-Made Lakes and Conversion of Floodplain Areas

According to Rigó and Barina (2021), the standing and running freshwater ecosystems in Albania are represented by 247 natural lakes, from which 103 have been glacial lakes, and about 800 artificial water reservoirs. Recent observations show that 19 glacial lakes in Albania, due to the global warming and the low level of snow cover in the winter (Shuka et al. 2017; Špoljar et al. 2021), are dried or sustain water only during the winter and spring time. So, currently in Albania there are 84 glacial lakes with a surface more than 0.5 ha (Shumka et al. 2024). The Albanian lakes and man-made reservoirs differ according to their formation origin, altitudinal elevation, and the type of their water supply and most importantly, climate and the human activities.

The great human impact on water flow modification and climate change has also affected the Albanian artificial water bodies. Currently, there are 630 reservoirs in the country, which is 21% fewer than before 1995. Of these, 65% of the existing dams are in a technically dangerous condition, and 524 of them have a depth ranging from 15 to 30 m (Anonymous 2003). The main reason of the man-made water bodies' reduction or water volume decrease, is destruction of their dams and lack of their maintenance. The implementation of new management strategies (Muir et al. 2012) in terms of the conservation of freshwater habitat types (Piczak et al. 2023) is important to reduce biodiversity loss (Lynch et al. 2023), stabilizing ecological function and guarantee ecosystem services for a long time (Ge et al. 2018; Grzybowski and Glińska-Lewczuk 2019).

2.5 Examples of Changing Land Use

Significant changes have been reported during 30-years in land use and Albanian freshwater ecosystems, including the man-made water reservoirs and ponds and their watershed (Shuka et al. 2005, 2017; Zakkak et al. 2018). Land use changes and urbanization have posed serious threats to the lake shorelines of Shkodra and Ohrid, negatively impacting their habitats and conservation status. The overexploitation of the natural resources, e.g. forest cutting and overgrazing or using of the waters for irrigation and electricity production, has drastically modified the lakeshore and water volume of the lakes in Valamara, Martaneshi, Kacnia and Lura areas and almost all the karstic lakes of Belshi municipality (Zakkak et al. 2018; Shumka et al. 2024). The high cover vegetation of tall floating and free-floating species in Lesser Prespa, Dushku and Shkodra Lakes, as well as in the lakes of Belshi area and Roskoveci swamp is a reflection of the high level of nutrients discharged into these water bodies. Many of the changes and modifications that already have been occurred in some of the lakes and reservoirs are irreversible e.g. Lesser Prespa Lake, Black and Kashta swamps and some lakes in Martaneshi area.

2.6 Characteristics, Main Impacts, and Stewardship of Natural and Artificial

The biggest lakes of Albania; Shkodra, Ohrid and Prespes, that are also considered as transboundary, lie respectively between Albania and Montenegro, Albania and North Macedonia and Albania and Greece. They are the largest lakes in Balkans, with tectonic ore tectonic-karstik origin. The largest number of karstic lakes in Albania is found in the Dumre plate c. 84, currently Belshi municipality. The past and current karst phenomena that occur in this area is one of the causes of the change in water level and their morphology. The second largest group of natural lakes are of glacial origin, 86.6% of them or 59 lakes are found in the catchments of ultramafic rocks, 22% in granitic rocks and only 9.3% of them in limestone rocks. The poor catchment substrate and low vegetation cover of the ultramafic bedrocks determine the good quality of waters and also poor floristic richness of the lakes in these substrates.

The main activities and threats that impact freshwater habitats in Albania lakes and reservoirs are: urbanization and construction of residential centers along the lake shorelines and reservoirs; alteration in flow water regime and disconnection between surface water and ground waters; decreased and unstable water use for agriculture and electricity produce; developing of aquaculture, urban and agricultural pollution; vegetation succession, invasive alien species and burning the drainage ditches; overexploitation and overgrazing on watershed and forest catchment and intensification of touristic activities (Shuka et al. 2009; Shumka et al. 2023).

The high level of water fluctuation as a result of climate change and also due to the use of lake waters for the electricity production and for irrigation, have increased the impact and modified the shoreline and riparian vegetation zone on the lakes and reservoirs. Riparian and shoreline vegetation act as a buffer strip and as a biological filter for the waters that are discharged into the lakes and reservoirs. The watershed cover vegetation slows runoff of surface waters and prevents erosion and deposition of sediment in the lakes, helping in the water quality and biodiversity conservation. Since lakes and reservoirs provide many benefits for the community, their conservation should be considered a more complex issue rather than simply the protection of the species and habitats.

3 Trends in Pollutants in Lake Systems

Many human-caused factors are threatening the diversity of life on Earth (Lacher and Roach 2018). Some have called the current situation the sixth mass extinction and a global annihilation (Ceballos et al. 2017). The urgency of taking immediate conservation actions to halt and reverse these decreases has been highlighted (Martin et al. 2012; Norris et al. 2020). However, the required course of action will rely on the gravity of the dangers as well as the possibility and viability of a successful intervention (Norris et al. 2020). The interplay of biological, social, economic, and governance elements can make even well-funded conservation efforts unsuccessful if they are not taken into consideration. Prioritizing species for conservation action and concentrating efforts are urgently needed. All lakes and other aquatic ecosystem types are facing serious different threats linked with accelerated eutrophication process, water level oscillation and inappropriate management.

3.1 Nutrients

Lake Prespa, that contributes approximately 50% of the total catchment area of Lake Ohrid, has a total phosphorus concentration (TP) of 31 mg/m^3, compared to 4.5 mg/m^3 in Lake Ohrid (Matzinger et al. 2006a). The high TP is addressed to intensified agriculture, villages' waste disposal sites scattered along the coastal area of Lake Prespa, increased use of P-containing detergents in the lake's coastal area, and the lack of sewage treatment (Shuka et al. 2005; Eftimi et al. 2021). In addition, the TP increase in Lake Prespa is related to the lake water level decline occurred since 1965 (Eftimi and Zojer 2015). The high total phosphorus (TP) in the waters of Lake Prespa serves as a significant warning for preserving the oligotrophic state of Lake Ohrid, given that there is an intensive water discharge from Prespa Lake to the oligotrophic Ohrid Lake.

Such a great contrast in trophic states between two lakes promoted Matzinger (2006) to raise the question much of the 65% of TP leaving Lake Prespa and entering the karst aquifer is retained in karst conduits, considering the fast flow rates and the limited contact between the water and limestone. One explanation proposed by Fetter (1993) is that phosphorus reacts with calcium carbonate to form a precipitate of hydroxyapatite. Coxon (1999) also concluded that phosphate may indeed be retained within the aquifer, even in a conduit flow system like the Zaver swallow hole.

On the contrary, Eftimi and Zojer (2015) suggest that the polluted waters of Prespa Lake seeping into the Mali Thate–Galičica karst massif could pollute both the karst groundwater and the water of Lake Ohrid. This is because the rapid movement of groundwater through big conduits reduces necessary time for self-purification and for microorganisms to die off (Drew and Hötzl 1999; Coxon 1999).

There is a strong relationship between chlorophyll, phosphorus and trophic state of lakes around the world. The content of Chl "a" in waters of Prespa Lake resulted to be four times higher compared to the Ohrid Lake, ranging from 4.00 to 4.20 $\mu g\ L^{-1}$ and from 0.6 to 1.40 $\mu g\ L^{-1}$, respectively (Shehu et al. 2017). These values correspond to trophic state indexes (TSI), of 44.19–44.68 and 25.58–33.90 for the Prespa and Ohrid Lake, respectively. Such TSI values confirm the mesotrophic of the waters in Lake Prespa and the oligotrophic state of Lake Ohrid (Shehu et al. 2017).

The lakes show differences in dissolved oxygen levels which decrease at varying rates with depth in the water column. The lowest concentration of dissolved oxygen (5.48 mg/L) was recorded in the Prespa Lake, at 10 m depth. Intensive growth of aquatic vegetation has led to severe de-oxygenation of the water (Shehu et al. 2017).

Higher concentrations of nutrients, BOD and COD in the Prespa Lake are primarily attributed to the discharge of untreated or insufficiently treated waste water into the lake and its tributaries from the Northern Macedonian sub-catchment (Shuka et al. 2005; GIZ 2017). The increased phosphorus levels indicate that Lake Prespa in Northern Macedonia is experiencing eutrophication, as the reduced water volume has diminished the lake's ability to dilute organic loads. During the summer period, at a 15 m depth (the bottom of the lake) the concentration of dissolved oxygen was extremely low (anoxia), and the concentration of phosphorus was exceptionally high

(> 87 mg/L) (GIZ 2017). The Trophic State Index (TSI) for the littoral zone of Lake Prespa indicates meso-eutrophic conditions.

An increasing concern is also linked with impacts of considerable period of water level oscillation and dropping on Prespa Lakes trophic status, nutrients and biodiversity. The bioassay of Maliaka et al. (2018) and generated results indicate that Lake Lesser Prespa is susceptible to additional N enrichment. Since cyanobacterial blooms are already occurring in the waterways, steps must be taken to lower current in-lake nutrient concentrations as well as stop additional enrichment. It is likely that a co-limitation of N and P characterizes large portions of Lake Lesser Prespa. Furthest to this, after Cvetkovska et al. (2014) presence of palaeo-shorelines and historical communities surrounding the lake suggests that Lake Prespa has previously seen significant lake-level variations. With an increase in total phosphorus (TP) input from a historic mean of about 20 to 31 mg P m^{-3} in 2003, there has been recent acceleration of anthropogenic eutrophication. On a historical contexts (Panagiotopoulos et al. 2014) based on specific approaches emphases the fact that the percentages of pollen from aquatic and semi-aquatic vascular plants in Prespa can be used as a stand-in to deduce variations in lake levels.

3.2 Contaminants

Shehu et al. (2017) used boxplot graphs to evaluate spatial variation of element concentration and found that the mean concentrations of Fe, Ni, Cr and Cd are higher in waters of Ohrid Lake. Among the studied metals, Fe shows the highest concentration in waters of both lakes, ranging from 9.34 to 54.63 and 1.63 to 16.92 $\mu g\ L^{-1}$ in waters of Ohrid and Prespa Lake, respectively (Fig. 3).

The authors found that Chromium content varied from 0.98 to 17.93 in Ohrid Lake to 0.9–2.6 $\mu g\ L^{-1}$ in Prespa Lake, being about 8 times higher in Ohrid water, probably due to intensive former and actual mining activities along the slopes and ultramafic geoecosystem composition of the watershed of Ohrid Lake, dominated by heavy metals (Shumka et al. 2024). For the same reason, Nickel content is twice its content in Prespa Lake, ranging from 4.93 to 12.26 in Ohrid water, and from 3.64 to 5.81 $\mu g\ L^{-1}$ in the Prespa water environment. An important role in the filtering process and preventing the discharge of heavy metals in the lake's waters is also played by the cover vegetation that is clearly larger in the watershed of Prespe Lakes (Shuka et al. 2008), compared to the cover vegetation of the ultramafic substrate of Lake Ohrid. Copper and Lead contents showed no significant differences between the waters of both lakes. Mean concentration of cadmium in both lakes didn't exhibit significant differences, showing values near or below the detection limit.

3.3 Anthropogenic Waste and Management Needs

Livestock

Cattle, sheep, goats, pigs, horses, donkeys, and poultry are among the types of livestock that graze the grazing land in the Albanian side of the Ohrid Region. Livestock on the Albanian side of the basin generate about 30 tons of trash. Similar to the Albanian side, a variety of cattle are housed on the pasture plains in North Macedonia (Watzin et al. 2002).

In Albania, livestock, which is mostly raised on small farms run and owned by the older generation, is a significant source of revenue for the local population. The majority of the animal grazing area is located in Prespa National Park. In Albania, agriculture accounts for 35% of the region's annual income, with these farms providing 65% of that total. Farmers employ a portion of their agricultural output to feed their livestock (UNDP 2005; Kiri 2021).

In North Macedonia the Prespa National Park is used as a feeding area for the livestock. However, the number of hectares under cultivation is much smaller compared to Albania (UNDP 2005).

Sewage Systems

The water and sewage systems in the Albanian portion of the Ohrid Region were mainly insufficient to meet the real demands until 2004. Around 54,000 people live in Pogradec city and its suburbs on the Albanian beaches of Ohrid Lake, where a new sewage treatment facility was constructed in 2004 with money from KFW (Germany) and SECO (State Secretariat for Economic Affairs).

In North Macedonia, wastewater management was given the necessary importance to protect the Ohrid Lake. This system collects the wastewater from the shoreline communities and delivers it to the treatment station. After treatment, the water is discharged into the River Crn Drim. The capacity of this system enables the collection, transport and treatment of only 65% of the wastewater produced in Ohrid–Struga Region (Watzin et al. 2002). In Pustec village (Prespa Region, in Albania), a sewage treatment facility was installed in 2004, financed by KFW. Other villages do not have such a sewage system, instead, septic tanks are used. In some areas, villages continue to dump sewage directly into the lake. In contrast, on the Greek part, almost all the villages have sewage systems using evaporation ponds.

Six settlements in the Prespa Region in North Macedonia have sewage treatment facilities connected to them. The facility is not set up to handle industrial garbage, even though this development is a positive move. The industrial solids consequently enter the lake and river immediately. Many settlements in Albania and North Macedonia still use septic systems, which will eventually degrade the quality of the lake's water and groundwater.

3.4 Albanian Lakes in the Middle of Climate Changes

The Zaver ponor (swallow hole) could provide us with some interesting data about the influence of climate change in the region, as it is the biggest and most important swallow hole into which the water of Prespa Lake disappears. Other swallow holes, not easily visible but nevertheless important, are located near the village of Golomboç (Globina). When the water level is high, the Zaver swallow whole makes it look like the lake having length about 600 m. In 2002, the level of the Lake Prespa decreased to about 844.5 m a.s.l. and a mysterious "road" appeared linking both sides of the channel of Zaver swallow hole. The "road", being about 200 m long, 2.0 m wide and 0.5–3.0 m high, was constructed using stones of different dimensions, from about 15–20 cm to more than 70 cm, and it is constructed like a dam. The water level is different on two sides of the road; being higher in the Prespa Lake than in the channel of the Zaver swallow hole; this led researchers to believe that it was used by local fishermen to preserve the water volume of the lake during periods of critical water level decline, particularly during the hottest historical weather events (Eftimi et al. 2021). Interesting data about the past climate have been obtained through detailed multi-method geochemical investigations, including stable isotope data from carbonates (δ18O, δH, δ13C from both calcite and siderite) of core samples of the lake bottom performed by Leng et al. (2013). High δ18O values that were obtained from these samples support the conclusion that lake levels were low as result of a significant arid phase. The last dry climate phases, accompanied by extremely low levels of the lake occurred around 1 ka ago. Rapid reversal was established to have occurred in the last 0.5 ka and is thought to correspond with the ruin of several buildings at 849–842 m a.s.l. (Sibinović 1987). It is logical to think that the Zaver dam was constructed to protect the excessive Prespa Lake drop, about 1–0.7 ka ago, when the climate was hotter than today. The high sensitivity of Lake Prespa to climate changes was emphasized also by Wagner and Wilke (2011).

Given the aforementioned considerations, it would seem that the Zaver dam was built during a lengthy, arid climatic cycle that coincided with a significant drop in the Great Prespa Lake's level. In order to protect the lake fauna, one of the most vital food sources, the dam was built to maintain the level as high as feasible.

Consequently, it is natural to wonder whether the recent catastrophic decrease of the Prespa Lake level in the last 50 years could be the result of climate change, similar to those the area experienced 1000–500 years ago (Eftimi et al. 2021). For this reason, it is very important to establish at which rate the climate change will affect the decrease of the water level of Lake Prespa.

Building freshwater resilience and changing water use practices are essential strategies for mitigating the impacts of climate change.

Lake Prespa

Point Sources of Pollution

The Albanian part of the Lake Prespa watershed does not have any notable point sources of pollution. Even though the marginal wetlands surrounding the lake serve as a buffer and filter, the Ezerani waste water treatment facility is a major point source of pollution to Lake Prespa in Northern Macedonia, where poorly treated wastewater and storm water outfalls from separate and combined sewerage systems clearly have a negative impact on water quality. 16,825 people who live in 44 places within the watershed in Northern Macedonia are the source of the waste water pressure. This number swells during the summer, when approximately 4000 tourists stay in the catchment, accommodated in private houses, hotels and holiday camp facilities. Industrial pollution arises from several small-to-medium sized enterprises, dealing with food processing, poultry farming, textiles, metal processing, wood processing, civil construction, ceramics and chemicals. Only 25% of the storm water network has been completed.

Significant Diffuse Sources of Pollution

In Albania: Agriculture and untreated waste water from communities, particularly from the villages of Kallamas, Gorica, Gollomboc, and Pustec (Liqenas), have an impact on the water quality of Lake Prespa. About 230 kg COD is the pollutant burden. Extensive agriculture has a less harmful impact on the lake. Farmers use very little inorganic fertilizers (only for wheat), relying primarily on livestock manure. Considering the high soil erosion factor of 31.7 tonnes ha/yr and the low absorptive capacity of the soil, up to 50% of the applied agrochemicals may be transferred to the lake and its sediments. Furthermore, illegal landfills and dumps in the villages especially near the shoreline are potential sources for diffuse pollution.

In Northern Macedonia: Agricultural activities result in higher diffuse source nutrient loading to Prespa Lake. In total, some 920 tonnes nitrogen and 477 tonnes phosphorus are applied each year in the Northern Macedonian watershed of Lake Prespa (UNDP 2012). Depending on the geology, soil type, land slope, and vegetation, between 15 and 30% of nitrogen fertilizer is typically discharged to surface and ground waters. Similarly, between 1 and 5% of phosphorus is exported to surface waterways, primarily due to vegetation, drainage density, and land slope. It is anticipated that around 64 tons of insecticides are utilized annually. Large amounts of hazardous solid waste from agricultural operations (pesticide packaging) and mostly organic garbage (apples and yard waste) are dumped in the riparian corridor and river channel. Surface water bodies, ground waters, soil, and particularly the Golema Reka water environment are all severely harmed by this (UNDP 2012). Besides agricultural activities, coastal and other villages in the watershed (Stenje, Krani, Ljubojno, Dolno Dupeni) also contribute to the overall load of diffuse source-derived pollution. Most villages, especially those in River Golema sub-catchment are not connected to the sewerage system and represent significant diffuse source pollution. Some of these (Konjsko, Stenje, Brajchino) have high tourist capacities, so are thought to represent relatively important diffuse sources of pollution.

Water Abstraction

In Albania: Five communities that use lake water for domestic and agricultural purposes are situated along the shores of Prespa Lake. About 3500 people live in these communities; the majority of them are farmers. This amounts to 700 m^3/day for personal use or a maximum abstraction rate of 20 m^3/day per family (including irrigation). Regular state monitoring and proper documentation of water extraction from the lake does not exist. The amount of abstracted water is smaller in winter when a substantial proportion of the population moves away from the villages to other locations.

In Northern Macedonia: In the northern part of the Prespa basin, numerous artesian wells have been drilled to depths of 15–200 m, with maximum flow rates of 6 L/s. For water supply, the town of Resen uses water from 2 karst springs at Krushje locality (Bigla Mountain), providing a flow rate of 35–160 L/s and three production wells (alluvial aquifer at Carev Dvor village), each 30 m deep, with total capacity of 50 L/s. Numerous wells and springs are used for water supply by surrounding villages (UNDP 2012).

In Prespa's three neighboring states, the agricultural industry continues to be one of the main sources of income for the local populace (Greece, Albania and North Macedonia). Therefore, future climate change will have a negative impact on local bean farming on the Greek side. For climate-proof agriculture, they suggested that bean varieties that can withstand heat and drought and don't require a lot of irrigation be given priority and promoted. In the future, sustainable watering techniques like drip irrigation and the utilization of reclaimed wastewater should be chosen (van der Schriek et al. 2020; Trajkovski and Apostolova 2024).

Lake Ohrid

Significant Point Sources of Pollution

In Albania: The most significant point source is the discharge (4500–5000 m^3/day) from Pogradec WWTP, designed to cater for a population equivalent of 40,000. The BOD removal efficiency is 90% (outflow concentration = 15 mg/L), and 87% for COD (outflow concentration = 40 mg/L). The treatment plant cleans an estimated 70%of the waste water produced in Pogradec. Other point sources are discharges of untreated waste water from villages (Lin Udemisht, Mëmëlisht, etc.) and settlements which are not connected to a central sewerage system.

In Northern Macedonia: The impact of point source pollution has probably decreased as the majority (ca. 80%) of the settlements in the watershed of the lake are now connected to sewerage systems and treatment plants. Industrial pollution of the lake and the waterbodies originates from several small to mid-sized enterprises, dealing mainly with food processing, civil construction, chemicals (Watzin et al. 2002). However, 2002 was a long time ago in water management terms, and a revised inventory of pollution sources and activities is required.

Significant Diffuse Sources of Pollution

In Albania: Water quality of Lake Ohrid is affected by agriculture and industrial activities, especially the metallurgical, chemical and mining industries. Several mines are located close to the lake (2.5 km), whereas four other coalmines (Alarup, Petrush, Vërdovë and Dardhas) can be found within a distance of 10 km from the lake shore. Chromium mining and naturally existing serpentine soil from the Pojskë (Pogradec) area are other variables that pose a threat to the quality of the water. The latter is potentially toxic due to its high content of heavy metals and high nickel availability. Agricultural land use in the immediate proximity of the lake has decreased from 460 to 130 ha. About 10% of waste waters are directly discharged into the lake without being treated due to constructions distributed along the shore without any planning, especially touristic buildings (hotels and restaurants). In the villages along the lake shore, there is no sewage system except for Lin village which discharges the sewage into a wetland located north of it. Furthermore, illegal landfills and dumps are potential sources of diffuse pollution. Especially Çerrava landfill has an elevated risk potential.

In Northern Macedonia: Agriculture is concentrated in the plains north of the lake and along the watersheds of its main tributaries. Fertilizers, soil particles and pesticides are washed with the erosion processes. The pollution load in the rivers represents a potential pressure on the lake. The diverted River Sateska and other tributaries rivers and springs passing through cropland are significant sources of organic and chemical pollution and discharge high sediment loads into the lake.

Water Abstraction

In Albania: Water is abstracted for the irrigation of cropland south-west and south of the lake, and for domestic purposes in rural areas. The volume of water abstraction is higher during summer owing to increased consumption by tourist facilities. Official figures, however, are not available.

In Northern Macedonia: Ground water abstraction in Ohrid-Struga region.

This area is located in the western part of Macedonia and encompasses Ohrid-Struga valley and a fringe of mountains (Jablanica, Galichica, Karaorman and others). Main urban centres are Ohrid and Struga. The main recipients of the surface waters of Lake Ohrid is the Black Drin River. The rivers Koselska and Sateska, tributaries of Lake Ohrid, are also important. In the Black Drin sub-catchment, seven springs are abstracted for water supply. Following Preveling et al. (2015), the largest value is measured at St. Naum (> 10 $m^3\ s^{-1}$), Vevchani (1.5 $m^3\ s^{-1}$), Biljanini Izvori (0.2–1 $m^3\ s^{-1}$), Beli Vodi (0.3 $m^3\ s^{-1}$) and Bel Bunar (0.04–0.1 m^3 s). Ground water is used for public water supply in Ohrid and Struga, with boreholes supplying individual dwellings in rural areas.

Agricultural Water Use

The total area of arable land in Macedonia is approximately 667,000 ha, about 60% of which is currently (or planned to be) under irrigation. Sprinkler irrigation is the predominant technique, covering 61% of the irrigated area. Other types of surface irrigation are applied to the remainder of the land. It is estimated that 15,205 ha of arable land in the Ohrid-Prespa region will be irrigated by 2020, accounting for 75% of the total water demand. Two pumping stations, one in Asamati and the other one in Sirhan, have been used to supply irrigation systems east and west of Lake Prespa on Macedonian territory. Presently, Lake Prespa and its tributaries, as well as ground water resources, are all abstracted for irrigation purposes. Wells combined with drip-irrigation systems have become the predominant method of irrigation in the region due to the unreliability of channel irrigations systems. Some 8000–10,000 wells have been drilled, covering an area of at least 3000 ha. Besides wells, a number of irrigation water intakes exist in rivers in the catchment. Some of them use remnants of old irrigation systems but a significant number are newly constructed systems of low efficiency, with high water losses. These systems are largely unregulated and beyond the control of water authorities and may have negative effects on ground water and lake quality.

References

Albrecht C, Wilke T (2008) Lake Ohrid: biodiversity and evolution. Hydrobiologia 615:103–140. https://doi.org/10.1007/s010750-008-9558-y

Aliaj S (2012) Neotectonics of Albanian. KLEAN, Tirana, p 292 (in Albanian)

Amataj S, Anovski T, Benischke R, Eftimi R, Gourcy LL, Kola L, Leontiadis I, Micevski E, Stamos A, Zoto J (2007) Tracer methods used to verify the hypothesis of Cvijić about the underground connection between Prespa and Ohrid Lake. Environ Geol 51:749–753. https://doi.org/10.1007/s00254-006-0388-9

Andreychouk V, Eftimi R, Nita J, Klimchouk A (2021) Geomorphology and hydrogeology of an exposed evaporite dome: the Dumre karst area, Central Albania. Geol Q 65(4)

Anonymous (2003) Vlerësimi i rreziqeve nga fatkeqësitë në Shqipëri. Raport studimi. UNDP, DFID, Tirana, 106 pp

Anovski T, Andonovski B, Minceva B (1991) Study of the hydrologic relationship between Ohrid and Prespa lakes. In: Proceedings of IAEA international symposium, 11–15 Mar 1991, IAEA-SM-319, 62 pp

Catsadorakis G, Malakou M (1997) Conservation and management issues of Prespa National Park. Hydrobiologia 351:175–196. https://doi.org/10.1023/A:1003093330400

Ceballos G, Ehrlich PR, Dirzo R (2017) Biological annihilation via the ongoing sixth mass extinction signaled by vertebrate population losses and declines. Proc Natl Acad Sci USA 114:E6089–E6096. https://doi.org/10.1073/pnas.1704949114

Chavkalovski I (1997) Hydrology of Prespa Lake. In: International symposium "towards integrated conservation and sustainable development of transboundary macro and micro Prespa Lakes", Korça, Albania, 24–26 Oct 1997, pp 37–39

Coxon C (1999) Agriculture induced impacts. In: Drew D, Hötzl H (eds) Karst hydrogeology and human activities. A.A Balkema, Rotterdam, pp 37–80

Cvetkovska A, Levkov Z, Reed MJ, Wagner B (2014) Late Glacial to Holocene climate change and human impact in the Mediterranean: the last ca. 17 ka diatom record of Lake Prespa (Macedonia/Albania/Greece). Palaeogeogr Palaeoclimatol Palaeoecol 406(2014):22–32. https://doi.org/10.1016/j.palaeo.2014.04.1010

Cvijić J (1906) Fundamental of geography and geology of Macedonia and Serbia (in Serbian), spec. ed. books 1, 2, VIII + 688. Serbian Royal Academy, Belgrade

Dilek Y, Furnes H, Shallo M (2008) Geochemistry of the Jurassic Mirdita Ophiolite (Albania) and the MORB to SSZ evolution of a marginal basin oceanic crust. Lithos 100(1–4):174–209
Dill WA (1993) Inland fisheries of Europe. EIFAC technical paper. No. 52 suppl. FAO, Rome, 281 pp
Drew D, Hötzl H (1999) Introduction. In: Drew D, Hötzl H (eds) Karst hydrogeology and human activities. International contribution to hydrogeology, vol 23. Balkema, Rotterdam, pp 3–12
Eftimi R (2019) The catastrophic decrease of Prespa Lake level: there are natural or anthropogenic reasons? In: Proceedings of 4th IAH-CEG conference (Central European Group) on "towards sustainable management of groundwater resources". Donji, pp 20–24
Eftimi R, Zojer H (2015) Human impact on karst aquifers of Albania. Environ Earth Sci 7:57–704
Eftimi R, Zoto J (1997) Isotope study of the connection of Ohrid and Prespa lakes. In: International symposium "towards integrated conservation and sustainable development of transboundry macro and micro Prespa Lakes", symposium held in Korça, Albania, pp 32–37
Eftimi R, Amataj S, Zoto J (2014) Groundwater circulation in two transboundary carbonate aquifers of Albania; their vulnerability and protection. In: Selected papers on hydrogeology, vol 11. Taylor & Francis Group, London, UK, pp 199–212
Eftimi R, Stevanović Z, Stojov V (2021) Hydrogeology of Mali Thate–Galičica karst massif related to the catastrophic decrease of the level of Lake Prespa. Environ Earth Sci 80:708. https://doi.org/10.1007/s12665-021-10006-z
Eftimi R, Parise M, Liso IS (2022) Karst brackish springs of Albania. Hydrology 9:127. https://doi.org/10.3390/hydrology9070127
Fetter CW (1993) Contaminant hydrogeology. Macmillan Publishing Company, New York, 458 p
Ford D, Williams P (2007) Karst hydrogeology and geomorphology. Wiley, Chichester, p 562
Ge Y, Zhang K, Yang X (2018) Long-term succession of aquatic plants reconstructed from palynological records in a shallow freshwater lake. Sci Total Environ 643:312–323. https://doi.org/10.1016/j.scitotenv.2018.06.203
GIZ (2017) Fish and fisheries Prespa Lake. Implementing the EU water framework directive in South-Eastern Europe. Pegi Sh.p.k., Tirana, Albania, 120 pp
Goldscheider N, Drew D (2007) Methods in karst hydrogeology. In: Goldscheider N, Drew D (eds) International contribution to hydrogeology, vol 26. Taylor and Francis, London, p 264
Grzybowski M, Glińska-Lewczuk K (2019) Principal threats to the conservation of freshwater habitats in the continental biogeographical region of Central Europe. Biodivers Conserv 28(14):4065–4097. https://doi.org/10.1007/s10531-019-01865-x
Hoffmann N, Reicherter K, Fernandez-Steeger T, Grutzner C (2010) Evolution of ancient Lake Ohrid: a tectonic perspective, Biogeosciences 7:3377–3386. https://doi.org/10.5194/bg-7-2277-2010
Kiri E (2016) GIS: a useful tool for managing of transboundary aquifers: a case study from Prespa Lake Basin. Bull Geol Soc Gr 50(2):750–759
Kiri E (2021) Hydrodynamic and hydrochemical investigation of the transboundary aquifer system in Prespa–Ohrid watershed. Ph.D. thesis, 370 pp
Kolaneci M (2003) IAEA regional project RER/8/008 study of Prespa Lake using nuclear and related techniques
Kresic N, Stevanovic Z (eds) (2010) Groundwater hydrology of springs. Elsevier, Amsterdam, p 573
Lacher T, Roach N (2018) The status of biodiversity in the Anthropocene: trends, threats, and actions. In: DellaSala DA, Goldstein MI (editors-in-chief), Lacher TE Jr, Pyare S (eds) The encyclopedia of the Anthropocene, vol 3. Oxford, UK, Elsevier, pp 1–8
Leng MJ, Wagner B, Bohem A, Panagiotopoulos K, Vane CA, Snelling A, Haidon C, Woodly E, Vogel H, Zancheta I, Baneschi I (2013) Understanding past climatic and hydrological variability in the Mediterranean from Lake Prespa sediment isotope and geochemical record over the Last Glacial cycle. Q Sci Rev 66:123–136. https://doi.org/10.1016/j.quascirev.2012.07.015

Lynch AJ, Hyman AA, Cooke SJ, Capon SJ, Franklin PA, Jähnig SC et al (2023) Future-proofing the emergency recovery plan for freshwater biodiversity. Environ Rev. https://doi.org/10.1139/er-2022-0116

Malatesta L, De Sanctis M, Ammann E, Attorre F, Buffi F, Cambria VE, Shuka L, Fanelli, G (2023) Bioregionalization of Albania: mismatch between the flora and the climate suggests that our models of Southern European bioregions are in need of a revision. Folia Geobot 58(1):71–87. https://doi.org/10.1007/s12224-023-09432-7

Maliaka V, Faassen EJ, Smolders AJ, Lürling M (2018) The impact of warming and nutrients on algae production and microcystins in Seston from the iconic Lake Lesser Prespa, Greece. Toxins 10(4):144. https://doi.org/10.3390/toxins10040144

Martin TG, Nally S, Burbidge AA, Arnall S, Garnett ST, Hayward MW, Lumsden LF, Menkhorst P, McDonald-Madden E, Possingham HP (2012) Acting fast helps avoid extinction. Conserv Lett 5:274–280. https://doi.org/10.1111/j.1755-263X.2012.00239.x

Matzinger A (2006) Is anthropogenic nutrient input jeopardizing unique Lake Ohrid?—mass flux analysis and management consequences. Doctoral sciences thesis, p 130

Matzinger A, Spirovski Z, Patceva Wüest S (2006a) Sensitivity of ancient Lake Ohrid to local anthropogenic impacts and global warming. J Great Lakes Res 32(1):158–179

Matzinger A, Jordanoski M, Veljanoska-Sarafiloska E, Sturm M, Müller B, Wuest A (2006b) Is Lake Prespa jeopardizing the ecosystem of ancient Lake Ohrid? Hydrobiologia 553:89–109. https://doi.org/10.1007/s10750-005-6427-9

Meçe S, Aliaj S (2000) Geology of Albania. Gebruder Borntraeger, Berlin, Stuttgart, 246 pp

Melo V (2001) Pogradec (Ohrid Lake) and Plloca tectonic doorstep (morphostructural and cinematic analyses). Albanian J Nat Tech Sci 1:123–136

Miho A, Kashta L, Beqiraj S (2013) Between the land and the sea—ecoguide to discover the transitional waters of Albania. Julvin 2, Tirana, pp 1–462

Miho A, Marka J, Krasniqi Z (2023) Importance of EU integration for biodiversity and nature conservation in transboundary protected areas (TPAs) in the Western Balkan. Hydrobiology 2:235–243. https://doi.org/10.3390/hydrobiology2010015

Milevski J, Ristevski P, Tanusevska D (1997) In: International symposium "towards integrated conservation and sustainable development of transboundary macro and micro Prespa Lakes", Korça, Albania, 24–26 Oct 1997, pp 52–58

Muir MC, Spray CJ, Rowan JS (2012) Climate change and standing freshwaters: informing adaptation strategies for conservation at multiple scales. Area 44(4):411–422. https://doi.org/10.1111/j.1475.2012.01130.x

Norris K, Terry A, Hansford JP, Turvey ST (2020) Biodiversity conservation and the earth system: mind the gap. Trends Ecol Evol 35:919–926. https://doi.org/10.1016/j.tree.2020.06.010

Panagiotopoulos K, Böhm A, Leng MJ, Wagner B (2014) Climate variability over the last 92 ka in SW Balkans from analysis of sediments from Lake Prespa. Clim Past 10:643–660. https://doi.org/10.5194/cp-10-643-2014

Pano N (1984) Hydrology of Albania. Institute of Hydrometeorology, Academy of Science, Tirana, 441 pp (in Albanian)

Parise M, Closson D, Gutiérrez F, Stevanović Z (2015) Anticipating and managing engineering problems in the complex karst environment. Environ Earth Sci 74:7823–7835. https://doi.org/10.1007/s12665-015-4647-5

Parise M, Andreychouk V, Liso IS, Trocino A, Eftimi R (2024) Karst geomorphology, hydrogeology and caves in the areas of the Prespa Lakes, at the Albania-Macedonia-Greece border. Geophys Res Abstr. https://doi.org/10.5194/egusphere-egu24-11841

Piczak ML, Perry D, Cooke SJ, Harrison I, Benitez S, Koning A et al (2023) Protecting and restoring habitats to benefit freshwater biodiversity. Environ Rev. https://doi.org/10.1139/er-2023-0034

Poff NL, Olden JD, Merritt DM, Pepin DM (2007) Homogenization of regional river dynamics by dams and global biodiversity implications. Proc Natl Acad Sci USA 104:5732–5737. https://doi.org/10.1073/pnas.0609812104

Popov V, Anovski T (2009) Determination of volume of water reservoir by using tritium concentrations. Water Resour Manag 125. https://doi.org/10.2495/WRM090411

Popov V, Anovska E, Arsov M, Amataj S, Kolaneci M, Stamos A, Arsov L, Anovski T (2009) Study of Prespa-Ohrid lake system using tracer experiments and the lake's water balance. WIT Trans Ecol Environ 125:75–84

Popovska C, Bonacci O (2007) Basic data on the hydrology of Lakes Ohrid and Prespa. Hydrol Process 21(5):658–664. https://doi.org/10.1002/hyp.6252

Preveling R, Bramick U, Densky H, Parr B, Pietrock M, Adhami E et al (2015) Initial characterization of Lakes Prespa, Ohrid and Shkodra/Skadar. Implementing the EU water framework directive in South-Eastern Europe. Deutsche Gesellschaft fur Internationale Zusammenarbeit, Bonn, Eschborn

Rigó A, Barina Z (2021) The floristic composition of irrigation ponds and water reservoirs in Albania after the long persistent drought of 2016–2017. In: 1st international electronic conference on biological diversity, ecology and evolution, vol 10. MDPI, Basel, Switzerland. https://doi.org/10.3390/BDEE2021-09509

Selenica A, Kolaneci M (1997) The catastrophic falling of the water level in Prespa Lake. In: Gjiknuri L, Miho A (eds) International symposium "towards integrated conservation and sustainable development of transboundary macro and micro Prespa Lakes", Korça, Albania, 24–26 Oct 1997, pp 49–51

Shehu A, Vasjari M, Duka S, Vallja L, Broli N (2017) Comparative study of the environmental state of Ohrid and Prespa Lakes, Albania. Eur Water 58:237–244

Shuka L, Hallaçi B (2010) Is determined flora and vegetation of Mirusha (Kosovo) and Kolshi (Albania) area from the serpentine substrate? In: BALWOIS 4th conference-Ohrid, pp 25–29

Shuka D, Tan K (2024) On the identity of *Scorzonera doriae* (Asteraceae) and *Athamanta densa* (Apiaceae), two taxa collected by Antonio Baldacci in South Albania. Phytol Balc 30(3):361–368. https://doi.org/10.7546/PhB.30.3.2024.9

Shuka L, Malollari I, Veizaj D (2005) Assessment of phosphates and nitrates effects in water's, quality of Ohrid and Prespa Lakes. J Environ Prot Ecol 6(4):761–770

Shuka L, Malltezi J, Mersinllari M, Vardhami I (2008) Dynamics of vegetation cover of Prespa Lakes and its watershed (Albanian side). In: Proceedings of the third international scientific conference on Balkan water observation and information system (BALWOIS), pp 27–31

Shuka L, Çullaj A, Shumka S, Miho A, Duka S, Bachofen R (2009) Response of drinking-water reservoir ecosystems to anthropogenic impacts in Albania: trends of interrelationship. J Int Environ Appl Sci 4(4):478–486

Shuka L, Mullaj A, Hoda A, Kashta L, Miho A (2017) Overview on the flora and vegetation of the Albanian Alps—the degree of conservation and threats. In: Conference: 37th meeting of Eastern Alpine and Dinaric society for vegetation ecology (EADSVE), Prizreni, Kosovo, pp 13–16. https://doi.org/10.13140/RG.2.2.23064.70400

Shuka L, Shuka D, Diku A (2021) Bimët Endemike dhe ato me Përhapje të Kufizuar, Parku Kombëtar "Prespa". Gent Grafik, Tiranë, 150 pp

Shumka S, Lalaj S, Šanda R, Shumka L, Meulenbroek P (2023) Recent data on the distribution of freshwater ichthyofauna in Albania. Croat J Fish 81:33–44. https://doi.org/10.2478/cjf-2023-0004

Shumka S, Shumka L, Špoljar M, Shuka L (2024) Evidence of climate Change and the conservation needed to halt the further deterioration of small glacial lakes. Climate 12:124. https://doi.org/10.3390/cli12080124

Sibinović M (1987) Ezera Prespansko i Ohridsko. The SRC & Agency of Water Resources of Macedonia, Skopje

Soria J, Apostolova N (2022) Decrease in the water level of Lake Prespa (North Macedonia) studied by remote sensing methodology: relation with hydrology and agriculture. Hydrology 9:99

Špoljar M, Shumka S, Tasevska O, Tomljanović T, Ostojić A, Balkić A, Lajtner J, Dražina T, Ternjej I (2022) Small standing-water ecosystems in the transitional temperate climate of the Western

Balkans. In: Pešić V, Milošević D, Miliša M (eds) Small water bodies of the Western Balkans. Springer, Cham. https://doi.org/10.1007/978-3-030-86478-1_2

Stamos A, Batsi A, Xantopoulou A (2014) Karstic aquifer systems and relations of hydraulic communication with the Prespa Lakes. In: The tri-national Prespa basin, Kalavrita, pp 527–534

Stevanović Z (2015) Karst aquifers: characterization and engineering. In: Stevanović Z (ed) Professional practice in earth sciences series. Springer, Heidelberg, p 687

Stevanović V, Tan K, Iatrou G (2003) Distribution of the endemic Balkan flora on serpentine I. Obligate serpentine endemics. Plant Syst Evol 242:149–170

Stojov V (2011) Climatic and anthropogenic impacts on water reserves on the tectonic lakes. M.S. thesis, University of Ss. Cyril and Methodius, Faculty of Civil Engineering, Skopje, Macedonia. https://www.moepp.gov.mk/wp-content/uploads/2020/12/prespa_klima_antropogeni_vlijanija_20201129.pdf

Stojov V (2020) Climate and anthropogenic impacts on the water reserves of the Prespa Lake (Климатски и антропогени влијанија врз водните резерви на Преспанското Езеро). Accessed 23 Jan 2021. https://www.moepp.gov.mk/prespa_klima_antropogeni_vlijanija_20201129.pdf

Trajkovski D, Apostolova N (2024) The catastrophic water loss of ancient Lake Prespa: a chronicle of a death foretold. Hydrology 11(12):199. https://doi.org/10.3390/hydrology11120199

Trocino A, Parise M, Rizzi A (2010) Ricerche speleologiche in Albania: primi dati sulle cavità nei pressi del lago di Prespa. In: XII reg. meeting speleology "Spelaion 07", pp. 246–259

UNDP (2005) Integrated ecosystem management in the Prespa Lakes Basin of Albania, FYR Macedonia and Greece. UNDP project report, 132 pp

UNDP (2012) Prespa Lakes Basin, strategic action programme (integrated ecosystem management in the Prespa Lake Basin). Report, 131 pp

van der Schriek T, Giannakopoulos C (2017) Determining the causes for the dramatic recent fall of Lake Prespa (Southwest Balkans). Hydrol Sci J 62(7):1131–1148. https://doi.org/10.1080/02626667.2017.1309042

van der Schriek T, Giannakopulos C, Varotsas KV (2020) The impact of future climate change on bean cultivation in the Prespa Lake catchment, northern Greece. Euro-Mediterr J Environ Integr 5(1). https://doi.org/10.1007/s41207-020-0151-8

van der Schriek T, Giannakopoulos C, Koutseri I, Malakou M, Kitsara G (2023) Assessing future fire and drought risk to the lake-wetland system in the Prespa National Park (Greece). In: EMS annual meeting 2023, EMS 2023-57, Bratislava, Slovakia, 4–8 Sept 2023. https://doi.org/10.5194/ems2023-57

Wagner B, Wilke T (2011) Evolutionary and geological history of the Balkan lakes Ohrid and Prespa. Biogeosciences 8:995–998. https://doi.org/10.5194/bg-8-995-2011

Wagner B, Vogel H, Zanchetta G, Sulpizio R (2010) Environmental change within the Balkan region during the past ca. 50 ka recorded in the sediments from lakes Prespa and Ohrid. Biogeosciences 7:3187–3198. https://doi.org/10.5194/bg-7-3187-2010

Watzin MC, Puka V, Naumovski TB (2002) Lake Ohrid and its watershed. State of the environment report, 134 pp

Xhulaj M, Shuka L (2007) Important planta areas (IPAs) in biodiversity enabling activity. Ministry of Environment, Forestry and Water Administration, Tirana, pp 231–276

Zakkak S, Radović A, Panitsa M, Vassilev K, Shuka L, Kuttner M, Schindler S, Kati V (2018) Vegetation patterns along agricultural land abandonment in the Balkans. J Veg Sci 29(5):877–886. https://doi.org/10.1111/jvs.12670

Current Status of Greek Lakes and Future Perspectives: Results from WFD Monitoring

Vasiliki Tsiaoussi, Efpraxia Mavromati, Kleoniki Demertzi, Victoria Giourieva, Sotiria Katsavouni, Dimitra Kemitzoglou, Valentini Navrozidou, Iliana Nikolopoulou, Dimitrios Papadimos, Athina Patsia, Miltiadis Seferlis, Maria Skopa, Eirini Varsami, Ioanna Zerva, and Marianthi Zioga

Abstract Lakes host rich biodiversity and provide valuable ecosystem services, including water for human and economic use. However, knowledge of lake ecosystems in Greece was previously limited. This gap has been addressed by the substantial increase in data from the National Lake Monitoring Network, which has been operational since 2012. The EU Water Framework Directive (WFD) has set the guidelines for lake monitoring and assessment systems. In this chapter, we present the key

V. Tsiaoussi (✉) · E. Mavromati · K. Demertzi · V. Giourieva · S. Katsavouni · D. Kemitzoglou · V. Navrozidou · I. Nikolopoulou · D. Papadimos · A. Patsia · M. Seferlis · M. Skopa · E. Varsami · I. Zerva · M. Zioga
Greek Biotope/Wetland Centre, National Museum of Natural History Goulandris, Thermi, Greece
e-mail: vasso@ekby.gr

E. Mavromati
e-mail: emavromati@ekby.gr

K. Demertzi
e-mail: kldemertzi@ekby.gr

V. Giourieva
e-mail: victoria@ekby.gr

S. Katsavouni
e-mail: sotiria@ekby.gr

D. Kemitzoglou
e-mail: dimitra@ekby.gr

V. Navrozidou
e-mail: vnavrozidou@ekby.gr

I. Nikolopoulou
e-mail: inikolop@ekby.gr

D. Papadimos
e-mail: papadimos@ekby.gr

A. Patsia
e-mail: athina.patsia@ekby.gr

K. Stefanidis et al. (eds.), *Lakes of the Eastern Mediterranean and Balkans*,
https://doi.org/10.1007/978-3-032-02168-7_5

characteristics, ecological status, and trends of 49 Greek lakes between 2016 and 2022, as assessed according to the WFD. Overall, 55% of the lakes were classified as having good or better ecological status/potential. Of the remaining lakes, 33% were rated as moderate, 10% were classified as poor, and only 2% were rated as having bad ecological status. Specifically, the majority of reservoirs (85%) were assessed as having good or above ecological potential, while only 22% of natural lakes were rated similarly. The majority of natural lakes were classified as having moderate (52%), poor (22%), or bad ecological status (4%). Monitoring activities have played a crucial role in improving our understanding of the structure and functioning of lake ecosystems across different spatial and temporal scales. Additionally, the national monitoring network has created a valuable time series of data documenting the status and trends of Greek lakes, available to all interested parties. Authorities and other stakeholders can use these data to prioritize relevant conservation and restoration measures and assess their effectiveness.

Keywords Lakes · WFD · Biological indicators · Physicochemical parameters · Hydrology · Monitoring · Assessment · Ecological status · Greece

1 Introduction

Freshwaters are considered biodiversity hotspots, supporting approximately 10% of all known species and around one-third of vertebrate species (Strayer and Dudgeon 2010). Lakes larger than 0.002 km^2 cover 3.7% of Earth's non-glaciated land area, as counted by satellite-based remote sensing (Verpoorter et al. 2014). In addition to their intrinsic biodiversity value, lakes are widely recognized for providing essential ecological services, including being a primary source of water for human and economic use (Reynaud and Lanzanova 2017). Despite their importance, freshwater biodiversity is in decline, as reflected by the downward trends observed in many freshwater taxa (Reid et al. 2019). Scientists have identified current challenges (e.g., Reid et al. 2019) and proposed strategies to advance freshwater biodiversity research,

M. Seferlis
e-mail: seferlis@ekby.gr

M. Skopa
e-mail: mskopa@ekby.gr

E. Varsami
e-mail: eirini@ekby.gr

I. Zerva
e-mail: izerva@ekby.gr

M. Zioga
e-mail: marianthi@ekby.gr

aiming to inform conservation and restoration efforts (Harper et al. 2021; Maasri et al. 2022).

In Greek lakes, land use changes and population density of catchments play an important role in shaping lake water quality (Mavromati et al. 2018). In particular, intensive agriculture and urbanization seem to be the main driving forces of nutrient inputs from point and non-point sources, water abstraction and habitat degradation, leading to eutrophication and hydromorphological alterations (Latinopoulos et al. 2016; Mavromati et al. 2018). The sensitivity of lakes to nutrient enrichment—and their capacity to retain these nutrients—is also influenced by their morphometric characteristics, which also play a critical role in determining water retention time (Kagalou et al. 2021a). The climate crisis further exacerbates these challenges, as the Mediterranean region is especially vulnerable to global change (Giorgi and Lionello 2008).

To address human-induced pressures such as those mentioned above, the European Union has put in place the Water Framework Directive (WFD; European Commission 2000), which serves as the primary regulatory framework for the protection of surface waters and groundwater. In relation to lakes, the directive aims to achieve and maintain good ecological and chemical status. However, when water bodies are designated as heavily modified or artificial, the goal is to reach good ecological potential instead. The WFD defines ecological status as an expression of the quality of the structure and functioning of aquatic ecosystems in surface waters, classified according to Annex V. To meet the WFD's requirements for assessing ecological status/potential, EU Member States have developed standardized biological assessment methods based on a range of Biological Quality Elements (BQEs). In lakes, these BQEs include: phytoplankton, macrophytes, phytobenthos, benthic invertebrate fauna, and fish fauna. Regulatory thresholds, aligned with WFD definitions, are compared and harmonized across countries through a process called intercalibration. As a result of this exercise, Commission Decision (EU) 2024/721 sets the values for Member State monitoring system classifications that are applied across Europe. For Greek lakes, the following biological assessment systems have been developed for each BQE:

- **Phytoplankton**: New Mediterranean Assessment for Reservoirs (NMASRP), Hellenic Lake Phytoplankton Assessment Method (HeLPhy), Greek Phytoplankton Trophic Index (GR-PTI)
- **Macrophytes**: Hellenic Lake Macrophytes Assessment Method (HeLM)
- **Phytobenthos**: Hellenic Assessment Method for Lake Diatoms
- **Benthic Invertebrate Fauna**: Hellenic Assessment Method for Lake Littoral Benthic Invertebrate Fauna (HeLLBI), Greek Lake Benthic Invertebrate Index (GLBiL)
- **Fish**: Greek Lake Fish Index (GLFI).

Notably, the phytobenthos index (Kelly et al. 2023) has not yet been included in the Commission Intercalibration Decision (2024), while the GLBiL was applied during the first revision of the River Basin Management Plans. Finally, for physicochemical quality elements, boundaries for Total Phosphorus (TP) have been established (Kagalou et al. 2021a).

The aim of this chapter is to present the status and trends of Greek lakes, based on the results of biological indicators and TP boundaries, as monitored and assessed according to the WFD. We begin by outlining the lakes included in the monitoring network, focusing on their key physicochemical and hydromorphological characteristics, with hydrology primarily addressing reservoirs. Next, we present the biological assessment methods developed and included in the Intercalibration Decision (2024), and outline the lake classification outcomes derived from their application. As the primary focus is on the current status of the lakes, the monitoring results cover the period from 2016 to 2022, unless otherwise specified. Finally, we provide an overall classification of the lakes' ecological status/potential and discuss the outcomes of our analyses.

2 The Monitoring Network

In Greece, the competent body responsible for monitoring natural lakes and reservoirs is The National Museum of Natural History Goulandris—Greek Biotope/Wetland Centre (NMNHG/EKBY), as designated by Joint Ministerial Decision 140384/2011. NMNHG/EKBY has operated the National Lake Monitoring Network continuously since 2012. The network includes 50 lake water bodies, both natural lakes and reservoirs, with their distribution shown in Fig. 1. Each lake typically has one sampling station, except for transboundary lakes (Doirani, Megali Prespa, and Mikri Prespa), which each have two. In total, there are 53 sampling stations under surveillance and operational monitoring. Since 2021, the network has expanded further, with monitoring now taking place at 80 sampling stations (JMD YPEN/DPDYP/107168/1444/2021).

Natural lakes in Greece are grouped into three types, primarily based on mean depth values and mixing regimes (Kagalou et al. 2021a; Perivolioti et al. 2025):

1. **Warm monomictic deep natural lakes** with a mean depth greater than 9 m (GR-DNL)
2. **Shallow polymictic natural lakes** with a mean depth between 3 and 9 m (GR-SNL)
3. **Very shallow natural lakes** with a mean depth of less than 3 m (GR-VSNL).

Reservoirs are divided into two categories based on depth:

1. **Deep reservoirs** with a mean depth greater than 15 m (GR-DR). This category refers to the calcareous and siliceous reservoir types (types LM 8 and LM 5/7 respectively) as described in the Intercalibration Decision (2024);
2. **Shallow reservoirs** with a mean depth of less than 15 m (GR-SR).

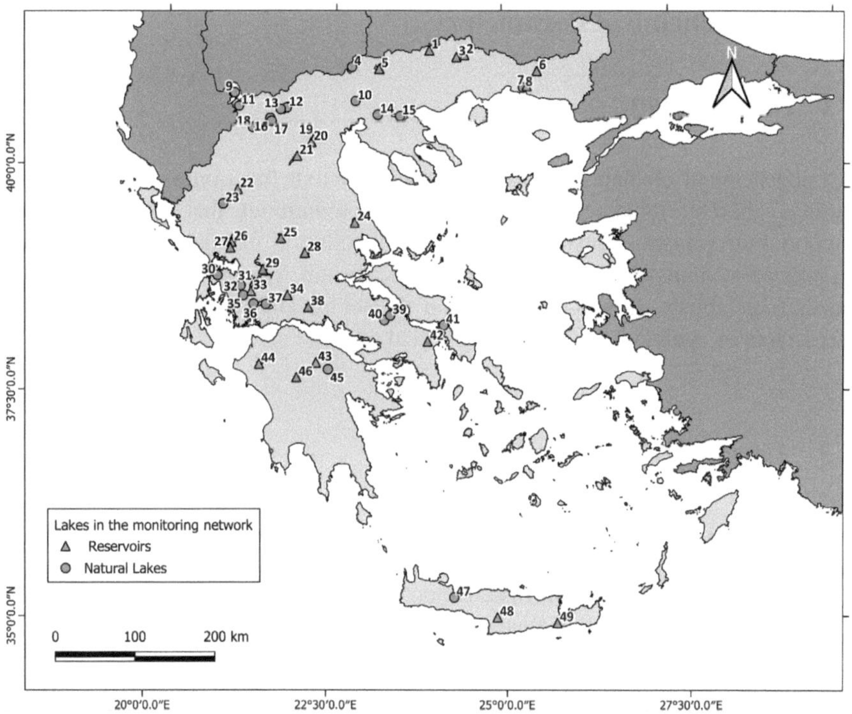

Fig. 1 Monitoring network for Greek lakes. Reservoirs: (1) T.L. Leukogeion, (2) T.L. Platanovrysis, (3) T.L. Thisavrou, (5) T.L. Kerkinis, (6) T.L. Gratinis, (7) T.L.N. Adrianis, (19) T.L. Asomaton, (20) T.L. Sfikias, (21) T.L. Polyfytou, (22) T.L. Pigon Aoou, (24) T.L. Karlas, (25) T.L. Tavropou, (26) T.L. Pournariou, (27) T.L. Pournariou II, (28) T.L. Smokovou, (29) T.L. Kremaston, (31) T.L. Kastrakiou, (33) T.L. Stratou, (34) T.L. Evinou, (38) T.L. Mornou, (42) T.L. Marathona, (43) T.L. Feneou, (44) T.L. Pineiou, (46) T.L. Ladona, (48) T.L. Faneromenis (49) T.L. Bramianou. Natural Lakes: (4) Doirani, (8) Ismarida, (9) Megali Prespa, (10) Pikrolimni, (11) Mikri Prespa, (12) Vegoritida, (13) Petron, (14) Koroneia, (15) Volvi, (16) Zazari, (17) Cheimaditida, (18) Kastoria, (23) Pamvotida, (30) Voulkaria, (32) Amvrakia, (35) Ozeros, (36) Lysimacheia, (37) Trichonida, (39) Paralimni, (40) Yliki, (41) Dystos, (45) Stymfalia, (47) Kourna

Lake Pikrolimni, a natural lake, is an exceptional case as its characteristics resemble those of "soda lakes", as suggested by Kagalou et al. (2021b). Due to its extreme physicochemical parameters, it is often excluded from statistical analyses.

Finally, Lake Saltini was previously included in the Lake Monitoring Network, but it was removed from the list of sampling stations in 2021, as it has been classified as a lagoon. Therefore, in the following chapters, we present analyses for a total of 49 lakes.

3 Physicochemical Parameters

3.1 Introduction

Physicochemical elements are essential to ecosystem functioning and its habitat settings, thereby influencing the biological communities that lake ecosystems support. Eutrophication, resulting from nutrient pollution, exerts significant pressure on lake ecosystems and the communities they sustain. The WFD requires Member States to establish nutrient concentration thresholds and use these as part of the assessment of ecological status (Poikane et al. 2022).

3.2 Materials and Methods

Physicochemical parameters were measured seasonally over the period from 2016 to 2022. Transparency was measured using a Secchi disc. Electrical conductivity was measured with a portable multiparameter instrument (Orion Star A320, Thermo Fischer Scientific Inc., US). Water samples for further laboratory analysis were collected from the euphotic zone (2.5 × Secchi depth). TP was determined using the ascorbic acid method (APHA 4500, P-E, 24th ed., 2023). Anion (Cl^-, SO_4^{3-}) and cation (Na^+, K^+, Mg^{2+}, Ca^{2+}) concentrations were measured using the methods ISO 10304-01:2007 and ISO 14911:1998, respectively, with an Ion chromatograph (Dionex Aquion IC system, Thermo Fischer Scientific Inc., US).

A Gibbs diagram (Gibbs 1970) was constructed as a scatter plot where $Na^+/(Na^+ + Ca^{2+})$ ratios were plotted against Total Dissolved Solids (TDS) of water samples. The major cations that characterize the end-members of surface waters are calcium for freshwaters and sodium for high-saline water bodies. Therefore, lake waters located closer to the right side of the plot are more saline. The concentration of TDS in lake waters reflects the relative influence of each of the natural processes of evaporation, precipitation and rock weathering.

Principal Component Analysis (PCA) was used to interpret the major variation patterns within the data. Highly correlated variables were excluded from the analysis. The Kaiser criterion (Kaiser 1960) was applied, and principal components (PCs) with eigenvalues greater than 1 were selected for further analysis. Statistical analyses were conducted using R statistical software (R Core Team 2019), while the Gibbs diagram was created using Excel.

3.3 Results

The Gibbs diagram (Gibbs 1970) is presented in Fig. 2.

According to Gibbs (1970), the natural processes that govern the chemistry of surface waters include evaporation, precipitation, and water–rock interaction. Lake waters with higher annual precipitation are examples of precipitation-dominated chemistry. In Greece, increased rainfall is observed in the western regions due to the effects of relief and the typical movement of weather systems from west to east. Conversely, lakes with evaporation-dominated chemistry are typically found in warm and arid regions.

The influence of these processes can be seen on the scatter plot shown on Fig. 2 where $Na^+/(Na^+ + Ca^{2+})$ ratios are plotted against TDS. Only two lakes (Lake Zazari and the Reservoir T.L. Leukogeion) are present on the rock weathering-precipitation zone (TDS < 100 mg/L), whereas the majority are located in the zone of rock weathering-dominance, as TDS values range from about 100 to 1000 mg/L. The positions of lakes within this grouping depend on relief, climate and bedrock material of each basin (Gibbs 1970). On the other hand, certain lakes show a saline character (e.g. Lakes Ismarida, Pikrolimni, Karla, Koroneia) as they have both high TDS values (> 1000 mg/L) and high $Na^+/(Na^+ + Ca^{2+})$ ratios.

The majority of lakes are located in the first half of the x-axis, which is expected, as lakes are typically characterized by low to moderate salinity. Additionally, deep natural lakes (DNL) and reservoirs (both deep and shallow) are positioned closer to the lower left part of the diagram, with $Na^+/(Na^+ + Ca^{2+})$ ratios of up to about

Fig. 2 Gibbs diagram showing the ratio between Na^+ and Na^+ and Ca^{2+} in water samples in relation to Total Dissolved Solids (TDS). Horizontal lines divide the graph into three zones; > 10,000 mg/L TDS: evaporation dominance; > 1000 mg/L TDS: evaporation-precipitation dominance; > 100 mg/ L TDS: rock weathering dominance; > 10 mg/L TDS: rock weathering and rainwater co-dominance. Typical seawater corresponds the upper right corner of the plot

0.3. Shallow natural lakes (SNL) exhibit a wider range of ratios, extending up to approximately 0.5. Very shallow natural lakes (VSNL) are associated with more saline conditions, as their ratios primarily range from 0.5 to 1.

Finally, Lake Pikrolimni stands out due to its considerably higher values in both TDS and the $Na^+/(Na^+ + Ca^{2+})$ ratio compared to the other lakes. A study by Kagalou et al. (2021b) highlights the unique characteristics of this lake, noting its strong resemblance to typical "soda lakes". These lakes are characterized by high pH values (pH > 9) and a high proportion of sodium and bicarbonate and carbonate ions, with little to no soluble magnesium or calcium. Soda lakes may also have high concentrations of sodium chloride and other salts, leading to saline conditions. Currently, no established nutrient thresholds exist for soda lakes that would support the attainment of good ecological status. Therefore, Pikrolimni was excluded from the subsequent physicochemical analysis.

To visualize the multivariate observations of our dataset within a two-dimensional frame, PCA was applied to both natural lakes and reservoirs, as shown in Fig. 3a, b, respectively. The eigenvalues of the first two principal components account for up to 83% (PC1: 47%; PC2: 36%) and 89% (PC1: 61%; PC2: 28%) of the total variance in the observations for natural lakes and reservoirs, respectively.

As expected, clear natural lakes with high Secchi Depth (SD) are separated from the more impacted lakes with low SD (Fig. 3a). Similarly, clear water reservoirs are grouped together in the central and lower-right quadrants of the plot (Fig. 3b). The relationship between SD and TP, two of the main driving factors, indicates a gradient from shallower, more impacted water bodies to less impacted ones. These differences can also be attributed to variations in population density, land use within the catchment area, and hydromorphological characteristics (Mavromati et al. 2018).

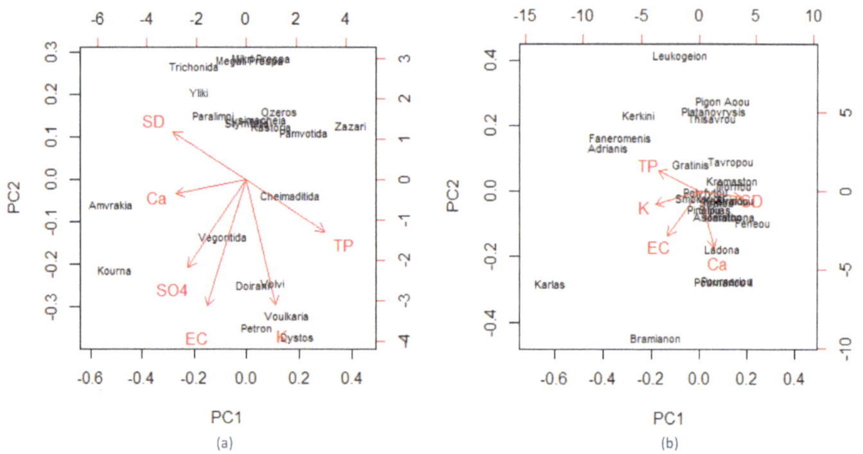

Fig. 3 PCA biplots of selected physicochemical variables in Greek **a** natural lakes and **b** reservoirs dataset

4 Characteristics of the Main Greek Reservoirs

4.1 Introduction

According to WFD, heavily modified water bodies are those whose physical characteristics or hydromorphological conditions have been altered to serve a specific purpose (e.g., irrigation or drinking water supply). These modifications often directly affect the structure and function of the water bodies, which is why heavily modified water bodies have distinct environmental objectives, namely the achievement of good ecological potential.

The construction of the first dam in Greece, the Marathon Dam, was driven by the need for a reliable drinking water supply, resulting in the creation of the Marathon Reservoir in 1929. In 1935, the need for flood control and irrigation led to the construction of the Lithotopos Dam and the formation of artificial Lake Kerkini in Northern Greece. After World War II, the demand for electricity generation led to the construction of three large dams in Greece during the 1950s and 1960s. Since then, hundreds of dams have been constructed across the country. Many reservoirs serve multiple functions; while they may be initially constructed for a primary purpose, they often fulfill additional roles throughout their operational lifespan, such as providing irrigation water. Today, the primary reasons for constructing dams include electricity generation, irrigation, drinking water supply, drought management, flood control, as well as fish farming and recreation.

Reservoirs in Greece are owned and managed by various authorities, including the Public Power Corporation (PPC), the Hellenic Ministry of Rural Development and Food, the Water Supply and Sewerage Company of the Capital Region (EYDAP), and other entities (e.g., municipalities). According to the International Commission on Large Dams (ICOLD 2013), large dams are defined as those exceeding 15 m in height from the deepest point of the foundation, or those with a capacity of 3×10^6 m^3 or greater, provided their height exceeds 5 m. In Greece, 138 dams meet the criteria of exceeding 15 m in height.

4.2 Data Acquisition

Twenty-six reservoirs are monitored as part of the Greek National Monitoring Network according to WFD. Data for 18 of these reservoirs are provided by the authorities that operate them (PPC and EYDAP).

4.3 Results

The main characteristics of the 26 reservoirs are presented in Table 1. The data span the periods 2012–2022 or 2016–2022, depending on the reservoir.

It is notable that the capacity of most reservoirs (16) does not exceed 300 hm^3, while another 7 reservoirs range between 400 and 900 hm^3. The only two reservoirs that stand out are Kremasta and Polifito, with capacities of 4495 hm^3 and 1939 hm^3, respectively. Greek reservoirs exhibit relatively short residence times, ranging from 0.1 to 4.6 years. This suggests that the characteristics of their aquatic ecosystems are primarily influenced by incoming water from the surrounding catchment rather than being sustained by self-regulating processes (Fussman et al. 2000; Rueda et al. 2006). These short residence times can partly be attributed to intense runoff, which is influenced by the seasonal fluctuations typical of the Mediterranean climate.

Kremasta is the largest reservoir in Greece, while Thisavros is the highest, not only in Greece but also in the Balkans (Melista 2015). The main operational purposes of the main reservoirs in Greece are irrigation, followed by power generation (65.38% and 57.69% of their usage respectively), as shown in Fig. 4. Additionally, three of these reservoirs supply drinking water to the Greek capital, Athens. These uses, combined with the climate, contribute to fluctuations in the reservoirs' water levels. Greek reservoirs are also strongly influenced by the seasonality of the Mediterranean climate, which causes significant variations in available water between the wet winter and dry summer seasons (Theodorakis et al. 2000). Consequently, their water level fluctuations range from 0.6 to 24 m (Table 1).

It is important to note that certain groups of reservoirs operate sequentially, with one reservoir following another. For example, Kremasta, Kastraki, and Stratos are reservoirs built along the Acheloos River, where the final reservoir downstream is primarily used for water abstraction while maintaining ecologically acceptable flow levels. A similar arrangement is found with the reservoirs along the Aliakmonas River, the longest river in Greece (reservoirs Polifitou, Sfikia, Asomaton). In contrast, artificial Lake Karla was constructed in place of a previously drained lake as part of an environmental project, serving flood protection purposes as well (Chamoglou et al. 2014; Perivolioti et al. 2024).

Table 1 Characteristics of 26 main Greek reservoirs

A/A	Reservoir name	Reservoir capacity (hm^3)	Surface (km^2)	Study period	Inflows (hm^3)	Extraction (hm^3)	Mean water level variance (m)	Mean residence time (years)	Purpose
1	Mornos	763.71	19.93	2012–2022	40.63	38.66	10.5	1.7	Drinking water supply
2	Marathonas	42.85	2.57	2012–2022	4.62	4.63	1.8	1.0	Drinking water supply
3	Evinos	137.63	3.6	2012–2022	25.52	25.72	24.2	0.5	Drinking water supply power generation
4	Kremasta	4495	80.6	2016–2022	257.81	249.76	17.7	1.6	Power generation
5	Kastraki	831	24.2	2016–2022	35.30	285.23	1.9	2.1	Drinking water supply irrigation power generation
6	Stratos	80	8.4	2016–2022	7.10	365.38	0.4	–	Irrigation power generation
7	Leukogia	12.99	1.192	–	–	–	–	–	Drinking water supply irrigation
8	Platanovrisi	73	3.3	2016–2022	12.24	121.97	1.7	0.5	Irrigation power generation
9	Thisavrou	705	20	2016–2022	81.51	90.83	18.9	0.8	Irrigation power generation flood protection
10	Gratini	12.8	0.85	2013–2022	–	–	4.1	–	Industrial water supply
11	Adriani	–	0.42	2012–2022	–	–	2.6	–	Irrigation
12	Polyfytou	1939	74	2016–2022	35.65	119.99	7.1	4.6	Power generation irrigation drinking water supply flood protection
13	Sfikias	99	4.3	2016–2022	6.71	132.65	2.3	1.2	Power generation
14	Asomaton	53	2.6	2016–2022	22.91	182.50	2.0	1.0	Irrigation power generation
15	Pournari I	730	20.6	2016–2022	128.26	160.70	10.8	0.5	Power generation

(continued)

Table 1 (continued)

A/A	Reservoir name	Reservoir capacity (hm^3)	Surface (km^2)	Study period	Inflows (hm^3)	Extraction (hm^3)	Mean water level variance (m)	Mean residence time (years)	Purpose
16	Pournari II	9	0.65	2016–2022	–	168.02	2.5	0.008	Power generation irrigation other
17	Pineios	420	19.895	–	–	–	–	–	Irrigation
18	Ladonas	49	4	2016–2022	30.06	33.28	10.5	0.2	Power generation
19	Feneos	5.3	0.509	–	–	–	–	–	Irrigation
20	Karla	198	38	2012–2022	–	–	0.6	–	Irrigation flood protection other
21	Tavropou	400	25.2	2016–2022	12.95	24.41	–	2.6	Power generation irrigation drinking water supply recreation
22	Smokovou	240	8.5	2016–2022	–	–	9.5	–	Irrigation power generation
23	Faneromeni (Crete)	19.679	1	2016–2022	–	–	11.9	–	Irrigation
24	Bramiania	16.4	1.05	2016–2022	–	–	3.7	–	Irrigation drinking water supply
25	Piges Aoou	260	11.5	2016–2022	8.79	8.61	5.4	2.8	Power generation
26	Kerkini	–	73	2020–2022	–	–	4.5		Irrigation flood protection

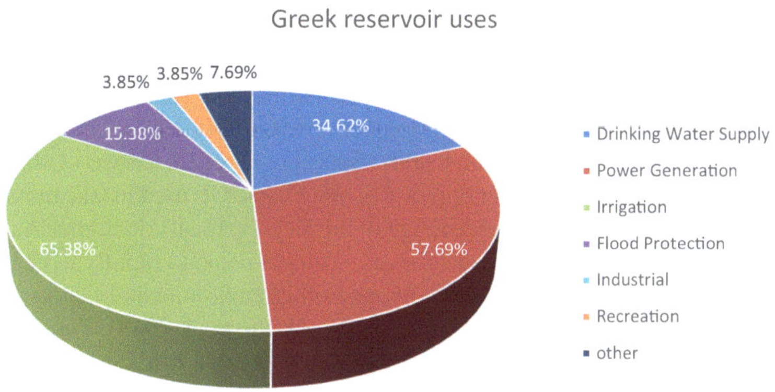

Fig. 4 Main operational purposes of the main reservoirs in Greece

5 Biological Monitoring and Assessment Systems

5.1 Introduction

WFD has fundamentally reshaped how freshwater ecosystems are assessed by placing BQEs at the center of ecological status classification (Free et al. 2024). The primary BQEs designated for lake monitoring under the WFD are phytoplankton, macrophytes, phytobenthos, benthic macroinvertebrate fauna and fish fauna, which are used to evaluate the degree to which human activities, particularly eutrophication and hydromorphological alterations, have disrupted the natural structure and functioning of aquatic ecosystems. In Greece, where lakes range from deep tectonic basins to shallow coastal reservoirs, a set of nationally tailored, WFD-compliant methodologies has been developed to address this ecological diversity. This chapter presents a comprehensive overview of these frameworks and their application across different lake types, offering insights into both ecological conditions and the pressures shaping them.

Phytoplankton is among the most widely used BQEs for assessing eutrophication in lake ecosystems due to its rapid response to nutrient enrichment and changes in water quality. The WFD specifies that assessments of phytoplankton should include evaluations of taxonomic composition and abundance, total biomass, and the frequency and intensity of planktonic blooms. These parameters are especially effective at detecting shifts in trophic state and early signs of ecological imbalance, making phytoplankton a sensitive and dynamic indicator of water body health (Birk et al. 2012; Carvalho et al. 2013).

Aquatic macrophytes, by contrast, provide a longer-term perspective on ecological conditions. These rooted plants are integral to the structural complexity and ecological functioning of lake habitats, influencing processes such as nutrient cycling, sediment stabilization, and the availability of shelter and food for aquatic fauna. Because they are closely linked to light availability, nutrient concentrations, and sediment conditions, macrophyte communities serve as reliable indicators of eutrophication and other environmental changes. Their distribution and composition can reflect both spatial patterns and long-term ecological trends, offering a high-resolution lens through which to assess ecosystem health (Bytyqi et al. 2020; Zervas et al. 2020).

Phytobenthos, particularly diatoms, are another key BQE used in lake monitoring, valued for their sensitivity to a wide range of physicochemical conditions. As microscopic algae inhabiting submerged surfaces, diatoms respond rapidly to changes in nutrient concentrations, pH, conductivity, and other environmental factors, making them highly effective indicators of water quality (Kelly et al. 2023). Their short life cycles and well-established ecological preferences enable precise assessments of both current conditions and long-term environmental trends. To date, phytobenthos-based lake assessment systems have been developed in numerous European countries and have undergone intercalibration in twelve of them (Kelly et al. 2023). In lake ecosystems, diatom-based indices are commonly used to detect eutrophication, acidification, and changes in substrate quality. The WFD mandates the use of taxonomic composition and relative abundance in phytobenthos assessments, enabling the detection of subtle ecological shifts that may not be apparent through other BQEs.

Benthic macroinvertebrates are another critical component of lake ecosystems and are widely employed in both lake and river monitoring due to their sensitivity to a broad range of pressures. These organisms inhabit the sediment–water interface and are affected by changes in nutrient levels, dissolved oxygen, acidity, and substrate structure. Traditionally, profundal and sublittoral macroinvertebrate communities have been used to detect pressures related to eutrophication and oxygen depletion, particularly in deep lakes (Rossaro et al. 2007). In line with this approach, a national assessment method, the Greek Lake Benthic macroinvertebrate Index (GLBiI) was developed, specifically targeting the profundal and sublittoral zones to address eutrophication pressures (Ntislidou et al. 2018). More recent research emphasizes the importance of littoral macroinvertebrate communities, which are directly influenced by shoreline alterations and combined stressors such as eutrophication and morphological change (Gabriels et al. 2010; Urbanič et al. 2012; Poikane et al. 2016). In response to these needs, the Hellenic Lake Littoral Benthic Invertebrate Index (HeLLBI) was developed, a method designed specifically to classify Mediterranean lakes according to the sensitivity of their littoral macroinvertebrate assemblages to these combined pressures (Mavromati et al. 2021).

Fish communities, as top-level consumers in aquatic ecosystems, provide an integrated view of environmental conditions over broad spatial and temporal scales. Fish are affected by a variety of anthropogenic pressures, including eutrophication, habitat fragmentation, flow regulation, and the presence of invasive species. As such, they are considered valuable indicators for assessing ecological integrity and cumulative impacts on lake ecosystems. Fish-based indices evaluate shifts in community composition, trophic structure, and species origin to detect signs of degradation and biotic homogenization (Pont et al. 2019). Their inclusion in ecological assessments enhances the robustness and ecological relevance of status classifications under the WFD.

Together, these five BQEs form the foundation of a comprehensive, multiscale assessment framework under the WFD. Each offers a unique window into the health of lake ecosystems and collectively they allow for a holistic understanding of the pressures, responses, and trajectories of aquatic environments. In Greece, the development of national methodologies tailored to Mediterranean lake typologies has ensured that these assessments remain ecologically valid, policy-relevant, and aligned with European intercalibration standards. This integrated biological approach is essential not only for meeting the regulatory requirements of the WFD but also for supporting the sustainable management and restoration of freshwater ecosystems in the face of growing environmental pressures.

5.2 *Materials and Methods*

Phytoplankton

Between 2016 and 2022, a total of 46 lakes were monitored for phytoplankton across 49 sampling stations in Greece. These included various lake types such as deep monomictic natural lakes, shallow polymictic lakes, very shallow polymictic lakes, and both deep (type LM 8) and shallow reservoirs. Most lakes were sampled at a single station located in the pelagic zone, with the exception of the transboundary lakes Megali Prespa, Mikri Prespa, and Doirani, which each had two stations. Phytoplankton samples were collected from the euphotic zone of the water column, defined as 2.5 times the Secchi disk depth.

Greece applies three phytoplankton assessment systems: the Hellenic Phytoplankton Assessment System for Natural Lakes (HeLPhy), the New Mediterranean Assessment System for Reservoirs Phytoplankton (NMASRP), and the Greek Phytoplankton Trophic Index (GR-PTI). They all respond to the eutrophication pressure.

The HeLPhy system is designed for natural lakes and aggregates four equally weighted parameters—Chlorophyll-a concentration, total biovolume, biovolume of cyanobacteria, and the Modified Nygaard Index—into a single multimetric index. The final score is derived by normalizing each metric's Ecological Quality Ratio (EQR) and averaging them, then assigning a status class from high to bad based on set thresholds (Tsiaoussi et al. 2017; Nikolopoulou et al. 2025).

The NMASRP method, intercalibrated within the Mediterranean GIG, applies to Greek, Cypriot, and Portuguese reservoirs. It incorporates similar parameters—Chlorophyll-a, biovolume, biovolume of cyanobacteria, and the Index Des Grups Algals (IGA)—which also reflect taxonomic composition and bloom occurrence (Pahissa et al. 2015; Tsiaoussi et al. 2016). As with HeLPhy, EQR values are normalized and combined into a final index, with classification thresholds adapted to reflect ecological potential in reservoirs.

For very shallow natural lakes and shallow reservoirs, the GR-PTI method is employed. This index includes a genera-based trophic score alongside Chlorophyll-a, total biovolume, and cyanobacteria biovolume. The PTI offers enhanced taxonomic resolution and consistency, especially for identifying responses to eutrophication pressures. The final lake score is calculated through normalization and averaging of metric EQRs, resulting in a status classification using the same five-tier scale (Kelly and Juggins 2023).

Macrophytes

Aquatic macrophyte monitoring between 2016 and 2022 was conducted in 24 lakes, including deep, shallow, and very shallow natural lakes, along with Feneos Reservoir, which—despite being artificial—was surveyed for macrophytes due to the fact that it has not been yet operational (it retains natural water level fluctuations). Sampling followed a belt transect-mapping protocol during the main growing season, covering five depth zones across 10–20 transects per lake. Species identification was carried out to species or subspecies level for higher plants and to genus level for macroalgae, with abundance recorded on a five-point DAFOR scale (Zervas et al. 2018).

The HeLM method is used to assess the ecological status of lakes based on aquatic vegetation and responds to eutrophication pressure (Zervas et al. 2018). It relies on two main metrics: the Trophic Index HeLM (TIHeLM) and the maximum colonization depth (Cmax). TIHeLM is adapted from European intercalibration procedures and is calculated per transect using species-specific trophic scores combined with the relative abundance of each species. These transect-level scores are then averaged to produce a lake-wide value (Zervas et al. 2018).

Cmax represents the deepest point at which rooted aquatic vegetation is observed and is determined over a three-year period to account for natural variability. Both TIHeLM and Cmax are converted into Ecological Quality Ratios (EQRs) through a normalization process. The final HeLM score is calculated as the average of these two EQRs. This score is then used to classify the lake's ecological status into one of five categories: high, good, moderate, poor, or bad, based on established threshold values (Zervas et al. 2018).

Phytobenthos

The first step of a new diatom-based method for the ecological assessment of Greek lakes is presented in Kelly et al. (2023); however, as this method has not been forwarded yet to the ECOSTAT working group and has not been incorporated into the Commission Decision on Intercalibration for the Water Framework Directive, it is not further discussed in this chapter.

Benthic Invertebrate Fauna

From 2018 to 2022, benthic macroinvertebrates were surveyed in 23 lakes, including 22 natural lakes and one artificial water body, Lake Karla (Patsia et al. 2023). In total 105 littoral zone sampling sites were selected based on lake morphology, habitat, and land use characteristics. Sampling involved three-minute kick or sweep procedures using 500 μm hand nets across multiple mesohabitats per site. Collected organisms were preserved, sorted, and identified primarily to family level.

The assessment method used, HeLLBI (Mavromati et al. 2021), incorporates three metrics: the relative abundance of Odonata, the Average Score Per Taxon (ASPT), and Simpson's Diversity Index. These metrics address community composition, sensitivity to disturbance, and diversity, respectively. Each metric's value is converted to a normalized EQR by comparison to a reference value. The final HeLLBI index is calculated as the mean of these normalized EQRs, and lakes are classified into ecological status classes based on this composite score. HeLLBI responds mainly to morphological alterations of lake shorelines (Mavromati et al. 2021). Multivariate analyses, including non-metric multidimensional scaling (NMDS) and analysis of similarities (ANOSIM), were used to explore patterns and differences in macroinvertebrate communities across lake types, implemented via PRIMER v7.0.5 software.

Fish

In 2022, four natural lakes were sampled for fish monitoring—three shallow and one deep—under the Greek National Monitoring Program (Petriki et al. 2023). Sampling adhered to European standards using stratified Nordic-type gillnets and, where applicable, electrofishing. All fish were identified to species level and measured for biomass and abundance.

According to Petriki et al. (2017), the Greek Lake Fish Index (GLFI) uses two key metrics: the relative biomass of omnivorous species ($OMNI_b$) and the relative abundance of introduced species ($Introduced_a$). $OMNI_b$ is sensitive to eutrophication as it correlates with total phosphorus levels, while $Introduced_a$ reflects habitat degradation, particularly shoreline modifications. Both metrics are transformed into EQRs using specific equations incorporating hindcasting adjustments. The GLFI score is the average of the normalized EQRs, and lakes are classified according to the standard five-class ecological status system (Petriki et al. 2017).

This comprehensive, multi-taxa monitoring framework provides a robust basis for assessing the ecological status of Greek lakes under the Water Framework Directive. The integration of various biological quality elements ensures that multiple facets of ecosystem health are considered, offering valuable insights into the pressures affecting these freshwater bodies.

5.3 Results and Discussion

Phytoplankton

The results from the application of the three assessment phytoplankton methods are given in Figs. 5, 6 and 7.

As shown in Fig. 5, five out of seven deep natural lakes are classified as having good or high ecological status (Lakes Kourna, Amvrakia, Yliki, Trichonida, and Megali Prespa). Lakes Megali Prespa and Vegoritida are near the 0.6 boundary, with Megali Prespa in good status and Vegoritida in moderate status. Lake Volvi lies closer

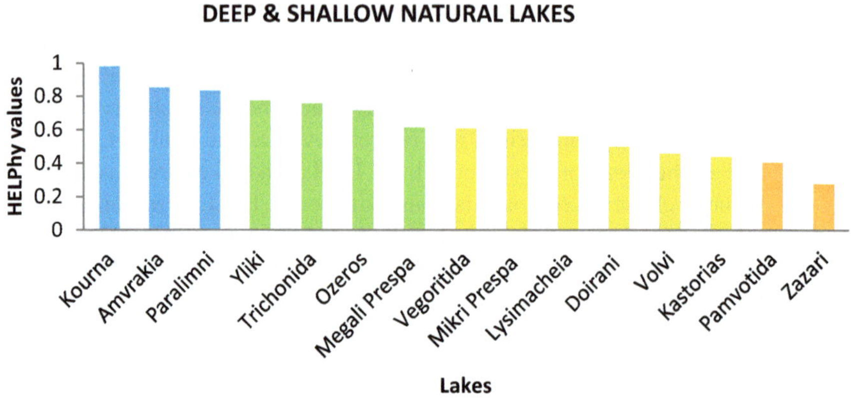

Fig. 5 Average ecological quality ratios of HeLPhy in deep and shallow natural lakes in Greece for the period 2016–2022

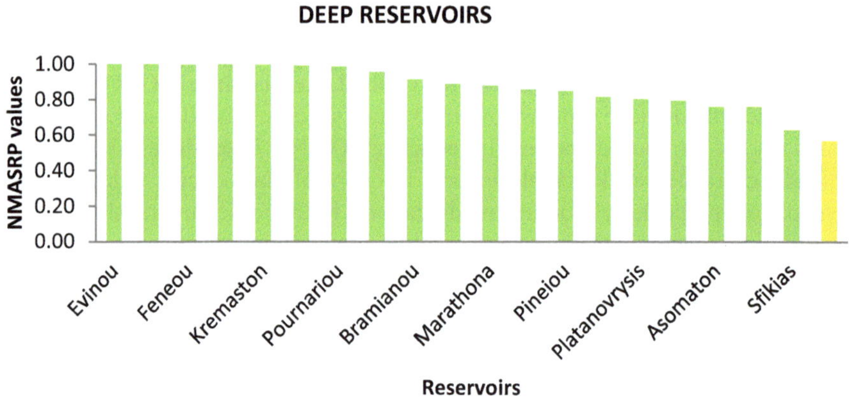

Fig. 6 Average ecological quality ratios of NMASRP in deep reservoirs in Greece for the period 2016–2022

to the 0.4 boundary. Shallow lakes show a wider range of Ecological Quality Ratios (EQRs). Of the eight shallow lakes, one is in high ecological status (Lake Paralimni), another in good status (Lake Ozeros), four are in moderate status, and two are in poor status. The EQR of Lake Mikri Prespa, classified in moderate status, is close to the 0.6 boundary. The classification of Lake Paralimni as high status is based on high EQR values in 2022, although the lake was previously classified as having good status between 2016 and 2021 (Nikolopoulou et al. 2025). Overall, 47% of deep and shallow natural lakes are in high or good status, 40% in moderate status, and 13% in poor ecological status.

The results from applying the NMASRP index to 20 deep reservoirs (type LM 8) in Greece are presented in Fig. 6. All but one of the reservoirs are classified

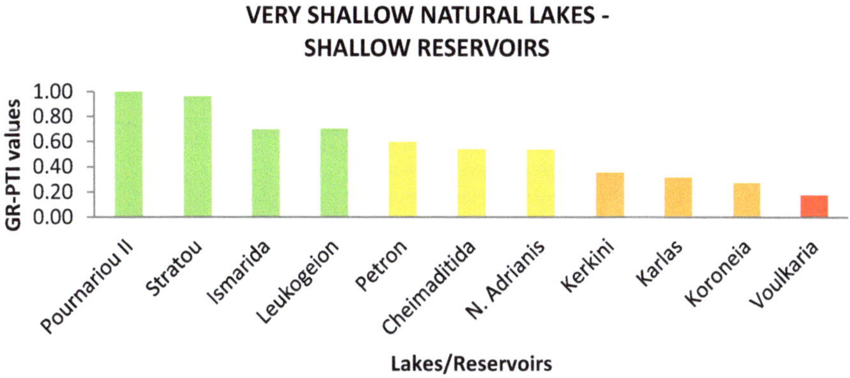

Fig. 7 Average ecological quality ratios of GR-PTI in very shallow natural lakes and in shallow reservoirs in Greece for the period 2016–2022

as having good ecological potential and above. The land cover in their catchment areas are predominantly natural and semi-natural (Mavromati et al. 2018). Previous phytoplankton data from the period 2013–2016 indicated low total phytoplankton and cyanobacteria biovolume in reservoirs located in river basins with a high proportion of natural and semi-natural land cover (Navrozidou et al. 2019).

The results from applying the GR-PTI assessment method to 13 lakes (7 very shallow natural lakes and 6 shallow reservoirs) are shown in Fig. 7. It is important to note that ecological potential is assigned to reservoirs. Lake Voulkaria is classified as having bad ecological status, making it the only lake in the entire phytoplankton dataset of Greek natural lakes and reservoirs to receive this classification.

Macrophytes

Regarding macrophytes, the HeLM assessment method was applied to determine the ecological status of lakes for two distinct three-year assessment periods: 2016–2018 and 2019–2021. The results are shown in Figs. 8 and 9. During the first sampling period (2016–2018), the majority of lakes (75%) were classified as high or good status. However, this percentage dropped to 68.75% during the 2019–2021 period. Specifically, two of the eight deep lakes were rated as high (Lake Kourna and Feneos Reservoir), while the remaining lakes were classified as good. In the GR-SNL type, one lake was classified as high (Lake Paralimni), two as good (Lakes Doirani and Kastoria), and three as moderate (Lakes Lysimacheia, Mikri Prespa, and Zazari). Lakes Ozeros and Pamvotida showed different ecological status between the two periods. Lake Ozeros, which was rated as good during the first period, saw a significant decline, being classified as bad during the second period. Conversely, Lake Pamvotida improved from poor to moderate status. Finally, Lake Paralimni was only evaluated during the first period (2016–2018), as no macrophytes were present during the second period due to an abrupt rise in its water level that led to macrophyte drowning (for further details on water level fluctuations, see wfd.ekby.gr).

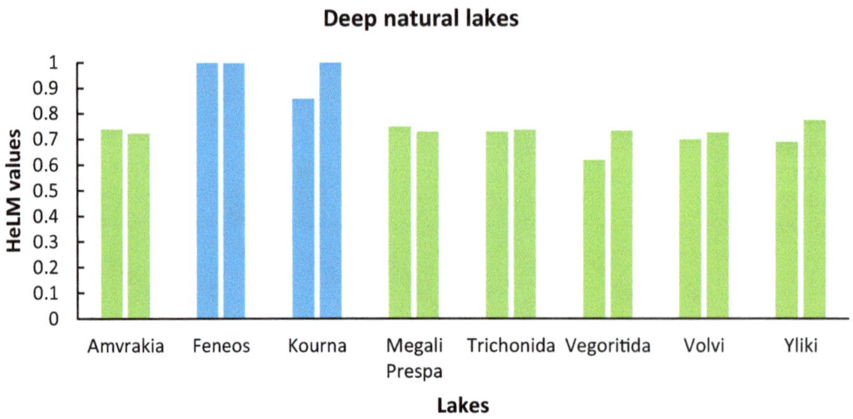

Fig. 8 Average ecological quality ratios of HeLM in deep natural lakes in Greece for the periods 2016–2018 and 2019–2021

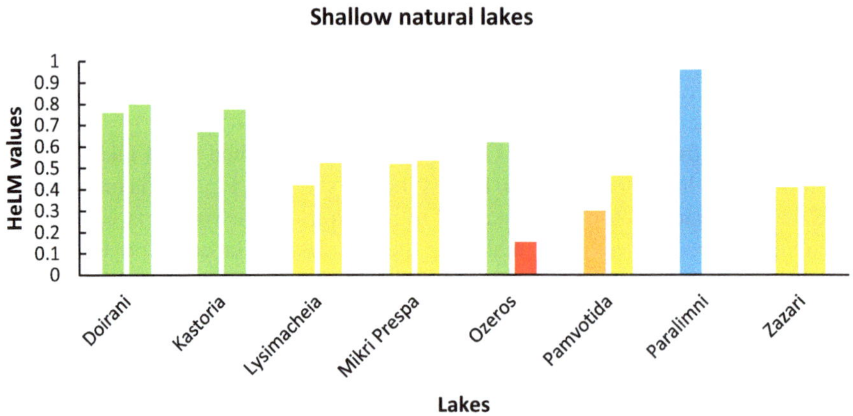

Fig. 9 Average ecological quality ratios of HeLM in shallow natural lakes in Greece for the periods 2016–2018 and 2019–2021

Figures 10 and 11 display the maximum values of Cmax for the period 2016–2022 for both deep and shallow lakes. Missing values for Lakes Ozeros (2019, 2020, and 2021) and Paralimni (2019, 2020, 2021, and 2022) indicate zeros, as no rooted submerged macrophytes were recorded in these years. Lakes Feneos (2020), Kourna (2022), and Pamvotida (2020) were not monitored during the respective years.

The maximum recorded Cmax value in deep natural lakes was 17.2 m, observed in Feneos Reservoir in 2018, while the minimum value was 1.4 m, recorded in Lake Yliki in 2016. In contrast, for shallow natural lakes, the highest Cmax value was 6.4 m, found in Lake Kastoria in 2020, and the lowest was 1 m, recorded in Lake Pamvotida in 2017.

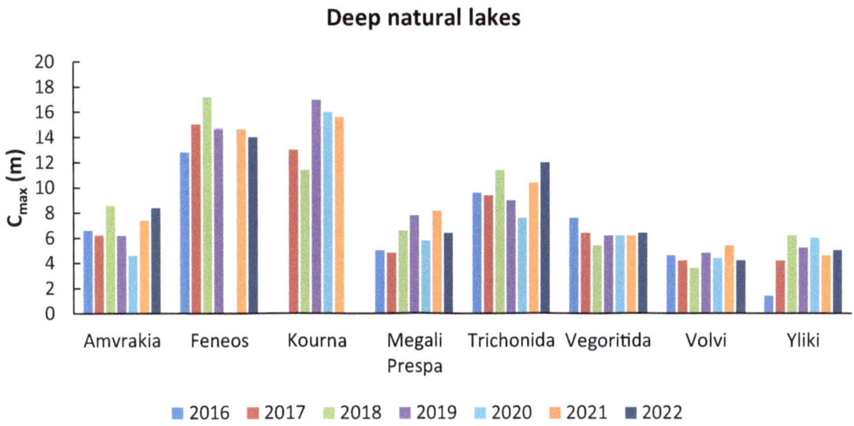

Fig. 10 Maximum values of Cmax in deep natural lakes in Greece for the period 2016–2022

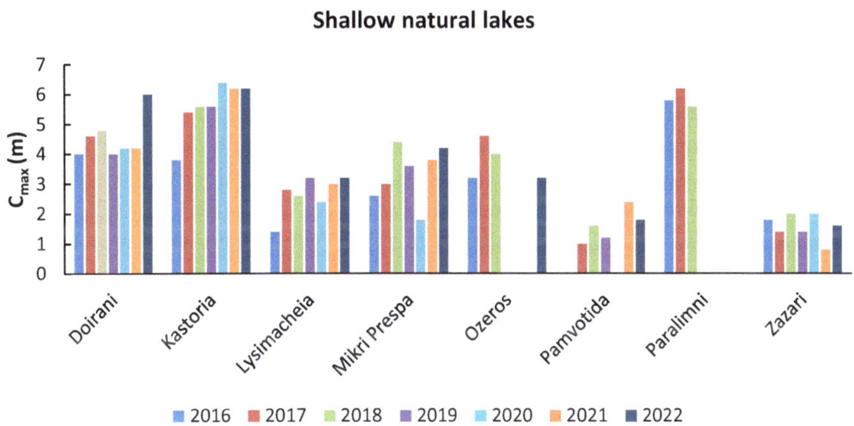

Fig. 11 Maximum values of Cmax in shallow natural lakes in Greece for the period 2016–2022

Regarding species richness, Lake Trichonida, from the deep natural lakes, and Lake Mikri Prespa, from the shallow lakes, exhibited the highest number of taxa, with 26 and 28 species, respectively. Common elodeids (rooted submerged macrophytes) observed in both deep and shallow lakes were *Myriophyllum spicatum*, *Vallisneria spiralis*, *Stuckenia pectinata*, and *Najas marina*. Among the various life-forms of aquatic vegetation, charids (primarily the genera *Chara* and *Nitella*), which are typically associated with high or good ecological status according to the WFD (Poikane et al. 2018), were found in high abundances in Lake Kourna and the Reservoir Feneos. The relative abundance values of different life-forms in very shallow lakes are shown in Fig. 12.

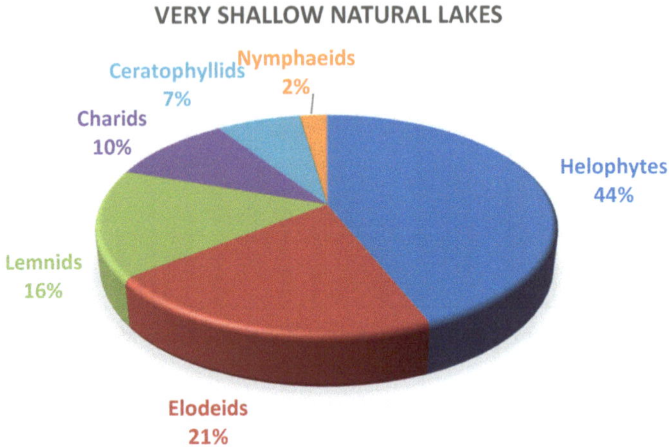

Fig. 12 Relative abundance values of different life forms of aquatic vegetation in very shallow lakes in Greece for the period 2016–2022

A total of 82 taxa were recorded in very shallow lakes, including the macroalgae *Cladophora glomerata* and *Enteromorpha* spp. Helophytes were the most abundant group, comprising 44% of the recorded taxa, with *Phragmites australis* (present in every lake) and *Typha* spp. (found in five lakes) being the primary representatives. The least abundant groups were ceratophyllids (7%) and nymphaeids (2%). Lake Cheimaditida had the highest number of taxa (47), while the lowest was recorded in Lake Koroneia, with only 8 taxa. Currently, there is no assessment method for aquatic macrophytes specific to very shallow lakes. In light of the new data, future research should focus on developing an index based on macrophytes for these water bodies.

Benthic Invertebrate Fauna

With respect to littoral benthic macroinvertebrates, a total of 76,374 individuals were collected, representing 72 identified taxa. A small number of taxa dominated the benthic invertebrate communities across the 23 lakes, with just five taxa making up 90% of the collected individuals: Corixidae, Gammaridae, Chironomidae, Oligochaeta, and Asellidae (55%, 13%, 12%, 7%, and 3%, respectively) (Fig. 13). The remaining 10% of the benthic communities consisted of 67 taxa. The highest mean number of taxa was observed in Lake Dystos (15), while the lowest was recorded in Lake Kourna (5.42). The fewest individuals were recorded in Lake Voulkaria (117.5), and the most were in Lake Karla (3704).

The HeLLBI assessment method was applied to classify the ecological status of 23 lakes, and the results are presented in Table 2. According to HeLLBI, the majority of lakes (57%) were classified as having moderate or poor status. Specifically, 2 lakes were evaluated as high (Lakes Dystos and Voulkaria), 8 lakes as good, 10 lakes as moderate, and 3 lakes as poor (Lakes Ozeros, Ismarida, and Zazari).

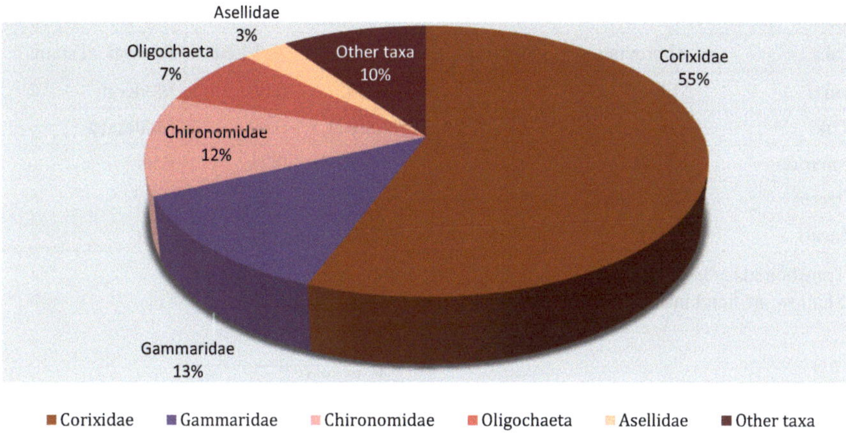

Fig. 13 Relative abundance (%) of benthic macroinvertebrate taxa at 105 sampling sites of 23 lakes in Greece for the period 2016–2022

Table 2 No of sampling sites, year of sampling, HeLLBI index and ecological status of the studied lakes in Greece for the period 2016–2022

Lake	Abbreviation	Sampling sites	Year	HeLLBI	Ecological status
Dystos	Dyst	3	2022	0.85	High
Voulkaria	Voulk	2	2021	0.88	High
Cheimaditida	Cheim	4	2021	0.76	Good
Doirani[a]	Doir	4	2022	0.80	Good
Koroneia	Kor	5	2022	0.68	Good
Kourna	Kourn	7	2018	0.62	Good
Lysimacheia	Lysim	6	2022	0.75	Good
Paralimni	Para	8	2018	0.78	Good
Pikrolimni	Pikro	3	2018	0.67	Good
Stymfalia	Stymf	1	2022	0.69	Good
Karla[b]	Karl	4	2021	0.56	Moderate
Kastoria	Kast	5	2021	0.57	Moderate
Megali Prespa	Meg Pres	4	2022	0.49	Moderate
Mikri Prespa	Mik Pres	3	2021	0.42	Moderate
Pamvotida	Pamv	6	2022	0.51	Moderate
Petron	Petr	3	2021	0.55	Moderate
Trichonida	Trich	13	2022	0.55	Moderate
Vegoritida	Veg	6	2021	0.43	Moderate

(continued)

Table 2 (continued)

Lake	Abbreviation	Sampling sites	Year	HeLLBI	Ecological status
Volvi	Vol	6	2021	0.60	Moderate
Yliki	Yli	6	2021	0.60	Moderate
Ismarida	Ism	1	2021	0.21	Poor
Ozeros	Ozer	2	2021	0.40	Poor
Zazari	Zaz	2	2021	0.34	Poor

[a] Transboundary lake
[b] Shallow artificial lake (restored)

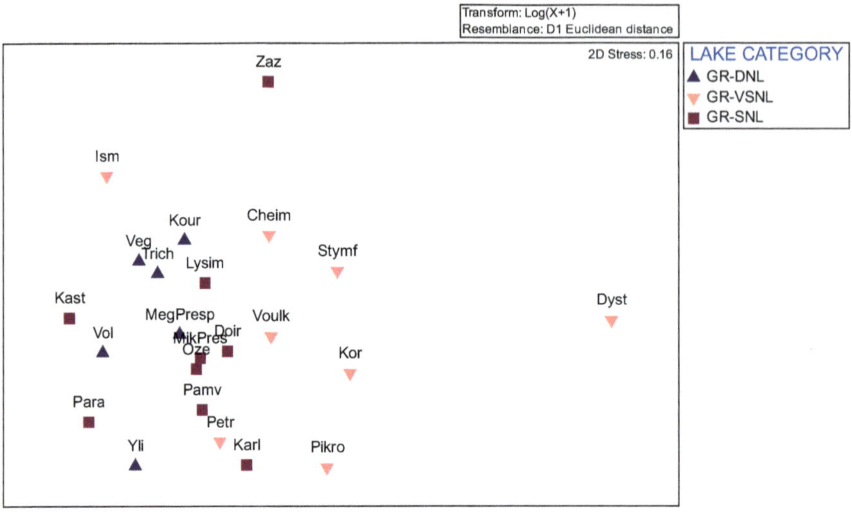

Fig. 14 Non-metric multidimensional scaling (NMDS) plot of the benthic macroinvertebrate communities among different lake categories, based on Euclidean distance among lakes

The results of the NMDS are presented in Fig. 14. The stress value of the NMDS (0.16) indicates that the ordination effectively captures the main compositional patterns of macroinvertebrates. The group of very shallow natural lakes (GR-VSNL) clustered together in the right part of the ordination plot, except for Lake Ismarida, which is positioned in the upper left part of the plot. Lake Ismarida, connected to the Thracian Sea by a 5 km channel, has a distinct saline character (Fig. 2). Its benthic community is dominated by Oligochaeta (98%) and it had the lowest EQR (0.21).

Very shallow lakes share common characteristics: extensive reedbeds, mostly *Phragmites australis*, and a high cover of aquatic macrophytes in the littoral zone. Substrate types rich in aquatic vegetation and habitat complexity seem to support distinct macroinvertebrate assemblages (Mavromati et al. 2023). In our results, the dense vegetation and extensive littoral zones contributed to more distinct benthic macroinvertebrate assemblages, with higher taxa richness compared to other lake types.

The shallow lakes grouped together in the lower left part of the ordination plot, showing a strong resemblance in their benthic communities. However, one lake, Lake Zazari, was slightly different, located in the upper left part of the plot. Lake Zazari, a shallow eutrophic lake with few macrophyte species, high phytoplankton biomass, and relatively high seasonal variability in water levels (Papadimos et al. 2022), exhibited a distinct benthic community dominated by Asellidae. Asellidae are typically associated with poor water quality, and Krepski et al. (2014) reported a significant correlation with nitrites in streams. Deep lakes are clustered together in the left part of the plot, forming the most homogeneous group, with all lakes classified as moderate, except for Lake Kourna, which was classified as good.

This relatively uniform taxonomic composition across lake types is further supported by the non-significant result of the ANOSIM analysis ($R = 0.094$, $p = 0.1$). The most abundant taxa were found across all three lake groups, covering the full range of quality classes, with the exception of the very shallow lakes, which exhibited greater species richness. It may be worthwhile to consider alternative sampling approaches that focus on well-defined meso-habitats (such as vegetated areas dominated by submerged macrophytes and/or stony areas) to investigate potential site-specific taxa that could be more sensitive to environmental changes.

Fish

Fish sampling has been undertaken during 2022 in four lakes (Lakes Vegoritida, Mikri Prespa, Paralimni, and Kastoria). Results from these samplings (Petriki et al. 2023), indicated the presence of a total of 22 fish species and one hybrid, representing 8 families, recorded across the four lakes (Table 3). The most abundant family was Leuciscidae, which included 10 species (excluding the hybrid). *Cyprinus carpio* was the only species found in all four lakes, while *Lepomis gibbosus*, *Pseudorasbora parva*, and *Tinca tinca* were present in three out of the four lakes (Lakes Vegoritida, Kastoria, and Mikri Prespa) (Table 4).

The ecological quality ratios (EQRs) for the metrics $OMNI_b$ (relative biomass of omnivorous species in the benthic gillnet catches) and $Introduced_a$ (relative abundance of introduced species in the benthic gillnet catches) and the final GLFI value were calculated (Petriki et al. 2023) and are presented in Table 4. The final ecological status for each lake is also presented.

Based on the GLFI index (Petriki et al. 2017, 2023), three lakes (Lakes Vegoritida, Mikri Prespa, and Paralimni) were classified as having good ecological quality, while Lake Kastoria was rated as moderate (Table 5). Lake Paralimni had the highest GLFI value (0.68) and marked the first ecological evaluation of the lake based on fish.

Lakes Vegoritida, Kastoria, and Mikri Prespa had also been evaluated by GLFI in the period 2010–2012 (Petriki et al. 2017). In the current fish sampling period in 2022, all three lakes showed higher GLFI values compared to the previous evaluation. Lakes Vegoritida and Mikri Prespa were classified as moderate in Petriki et al. (2017), while Lake Kastoria was classified as poor. The change in the classification of the ecological status of these three lakes based on GLFI appears to be due to differences,

Table 3 Families and species recorded in the four lakes studied (Petriki et al. 2023)

Family	Species	Lakes			
		Vegoritida	Kastoria	Mikri Prespa	Paralimni
Acheilognathidae	*Rhodeus meridionalis*	√			
Centrarchidae	*Lepomis gibbosus*	√	√	√	
Cyprinidae	*Barbus prespensis*			√	
	Cyprinus carpio	√	√	√	√
	Carassius gibelio			√	√
	Luciobarbus graecus				√
Gobionidae	*Pseudorasbora parva*	√	√	√	
Leuciscidae	*Alburnoides prespensis*			√	
	Alburnus thessalicus	√			
	Alburnus belvica			√	
	Chondrostoma prespense			√	
	Hybrid			√	
	Leucos basak			√	
	Leucos ylikiensis				√
	Rutilus rutilus	√	√		
	Scardinius graecus				√
	Squalius vardarensis		√		
	Squalius prespensis			√	
Percidae	*Perca fluviatilis*	√	√		
Siluridae	*Silurus glanis*		√		
	Silurus aristotelis				√
Tincidae	*Tinca tinca*	√	√	√	
Total		8	8	12	6

Table 4 Ecological quality ratios (EQRs) estimated for metrics $OMNI_b$ (relative biomass of omnivorous species in the benthic gillnet catches) and $Introduced_a$ (relative abundance of introduced species in the benthic gillnet catches), the final GLFI value and ecological status assessment for the four lakes in Greece[a]

Lakes	EQR			Ecological status
	EQR-$OMNI_b$	EQR-$Introduced_a$	GLFI	
Vegoritida	0.647	0.675	0.66	Good
Mikri Prespa	0.737	0.503	0.62	Good
Paralimni	0.752	0.613	0.68	Good
Kastoria	0.470	0.448	0.46	Moderate

[a] EQRs and GLFI values were estimated in Petriki et al. (2023)

Table 5 Thresholds for total phosphorus (TP, μg/L) Good-Moderate (GM) class boundaries for Greek Deep Natural Lakes (GR-DNL), Greek Shallow Natural Lakes (GR-SNL), and for Greek Reservoirs (GR-RES)—types LM 8 and LM 5/7

Lake type		GR-DNL	GR-SNL	GR-RES	
				LM 8	LM 5/7
TP (μg/L)	GM	32	41	43	46

albeit relatively small, in the number of species captured during the two samplings (Petriki et al. 2023). GLFI was able to detect these subtle changes, leading to different results and thus altering the ecological status classification.

5.4 *Total Phosphorus*

Physicochemical parameters, such as nutrients, play a supportive role to biological quality elements in the classification of ecological status according to the WFD. While moderate nutrient enrichment can be beneficial for maintaining or increasing biodiversity and aquatic food production (Jeppesen et al. 2001), excessive nutrient levels are typically associated with degradation in biotic integrity and water quality. Eutrophication leads to undesirable disturbances in lakes, including harmful algal blooms, loss of submerged vegetation, and hypoxic conditions. Nutrient loads from urban wastewater and conventional agriculture are often rich in phosphorus (P), which is typically the limiting factor for algal growth (Wurtsbaugh et al. 2019). The need to control eutrophication is undeniable, and phosphorus concentration is one of the key elements that must be regulated. To effectively manage nutrient concentrations and improve the ecological status of lakes, it is crucial to establish appropriate criteria and thresholds. Specifically, for TP (μg/L), thresholds supporting good ecological status have been set and are presented on Table 5. TP thresholds regarding Good-Moderate class boundary are 32 and 41 μg/L for Greek deep warm monomictic

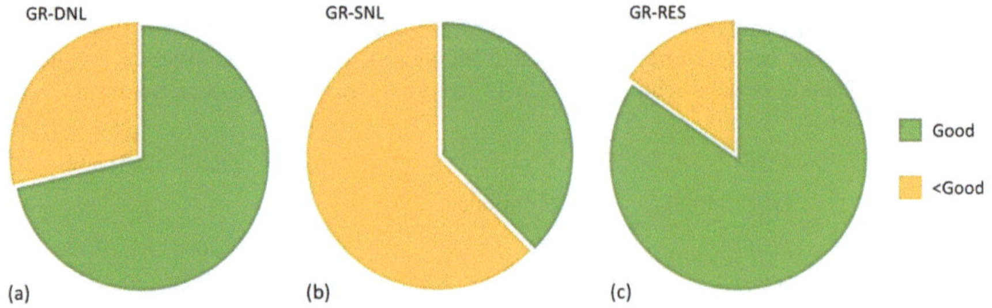

Fig. 15 Pie charts showing relative percentages of **a** Greek Deep Natural Lakes (GR-DNL), **b** Greek Shallow Natural Lakes (GR-SNL) and **c** Greek Reservoirs (GR-RES, corresponding to type LM 8) classified according to TP thresholds for good or < good status

and shallow polymictic natural lakes, respectively (Kagalou et al. 2021a), and 43 and 46 μg/L for reservoirs—types LM 8 and LM 5/7, correspondingly (Poikane et al. 2022). For the very shallow natural lakes and shallow reservoirs, TP thresholds have not been set yet (Kagalou et al. 2021a). Figure 15 presents the relative percentages of Greek lakes and reservoirs with a good or less than good status with regard to the above-mentioned TP thresholds. Most of the natural deep lakes (71%) and reservoirs (85%) have a good status, while most of shallow natural lakes (63%) have a less than good status.

Intensive agricultural activities increase TP loading in water bodies. A study by Mavromati et al. (2018) on Greek natural lakes and reservoirs highlights the positive relationship between TP levels, intensive agriculture, and population density. The study reports that natural lakes are considerably influenced by intensive agriculture. Additionally, urban lakes tend to experience higher nutrient loads and trophic status compared to non-urban lakes (Naselli-Flores 2008).

As a result, it can be argued that the water quality of reservoirs may differ from that of natural lakes. Reservoirs generally appear to be less affected by anthropogenic pressures than natural lakes. However, eutrophication remains a significant concern, as human activities disrupt natural processes in lake ecosystems. Effective management of both natural lakes and reservoirs is crucial due to the negative impacts of eutrophication on aquatic ecosystems and human well-being.

6 Overall Assessment and Discussion

In Greece, systematic monitoring of biological, physicochemical, and hydromorphological parameters in lakes has been ongoing since 2012 as part of the National Monitoring Network. This consistent, long-term dataset has facilitated the development of assessment methods for BQEs and TP, as detailed in the preceding sections. These methods encompass the underlying frameworks, specific methodologies and their practical applications in lake evaluation, with particular emphasis on spatial assessments conducted between 2016 and 2022.

These monitoring and assessment methods are employed to classify the ecological status of lake ecosystems in accordance with the provisions of WFD. The overall classification of ecological status or potential is based on various BQEs that represent key ecosystem components, further supported by physicochemical and hydromorphological quality elements. According to the WFD, hydromorphological quality elements are considered only in high status, while physicochemical quality elements are taken into account for classifications down to the Good/Moderate boundary. The "One Out–All Out" principle is applied, meaning that the worst status among any of the BQEs or supporting elements assessed determines the overall ecological status of the waterbodies.

Based on the application of the aforementioned assessment methods and following the "One-Out, All-Out" principle, Greek lakes were classified into ecological status/potential classes for the period 2016–2022. The overall results are presented in Figs. 16 and 17 for reservoirs and natural lakes, respectively. In total, 55% of all studied lakes were classified as having good and above ecological status/potential. The remaining lakes were classified as follows: 33% were assessed in moderate ecological status/potential, 10% in poor ecological status/potential and 2% in bad ecological status. Among the reservoirs, 85% were assessed with good and above ecological potential, with only 15% classified below good ecological potential. In contrast, 22% of natural lakes were classified with good ecological status, while 52% were rated as moderate. Of the remaining natural lakes, 22% were classified as poor and 4% as bad.

WFD defines ecological status as an expression of the quality of the structure and functioning of aquatic ecosystems in surface waters, classified according to Annex V. The objective of the WFD was to achieve at least good water status by 2015, defined as the worst status between ecological and chemical status. According to Article 4 of the WFD, Member States are required to prevent any deterioration in the

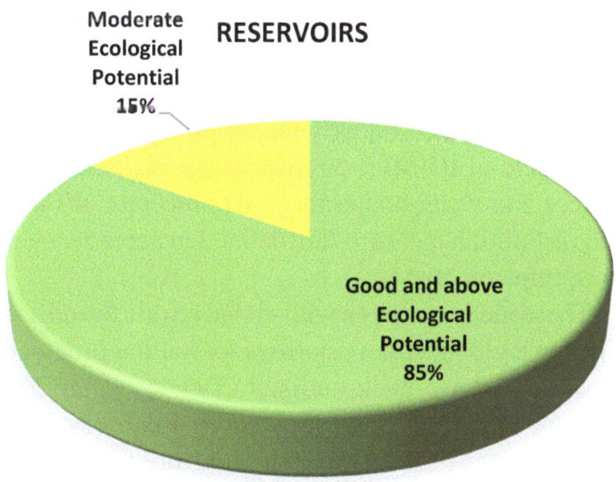

Fig. 16 Ecological potential of reservoirs in Greece for the period 2016–2022

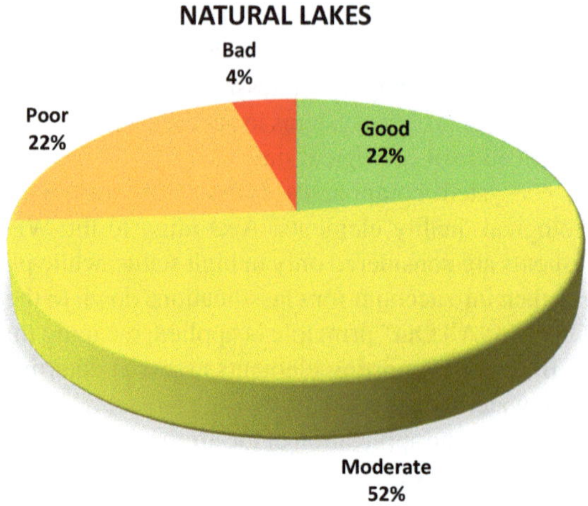

Fig. 17 Ecological status of natural lakes in Greece for the period 2016–2022

status of their surface waters, to protect, enhance, and restore all bodies of surface water, and to progressively reduce pollution from priority substances. Furthermore, the directive calls for the cessation or phase-out of emissions, discharges, and losses of priority hazardous substances.

The lake monitoring and assessment methods provide a comprehensive understanding of the structure of lake ecosystems across various spatial and temporal scales. Through the national monitoring network, Greece has developed time series of data that document the status and trends of its water bodies. These data serve as scientific evidence to track progress toward the achievement of the WFD objectives. As a result, the monitoring framework offers valuable insights for informed decision-making and the formulation of effective management strategies. Relevant authorities can prioritize conservation and management actions and assess their effectiveness over time.

A core aspect of the WFD is the use of the catchment scale as the management unit for river basins across Europe. This approach allows for the integrated consideration of all pressures affecting water bodies (Reyjol et al. 2014). Programmes of Measures are outlined in six-year River Basin Management Plans (RBMPs), targeting the pressures and their underlying causes that contribute to ecosystem degradation. The third cycle of RBMPs, covering the period 2022–2027, has just been completed in Greece. For water bodies that fail to meet the WFD objectives, specific conservation and restoration measures are defined in accordance with WFD guidelines and are set to be implemented.

These measures align with global efforts to address the loss of freshwater biodiversity. On a global scale, priorities for halting biodiversity loss and improving aquatic ecosystems have been outlined (Tickner et al. 2020; Birnie-Gauvin et al. 2023), with the upscaling of an Emergency Recovery Plan considered crucial to meeting the objectives of the Convention on Biological Diversity for 2030 and the Sustainable Development Goals (Tickner et al. 2020). Within the EU, the EU Biodiversity

Strategy for 2030 aims to set Europe on a path to biodiversity recovery. Protecting and restoring freshwater biodiversity and healthy freshwater ecosystems are key priorities, with the recognition that significant efforts are required to restore these ecosystems and the natural functions of rivers to meet the objectives of the WFD. The Nature Restoration Regulation [Regulation (EU) 2024/1991] provides a robust framework for the restoration of freshwater ecosystems, aiming for the achievement of favourable conservation status for river, lake, alluvial, and riparian habitats, as well as the species protected by the Habitats and Birds Directives (Directives 92/43/EEC and 2009/147/EC).

Acknowledgements EKBY's personnel conducted samplings and contributed to data analysis. Fish data used in this research came from contract 'Fish samplings in Greek Lakes', commissioned by EKBY to The Aristotle University of Thessaloniki, supervised by Dimitra Bobori. Data of 18 reservoirs presented in Table 2 were provided by the Public Power Corporation, the Water Supply and Sewerage Company of the Capital Region and the Administrative Region of Crete. Part of the phytoplankton analysis was commissioned by EKBY to the Laboratory Unit on Harmful Marine Microalgae of Aristotle University of Thessaloniki.

Funding This research was conducted in the frame of the Greek National Water Monitoring Network, according to the JMD YPEN/DPDYP/107168/1444/2021, implemented by The Goulandris Natural History Museum, Greek Biotope/Wetland Centre (EKBY). The network is supervised by the General Directorate for Waters of the Ministry of Environment and Energy. This research was financed by the European Union Cohesion Fund (Partnership Agreement 2014–2020, Act MIS 5001204 and Partnership Agreement 2021–2027, Act MIS 6005123).

References

APHA 4500-P E (2023) Standard methods for the examination of water and wastewater. In: Lipps WC, Baxter TE, Braun-Howland E (eds) Phosphorus-ascorbic acid method, 24th edn. American Public Health Association, American Water Works Association, Water Environment Federation

Birk S, Bonne W, Borja A, Brucet S, Courrat A, Poikane S, Solimini A, Van De Bund W, Zampoukas N, Hering D (2012) Three hundred ways to assess Europe's surface waters: an almost complete overview of biological methods to implement the water framework directive. Ecol Indic 18:31–41. https://doi.org/10.1016/j.ecolind.2011.10.009

Birnie-Gauvin K, Lynch AJ, Franklin PA, Reid AJ, Landsman SJ, Tickner DJ, Aarestrup K, Cooke SJ (2023) The RACE for freshwater biodiversity: essential actions to create the social context for meaningful conservation. Conserv Sci Pract 5(4):e12911. https://doi.org/10.1111/csp2.12911

Bytyqi P, Czikkely M, Shala-Abazi A, Fetoshi O, Ismaili M, Hyseni-Spahiu M, Ymeri P, Kabashi-Kastrati E, Millaku F (2020) Macrophytes as biological indicators of organic pollution in the Lepenci River Basin in Kosovo. J Freshw Ecol 35(1):105–121. https://doi.org/10.1080/02705060.2020.1745913

Carvalho L, Poikane S, Solheim AL, Phillips G, Borics G, Catalan J, De Hoyos C, Drakare S, Dudley BJ, Järvinen M, Laplace-Treyture C (2013) Strength and uncertainty of phytoplankton metrics for assessing eutrophication impacts in lakes. Hydrobiologia 704(1):127–140

Chamoglou M, Papadimitriou T, Kagalou I (2014) Key-descriptors for the functioning of a Mediterranean reservoir: the case of the new Lake Karla-Greece. Environ Process 1:127–135. https://doi.org/10.1007/s40710-014-0011-0

European Commission (2000) Directive 2000/60/EC of the European Parliament and of the Council of 23 October 2000 establishing a framework for community action in the field of water policy. Off J Eur Communities 327(43):1–72

Free G, Poikane S, Solheim AL, Bussettini M, Bradley C, Smith J, Caroni R, Bresciani M, Pinardi M, Giardino C, van de Bund W (2024) Climate change and ecological assessment in Europe under the WFD—hitting moving targets with shifting baselines? J Environ Manage 370:122884. https://doi.org/10.1016/j.jenvman.2024.122884

Fussman GF, Ellner SP, Shertzer KW, Hairston NG (2000) Crossing the Hopf bifurcation in a live predator–prey system. Science 290:1358–1360

Gabriels W, Koen L, Pauw N, Goethals P (2010) Multimetric macroinvertebrate index Flanders (MMIF) for biological assessment of rivers and lakes in Flanders (Belgium). Limnol Ecol Manag Inland Waters 40:199–207. https://doi.org/10.1016/j.limno.2009.10.001

Gibbs JR (1970) Mechanisms controlling world water chemistry. Science 170:1088–1090

Giorgi F, Lionello P (2008) Climate change projections for the Mediterranean region. Glob Planet Change 63:90–104. https://doi.org/10.1016/j.gloplacha.2007.09.005

Harper M, Mejbel HS, Longert D, Abell R, Beard TD, Bennet JR et al (2021) Twenty-five essential research questions to inform the protection and restoration of freshwater biodiversity. Aquat Conserv Mar Freshw Ecosyst 1–22. https://doi.org/10.1002/aqc.3634

ICOLD (2013) The dams of Greece, Greek Committee on large dams, p 112. https://eemf.gr/

ISO 10304-01 (2007) Water quality—determination of dissolved anions by liquid chromatography of ions—part 1: determination of bromide, chloride, fluoride, nitrate, nitrite, phosphate and sulphate

ISO 14911 (1998) Water quality—determination of dissolved Li^{+}, Na^{+}, K^{+}, NH^{4+}, Mg^{2+}, Ca^{2+}, Mn^{2+}, Sr^{2+} and Ba^{2+} using ion chromatography—method for water and waste water

Jeppesen E, Peder Jensen J, Søndergaard M, Lauridsen T, Landkildehus F (2001) Trophic structure, species richness and biodiversity in Danish lakes: changes along a phosphorus gradient. Freshw Biol 45(2):201–218. https://doi.org/10.1046/j.1365-2427.2000.00675.x

Kagalou I, Ntislidou C, Latinopoulos D, Kemitzoglou D, Tsiaoussi V, Bobori DC (2021a) Setting the phosphorus boundaries for Greek natural shallow and deep lakes for water framework directive compliance. Water 13(5):739. https://doi.org/10.3390/w13050739

Kagalou I, Ntislidou C, Latinopoulos D, Bobori D (2021b) Suggestions proposition for lakes' classification according to the new total phosphorus concentrations boundaries for each lake type. Technical report. Aristotle University, Thessaloniki, Greece

Kaiser HF (1960) The application of electronic computers to factor analysis. Educ Psychol Meas 20(1):141–151

Kelly M, Juggins S (2023) Revision of phytoplankton metrics for Greek lakes. Technical report

Kelly M, Juggins S, Moschandreou K, Kemitzoglou D, Tsiaoussi V (2023) Development of novel diatom metrics to assess ecological status of phytobenthos in Greek lakes. Ecol Ind 147:109974. https://doi.org/10.1016/j.ecolind.2023.109974

Krepski T, Pilecka-Rapacz M, Czerniawski R et al (2014) Analysis of benthic macroinvertebrate communities from the lower sections of large river in relation to different environmental factors. Cent Eur J Biol 9:1037–1047. https://doi.org/10.2478/s11535-014-0346-6

Latinopoulos D, Ntislidou C, Kagalou I (2016) Multipurpose plans for the sustainability of the Greek lakes: emphasis on multiple stressors. Environ Process 3:589–602. https://doi.org/10.1007/s40710-016-0152-4

Maasri A, Jähnig SC, Adamescu MC, Adrian R, Baigun C, Baird DJ et al (2022) A global agenda for advancing freshwater biodiversity research. Ecol Lett 25:255–263. https://doi.org/10.1111/ele.13931

Mavromati E, Kagalou I, Kemitzoglou D, Apostolakis A, Seferlis M, Tsiaoussi V (2018) Relationships among land use patterns, hydromorphological features and physicochemical parameters of surface waters: WFD lake monitoring in Greece. Environ Process 5:139–151. https://doi.org/10.1007/s40710-018-0315-6

Mavromati E, Kemitzoglou D, Tsiaoussi V, Lazaridou M (2021) A new WFD—compliant littoral macroinvertebrate index for monitoring and assessment of Mediterranean lakes (HeLLBI). Environ Monit Assess 193:745. https://doi.org/10.1007/s10661-021-09493-1
Mavromati E, Kemitzoglou D, Tsiaoussi V (2023) Does littoral substrate affect macroinvertebrate assemblages in Mediterranean lakes? Aquat Ecol 57:667–679. https://doi.org/10.1007/s10452-023-10037-7
Melista Eleni (2015) Earthen dams of Greece. Bachelor thesis, Patra, p 53
Naselli-Flores L (2008) Urban lakes: ecosystems at risk, worthy of the best care. In: Proceedings of TAAL 2007: the 12th world lake conference, pp 1333–1337
Navrozidou V, Apostolakis A, Katsavouni S, Mavromati E, Tsiaoussi V (2019) Water quality of Greek reservoirs: the impact of land cover and water resources management on phytoplankton. Paper presented at 11th world congress on water resources and environment (EWRA 2019) managing water resources for a sustainable future, Madrid, Spain, 25–29 June 2019
Nikolopoulou I, Mavromati E, Moschandreou K, Navrozidou V, Kemitzoglou D, Tsiaoussi V (2025) Lake phytoplankton status and trends: a case study from Greek lakes, Eastern Mediterranean. Environ Monit Assess 197, 733 (2025). https://doi.org/10.1007/s10661-025-14147-7
Ntislidou C, Lazaridou M, Tsiaoussi V, Bobori D (2018) A new multimetric macroinvertebrate index for the ecological assessment of Mediterranean lakes. Ecol Ind 93:1020–1033
Pahissa J, Catalan J, Morabito G, Dörflinger G, Ferreira J, Laplace-Treyture C, Gîrbea R, Marchetto A, Polykarpou P, de Hoyos CP (2015) Benefits and limitations of an intercalibration of phytoplankton assessment methods based on the Mediterranean GIG reservoir experience. Sci Total Environ 538:169–179. https://doi.org/10.1016/j.scitotenv.2015.08.042
Papadimos D, Demertzi K, Papamichail D (2022) Assessing lake response to extreme climate change using the coupled MIKE SHE/MIKE 11 model: case study of Lake Zazari in Greece. Water 14(6):921. https://doi.org/10.3390/w14060921
Patsia A, Kemitzoglou D, Mavromati E, Tsiaoussi V (2023) Benthic macroinvertebrate communities in the littoral zone of Greek lakes. Poster presentation in the 19th world lake conference, Balatonfüred, Hungary, 7–9 Nov 2023
Perivolioti TM, Zachopoulos K, Zioga M, Tompoulidou M, Katsavouni S, Kemitzoglou D, Terzopoulos D, Mouratidis A, Tsiaoussi V (2024) Monitoring the impact of floods on water quality using optical remote sensing imagery: the case of Lake Karla (Greece). Water 16:3502. https://doi.org/10.3390/w16233502
Perivolioti TM, Apostolakis A, Papadimos D, Tsiaoussi V (2025) Bathymetry and morphometric analysis of Greek natural lakes through a hybrid GIS-acoustic methodology. Inland Waters 1–45. https://doi.org/10.1080/20442041.2024.2447173
Petriki O, Lazaridou M, Bobori D (2017) A fish-based index for the assessment of the ecological quality of temperate lakes. Ecol Indic 78:556–565. https://doi.org/10.1016/j.ecolind.2017.03.029
Petriki O, Ntislidou C, Bobori D (2023) Ecological quality indicators of four lakes. Technical report, Aristotle University of Thessaloniki, p 14 (in Greek)
Poikane S, Johnson RK, Sandin L, Schartau AK, Solimini AG, Urbanič G, Arbačiauskas K, Aroviita J, Gabriels W, Miler O, Pusch MT, Timm H, Bohmer J (2016) Benthic macroinvertebrates in lake ecological assessment: a review of methods, intercalibration and practical recommendations. Sci Total Environ 543:123–134
Poikane S, Portielje R, Denys L, Elferts D, Kelly M, Kolada A, Mäemets H, Phillips G, Søndergaard M, Willby N, van den Berg MS (2018) Macrophyte assessment in European lakes: diverse approaches but convergent views of 'good' ecological status. Ecol Ind 94(1):185–197. https://doi.org/10.1016/j.ecolind.2018.06.056
Poikane S, Kelly M, Várbíró G, Borics G, Erős T, Hellsten S, Kolada A, Lukács B, Solheim A, López J, Willby N, Wolfram G, Phillips G (2022) Estimating nutrient thresholds for eutrophication management: novel insights from understudied lake types. Sci Total Environ 827:154242. https://doi.org/10.1016/j.scitotenv.2022.154242

Pont D, Valentini A, Rocle M, Maire A, Delaigue O, Jean P, Dejean T (2019) The future of fish-based ecological assessment of European rivers: from traditional EU water framework directive compliant methods to eDNA metabarcoding-based approaches. J Fish Biol 98(2):354–366

R Core Team (2019) R: a language and environment for statistical computing. R Foundation for Statistical Computing, Vienna. https://www.r-project.org/

Reid AJ, Carlson AK, Creed IF, Eliason EJ, Gell PA, Johnson PTJ, Kidd KA, MacCormack TJ, Olden JD, Ormerod SJ, Smol JP, Taylor WW, Tockner K, Vermair JC, Dudgeon D, Cooke SJ (2019) Emerging threats and persistent conservation challenges for freshwater biodiversity. Biol Rev 94:849–873. https://doi.org/10.1111/brv.12480

Reyjol Y, Argillier C, Bonne W, Borja A, Buijse AD, Cardoso AC et al (2014) Assessing the ecological status in the context of the European water framework directive: where do we go now? Sci Total Environ 497:332–344. https://doi.org/10.1016/j.scitotenv.2014.07.119

Reynaud A, Lanzanova D (2017) A global meta-analysis of the value of ecosystem services provided by lakes. Ecol Econ 137:184–194. https://doi.org/10.1016/j.ecolecon.2017.03.001

Rossaro B, Marziali L, Cardoso AC, Solimini A, Free G, Giacchini R (2007) A biotic index using benthic macroinvertebrates for Italian lakes. Ecol Ind 7(2):412–429. https://doi.org/10.1016/j.ecolind.2006.04.002

Rueda F, Moreno-Ostos E, Armengol J (2006) The residence time of river water in reservoirs. Ecol Model 191(2):260–274. https://doi.org/10.1016/j.ecolmodel.2005.04.030

Strayer D, Dudgeon D (2010) Freshwater biodiversity conservation: recent progress and future challenges. J N Am Benthol Soc 29(1):344–358. https://doi.org/10.1899/08-171.1

Theodorakis M, Margaris NS, Kainadas I (2000) Wetlands of PPC. University of the Aegean, Department of Environment, Kastaniotis Editions, p 127

Tickner D, Opperman JJ, Abell R, Acrema M, Arthington AH, Bunn SE et al (2020) Bending the curve of global freshwater biodiversity loss: an emergency recovery plan. BioScience 70:330–342

Tsiaoussi V, Kemitzoglou D, Mavromati E (2016) Report on the application of phytoplankton index NMASRP for reservoirs in Greece. Greek Biotope/Wetland Centre (EKBY) and Special Secretariat for Waters, Ministry of Environment, Thermi, Greece, 16 pp

Tsiaoussi V, Mavromati E, Kemitzoglou D (2017) Report on the development of the national method for the assessment of the ecological status of natural lakes in Greece, using the biological quality element "phytoplankton", 1st revision. Greek Biotope/Wetland Centre (EKBY) and Special Secretariat for Waters, Ministry of Environment, Thermi, Greece, 17 pp

Urbanič G, Petkovska V, Pavlin M (2012) The relationship between littoral benthic invertebrates and lakeshore modification. Fundam Appl Limnol 180:157–173. https://doi.org/10.1127/1863-9135/2012/0207

Verpoorter C, Kutser T, Seekell DA, Tranvik LJ (2014) A global inventory of lakes based on high-resolution satellite imagery. Geophys Res Lett 41:6396–6402. https://doi.org/10.1002/2014GL060641

Wurtsbaugh W, Paerl H, Dodds W (2019) Nutrients, eutrophication and harmful algal blooms along the freshwater to marine continuum. WIREs Water 6(5):e1373. https://doi.org/10.1002/wat2.1373

Zervas D, Tsiaoussi V, Tsiripidis I (2018) HeLM: a macrophyte-based method for monitoring and assessment of Greek lakes. Environ Monit Assess 190:326. https://doi.org/10.1007/s10661-018-6708-1

Zervas D, Tsiripidis I, Bergmeier E, Tsiaoussi V (2020) A phytosociological survey of aquatic vegetation in the main freshwater lakes of Greece. Veget Classif Surv 1:53–75. https://doi.org/10.3897/VCS/2020/48377

Satellite Remote Sensing (RS): A Valuable and Efficient Implement for Optically Diverse Greek Lakes' Water Quality (WQ) and Trophic Status Monitoring

Vassiliki Markogianni

Abstract Satellite remote sensing (RS) has emerged as a valuable tool for monitoring the water quality (WQ) and trophic status of optically diverse lakes, particularly in regions like Greece, characterized by varied aquatic ecosystems. Traditional in-situ methods for assessing WQ are often labor-intensive, costly, and limited in spatial coverage. RS provides a complementary approach, enabling frequent, wide-scale, and cost-effective monitoring. The optical diversity of lakes, influenced by factors such as varying chlorophyll-a (Chl-*a*) concentrations, suspended sediments, and dissolved organic matter, presents challenges in WQ assessment. RS techniques, leveraging spectral data from satellites, can effectively quantify WQ key indicators, which are crucial for determining trophic status. This chapter provides a comprehensive overview of specific RS applications across the National Lake Network Monitoring of Greece, conducted in accordance with the Water Framework Directive (WFD), focusing primarily on the empirical development of WQ quantitative models for Chl-*a* and total phosphorus concentrations, and Secchi Disk depths. Those same models' performance was further validated from the perspective of different atmospheric correction (AC) methods, since accurate AC is crucial for applications especially where small differences in surface reflectance (SR) are significant. The proven capability of RS to provide consistent and timely data supports sustainable water resource management through the continuous monitoring of Greece's optically diverse lake ecosystems.

Keywords Lake water quality (WQ) · Satellite remote sensing (RS) · Chlorophyll-*a* · Secchi disk depth (SDD) · Total phosphorus (TP) · WFD · Carlson's trophic state index (TSI) · Atmospheric correction (AC) · Landsat · Sentinel 2

V. Markogianni (✉)
Institute of Marine Biological Resources and Inland Waters, Hellenic Centre for Marine Research, Attica, Greece
e-mail: vmarkogianni@hcmr.gr

K. Stefanidis et al. (eds.), *Lakes of the Eastern Mediterranean and Balkans*,
https://doi.org/10.1007/978-3-032-02168-7_6

1 Introduction

Surface freshwater stands as one of the terrestrial ecosystem's vital resources and serves as the primary source of drinking water globally (Whyte et al. 2018). Lakes and reservoirs store a portion of this crucial resource, with over 14 million lakes (> 10 ha) documented worldwide (Meyer et al. 2020). Lake water fulfils both environmental and human needs and plays a pivotal role in the European and global economies. It is utilized for various purposes, including civil (e.g., irrigation), industrial (e.g., processing and cooling, energy production, fishery), and recreational activities. Particularly in landlocked nations, lakes hold significant value as they represent some of the most important water sources (Knoll et al. 2019).

Over the past two decades, limnological research indicates that ongoing global warming is exacerbating lakes' responses to cultural eutrophication. This includes the acceleration of hypolimnetic anoxia, nutrient release from lake sediments, intensified nutrient recycling, and by extent increased algal production (Salmaso and Tolotti 2021). Additionally, according to a report by the European Environment Agency (EEA 2018) examining the impact of nutrient loads on European surface water bodies, it was found that 60% of these bodies fail to meet the objectives of good water quality as defined by international WQ directives, such as the European Water Framework Directive (2000/60/EC).

The necessity for the sustainable management of water bodies underscores the fact that water resources are finite and possess limited resilience in the face of human-induced pressures such as ongoing drainage, conversion, and pollution. Therefore, continuous monitoring of both the quality of water bodies and their watersheds emerges as a pivotal aspect of sustainable water management (Gholizadeh et al. 2016). WQ monitoring stands out as the cornerstone of lake management, where the term "lake management" encompasses strategies aimed at preserving the long-term viability of lake ecosystems, which serve as habitats for aquatic and terrestrial life (Bonansea et al. 2015).

WQ is the primary indicator of a water body's ecological condition, necessitating continuous monitoring of mainly physico-chemical and biological elements (Fatima 2018; Nikolaidis et al. 2022). However, monitoring large water bodies is a challenging endeavor, requiring frequent and meticulous data collection and interpretation efforts. Achieving a comprehensive understanding of the spatial and temporal variations of common key WQ indicators requires exhaustive sampling fieldwork. This results in a crucial trade-off between the number of sampling stations and the imperative of keeping costs within reasonable limits (Strobel et al. 2000).

WQ parameters are conventionally assessed through a combination of in-situ measurements, water sample collection, and laboratory analysis (Li et al., 2016). While in-situ WQ monitoring offers high accuracy at specific locations and time, it is a time-intensive process and cannot guarantee a simultaneous WQ dataset on a regional or broader scale (Duan et al. 2013; Gholizadeh et al. 2016; Topp et al. 2020). Moreover, traditional point sampling methods fail to capture the spatial or temporal variations in WQ required for comprehensive assessment and management of water bodies. Additionally, the patchy distribution of elements such as nutrients, algal blooms, and Total Suspended Matter (TSM) renders these methods unsuitable for monitoring numerous water bodies at a regional or national level (Japitana and Burce 2019).

In recent years, the combined use of in-situ monitoring of lake WQ and satellite remote sensing (RS) has emerged as the latest scientific trend in many WQ monitoring programs worldwide (Japitana and Burce 2019; Neil et al. 2019; Topp et al. 2020). While RS has undeniable capabilities in assessing lake WQ, it is not sufficiently precise on its own and should be complemented with field water sampling (Gholizadeh et al. 2016). Consequently, the integration of point-specific WQ datasets, which lack spatiotemporal trends, with simultaneous RS datasets providing a synoptic view of ongoing Earth surface processes, facilitates the monitoring, assessment, and identification of WQ management strategies. Satellite RS proves to be an efficient and beneficial tool for evaluating spatial and temporal variations in WQ (Bonansea et al. 2015; Japitana and Burce 2019). RS technologies enable researchers to gain a comprehensive perspective of ecosystems. Viewing Earth from space, satellite data becomes invaluable in supporting lake management, particularly in the context of increasingly strict environmental regulations approved by governments worldwide, such as the Water Framework Directive (WFD; 2000/60/EC) and the European Marine Strategy Framework Directive (MSFD; 2008/56/EC) (Nikolaidis et al. 2022).

While RS approaches for inland WQ measurements have been utilized for nearly 50 years (Topp et al. 2020), the development of new instruments (platforms and sensors) for this purpose over the last two decades has been relatively slow compared to terrestrial and oceanic RS techniques (Pizani and Maillard 2022). The effectiveness of each RS application for WQ monitoring hinges on the selection of appropriate platforms and instruments (Pizani and Maillard 2022). Various categories of sensors are commonly used in WQ assessments, including airborne and unmanned aerial vehicles (UAV) sensors, satellite sensors with visible and infrared wavelengths, passive microwave radiometers (MWR), and synthetic aperture radar (SAR). According to Gholizadeh et al. (2016), Sagan et al. (2020), Topp et al. (2020), and Pizani and Maillard (2022), the most widely utilized sensors for inland WQ assessment are passive optical and thermal sensors. This preference is because water exhibits high absorptivity within the near and shortwave infrared spectrum, with the majority of water-leaving radiance occurring within the visible spectrum. However, in optically complex waters, sediment reflectance surpasses the absorptive properties of water in the near/shortwave infrared wavelengths (NIR/SWIR), where high absorption within the visible spectrum leads to a limited range of reflectance values. To address this limitation, there has been a significant improvement in spectral, spatial, and radiometric resolutions, revisit time, number of satellite bands, and free access to data (Topp et al. 2020).

Inland waters, particularly lakes, pose an additional challenge for current ocean color satellites due to their small size, making them undetectable. However, this limitation has been addressed through the utilization of multispectral images from satellites such as Landsat, Sentinel-2, and ASTER (Advanced Spaceborne Thermal

Emission and Reflection Radiometer). With their fine spatial resolutions ranging from 10 to 60 m, these satellites can effectively capture small freshwater lakes and rivers wider than a few hundred meters. Consequently, these images have become the preferred choice for freshwater lake mapping projects (Wang et al. 2020). Recent reviews of state-of-the-art-RS-based approaches by Gholizadeh et al. (2016) and Pizani and Maillard (2022) underscore the effectiveness of Landsat sensors, including TM (Thematic Mapper), MSS (Multi-Spectral Scanner), ETM (Enhanced Thematic Mapper), OLI (Operational Land Imager), and OLI-2 (Landsat 9), in assessing key WQ parameters such as Chlorophyll-*a* (Chl-*a*), Secchi Disk Depth (SDD), Total Phosphorus (TP), and by extension, trophic status.

1.1 Traditionally Used Approaches for Lake WQ Monitoring via Satellite RS

The predominant approach for monitoring the WQ of inland waters using RS involves the establishment of a standard linear regression between spectral band/band ratio values and temporally coincident in-situ WQ measurements (Topp et al. 2020).

According to Topp et al. (2020), there are three well-documented methodologies for estimating the concentration of WQ elements in inland waters: empirical, semi-empirical, and physical or analytical methodology (Table 1). Empirical methods aim to establish relationships between in-situ WQ measurements and water leaving radiance measured by the sensor without the need for prior understanding of the complex interactions between water and light. While these relationships often lead to effective data improvement, their transferability is limited (Austin and Petzold 1981). Additionally, empirical methods incorporate machine learning techniques, known for their robust ability to handle the complex non-linear relationships typical of WQ RS data (Sagan et al. 2020; Topp et al. 2020). Machine learning algorithms include artificial neural networks (ANN), genetic algorithms (GA), support vector machines (SVM), random forest regression trees (RFRT), and empirical orthogonal functions (EOF; Topp et al. 2020).

On the other hand, semi-empirical techniques combine spectral and physical knowledge of the properties of the studied WQ constituents, which are then correlated with in-situ concentrations. Physical or analytical approaches require the acquisition of certain biogeochemical parameter values (e.g., Chl-*a*, CDOM), as well as inherent (IOP) and apparent optical properties (AOP). These models are based on radiative transfer and are calibrated using field observations.

Although analytical methods such as fuzzy logic and Principal Component Analysis (PCA) have been extensively employed, empirical and semi-empirical predictive models remain widely utilized (Gholizadeh et al. 2016). The complexity of analytical methods in terms of theory and calculation difficulties (Gholizadeh et al. 2016), along with the unavailability of detailed spectral information of optically active water constituents (OACs; e.g. optical properties, radiometric quantities), have contributed

Table 1 Summary of the common approaches for RS of lake WQ monitoring (Markogianni 2023)

Modelling approach	Including	Advantages	Limitations
Empirical		Easily interpretable, no a priori assumptions required	Limited transferability (dependant on range of training data), uncapable of handling non-linear relationships
	Machine learning techniques	Capable of handling complicated non-linear Relationships, no a priori assumptions required	Computationally expensive, risk of overfitting
Semi-empirical		Easily interpretable	Uncapable of handling non-linear relationships
Physical/analytical		Theoretically generalizable	Computationally expensive, knowledge of optical properties and collection of in-situ Rrs spectra are required

to the continued utilization and development of empirical models. This trend is particularly evident in cases where machine learning models are employed, as they often reduce overall error and maximize model fit (Topp et al. 2020). However, it's important to note that empirical algorithms tend to be more specific to certain water types or regions, or optical conditions (Odermatt et al. 2012).

Moreover, it's worth mentioning that semi-analytical methods are considered superior to empirical ones in terms of result reliability and the fact that they do not require subsequent recalibration of the retrieval algorithm using in-situ data. However, these approaches necessitate the use of a spectroradiometer and the collection of in-situ-measured Reflectance (Rrs) spectra, including radiance from skylight, radiance from a standard gray board, and total upwelling radiance from the water (Jiang et al. 2019).

1.2 WQ Elements Measured by RS

RS offers a powerful means of monitoring various WQ parameters, both optically-active and non-optically-active, across diverse aquatic environments such as lakes, ponds, rivers, and coastal regions (Gholizadeh et al. 2016; Pizani and Maillard 2022). Optically-active parameters, including Chl-*a*, transparency, turbidity, total suspended matters (TSM), coloured dissolved organic matter (CDOM), true color, and temperature, can be directly monitored remotely. Non-optically-active parameters, such as nutrients, pH, and dissolved oxygen, can be indirectly assessed through their relationships with optically-active parameters, serving as proxies (Gholizadeh et al. 2016; Pizani and Maillard 2022). Chl-*a* concentration serves as a key indicator of phytoplankton abundance, trophic state, and overall water pollution indices related to public health, eutrophication, and ecosystem habitat degradation (Randolph et al. 2008; Ruiz-Verdu et al. 2008). RS techniques enable the direct quantification of Chl-*a* concentration, providing priceless insights into trophic levels, the presence of toxic algal blooms, and phytoplankton biomass (Randolph et al. 2008; Ruiz-Verdu et al. 2008).

Numerous studies have revealed that biological and chemical WQ parameters, such as Chl-*a*, exhibit distinctive spectral characteristics that can be captured using RS techniques. To translate spectral data into WQ parameters, researchers have developed various spectral indices based on empirical or semi-empirical relationships. However, these spectral indices have shown to be less reliable in diverse water bodies, including lakes, ponds, rivers, and streams in coastal regions (Yang et al. 2017; Sagan et al. 2020). This reduced reliability stems from the assumption of constant water and atmospheric conditions, leading to significant estimation errors, particularly when applied across time series. Consequently, spectral indices are suggested to be more suitable for identifying the spatial distribution of WQ rather than providing precise predictions (Sagan et al. 2020).

Concerning the assessment of nutrient concentrations via RS, challenges are presented due to weak optical characteristics and low signal-to-noise ratio, resulting in less statistically robust results compared to optically-active constituents (Gholizadeh et al. 2016). Nonetheless, RS has been effective in detecting the relationship between algae concentration and corresponding nutrients, providing valuable information about nutrient dynamics (Sagan et al. 2020). Nutrient models have generally not produced statistically robust results comparable to those of constituents with optical properties (Dewidar and Khedr 2001; Wu et al. 2010; Chen and Quan 2012; Isenstein and Park 2014). However, indirect methods can be employed to estimate nitrogen (N) and phosphorus (P) concentrations.

It's crucial, though, to pay special attention to which nutrient is growth limiting, as the correlation with Chl-*a* may vary depending on the water body. For instance, in one water body, the correlation with Chl-*a* might be with nitrogen, while in another, it might be with phosphorus (Topp et al. 2020).

Total phosphorus (TP) estimation via RS has been explored due to its high correlation with optically active constituents (Kutser et al. 1995; Wang et al. 2004; Wu et al. 2010), as it cannot be directly measured using optical RS instruments. The relationship between Chl-*a* and TP has been extensively investigated in individual lakes, demonstrating a strong and positive correlation among lakes (Healey and Hendzel 1979; Busse et al. 2006). Since TP is highly correlated with Chl-*a* concentration, and TSM usually reflects TP loading, TP is also closely related to SDD according to Carlson's findings (Carlson 1977).

TP is closely related to SDD, another important trophic state indicator, as increased algae and suspended inorganic matter reduce light penetration depth (Fuller et al. 2002). Two main methodologies are typically employed to retrieve Secchi Disk Transparency (SDT) based on RS data: the empirical approach, which estimates SD through regression analysis, and the semi-analytical approach, which retrieves SD based on underwater visibility theory (Jiang et al. 2019).

Furthermore, water clarity has been demonstrated to interact with nutrient availability and Chl-*a* concentrations within lakes (Song et al. 2012). As a result, RS studies frequently leverage water clarity assessments to evaluate the overall trophic status of lakes, categorizing them as oligotrophic, mesotrophic, or eutrophic based on their nutrient and algae levels.

Overall, RS provides esteemed insights into various WQ parameters, offering a comprehensive understanding of aquatic ecosystems while supporting management and conservation efforts.

1.3 Case 1-Case 2 Waters and RS

The categorization of waters into Case 1 (oceanic) and Case 2 (coastal regions, rivers, and lakes), as refined by Gordon and Morel (1983), holds significant importance in the application of RS techniques for monitoring WQ and trophic status. This classification scheme has notable implications for the interpretation and modelling of optical data.

In Case 1 waters, optical properties are primarily influenced by phytoplankton and co-varying substances, with Chl-*a* serving as a proxy for phytoplankton concentration. This understanding has facilitated the development and implementation of large-scale optical models and the creation of Chl-*a* predicting algorithms tailored to Case 1 waters (Markogianni et al. 2022).

In contrast to Case 1 waters, single-variable models are inadequate for Case 2 waters. It is widely recognized that Case 2 waters are more complicated in terms of their composition and optical properties. Monitoring the WQ of Case 2 waters presents a more complex challenge due to the independent variation of phytoplankton, suspended material, yellow substances, and potentially bottom reflectance. The primary difficulty arises from the small changes in optical signals and dissolved constituent concentrations, which often hinder the extraction of reliable information. Additionally, the optical signal may be influenced similarly by multiple substances, making it difficult to distinguish between them (Gholizadeh et al. 2016). Moreover, runoff and discharges from rivers/streams further complicate the retrieval process of water constituents in Case 2 waters. These inflows introduce various organic and inorganic particles, known as total suspended solids (TSS), which add to the complexity of the water composition (Gholizadeh et al. 2016).

Given the complexity of WQ monitoring in Case 2 waters, which represents a multi-variable, non-linear problem, it is more practical to develop a series of algorithms rather than relying on a single all-purpose model. By employing multiple algorithms, each designed to address specific variables and varying concentration ranges, a more comprehensive approach can be achieved (IOCCG 2000). This allows for a more exhaustive understanding of the diverse factors influencing WQ in Case 2 waters and improves the accuracy and reliability of WQ assessments.

The Case 1/Case 2 classification can significantly enhance RS products, particularly when combined with individual optical water types (OWTs). Coastal regions and inland waters exhibit remarkable optical diversity, making additional information about their variability in inherent optical properties (IOPs) and biogeochemical composition highly valuable. Various OWTs can be identified, such as hypereutrophic waters, turbid waters with high organic content, sediment-laden waters, CDOM-rich waters, or very clear blue waters.

To further discriminate these OWTs within and across Case 1 and Case 2 waters, several clustering techniques have been employed, including hierarchical, partitional, and hybrid methods (Spyrakos et al. 2018). By accurately classifying OWTs, the selection of appropriate constituent algorithms can be optimized, particularly when simpler approaches fail to provide reliable results.

The varying concentrations observed within Case 2 waters correspond to different trophic status classes. Carlson (1977) established a methodology for trophic status classification in inland waters, considering Chl-*a* and phosphorus concentrations, as well as SDD (m). These WQ elements are associated with three primary trophic classes: oligotrophic, mesotrophic, and eutrophic (McCullough 2012), which also may include transitional categories like ultra-oligotrophic and hypertrophic (Watanabe et al. 2020).

Based on this rationale, lakes with very clear water are classified as oligotrophic Case-2 rather than Case-1 waters (Gons et al. 2008). This classification is based on the fact that these lakes typically receive significant terrigenous input (Gons and Auer 2004), and their water clarity is primarily influenced by the concentration of Dissolved Organic Carbon (DOC) (Lisi and Hein 2019; Song et al. 2012).

The clarity of water, in turn, has wide-ranging effects on various chemical, physical, and biological processes, including thermal structure, light transmission for photosynthesis, attenuation of damaging ultraviolet light levels, vertical distribution of plants and animals, as well as the form and availability of toxic metals (Schindler et al. 1997; Williamson et al. 1999; Pérez-Fuentetaja et al. 1999; Gunn et al. 2001).

Considering the above, and based on the relevant literature, it has been noted that further algorithm development is needed, especially for oligotrophic water bodies. Additionally, the selection of appropriate wavelengths is of paramount importance (Markogianni 2023).

2 Estimating Chlorophyll-*a* of Inland Water Bodies in Greece Based on Landsat Data

This effort aims to explore different RS-derived Chl-*a* algorithms suitable for different types of lakes. The practical application of this proposed approach was assessed across a total of 50 lake water bodies (both natural and artificial) during the years 2013–2018. These lakes constitute the National Lake Network Monitoring of Greece, conducted in accordance with the Water Framework Directive (WFD; Fig. 1).

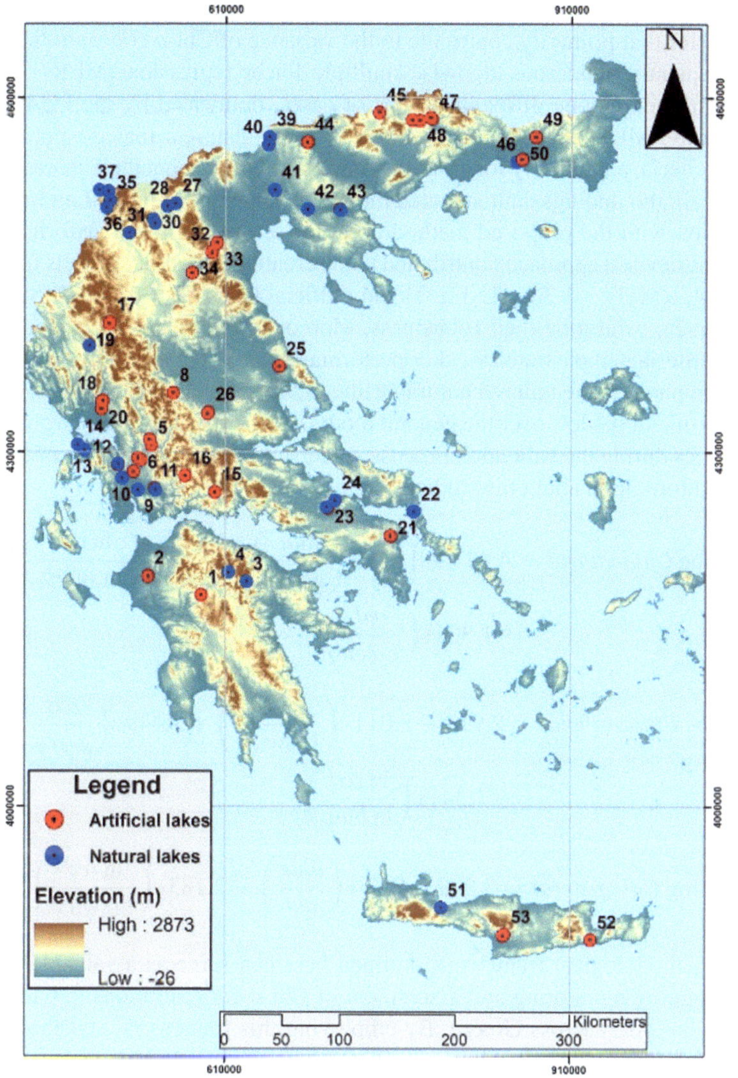

Fig. 1 National Lake Monitoring Network of Greece (numbers of sampling stations coincide with the numbers presented in Table 5 (Appendix)

From a statistical perspective, PCA was conducted to examine the interrelationships among Chl-*a* values and specific criteria of Greek lakes. These criteria include lakes' characteristics (artificial/natural), WFD typology, climatic type (based on the Köppen-Geiger climate classification; Peel et al. 2007), season of water samplings, and the time difference between sampling and satellite overpass. The PCA results

(Table 2) indicated that lake characteristics (natural/artificial) and WFD typology are the variables that primarily contribute to the variance of Chl-*a* concentration.

Subsequently, numerous stepwise multiple linear regression (MLRs) analyses were performed among different groups of cases, delineated by the PCA criteria. Additionally, MLR analysis was conducted using datasets that do not consider specific criteria, particularly for cases lacking information about their characteristics.

All in all, the findings demonstrated the effectiveness of utilizing Landsat data in conjunction with the proposed methodology for estimating log-transformed Chl-*a* levels. The devised approach contributed to the creation of distinct models for natural (R = 0.78; RMSE = 1.3 μg/L; Eq. 1) and artificial lakes (R = 0.76; RMSE = 1.29 μg/L; Eq. 2), exhibiting their robustness. Moreover, the model developed without specific criteria demonstrated weaker performance (R = 0.65; RMSE = 1.85 μg/L; Eq. 3) compared to the tailored natural/artificial lake models. This superiority of the natural/artificial models over the general model was further validated through statistical indices during the validation process, underscoring the importance of acquiring relevant information about the studied areas.

$$\log Chlanatural = 4.443 - 1.421*\left(\frac{blue}{green}\right) - 3.454*\left(\frac{\ln red}{\ln swir2}\right) + 1.304*\left(\frac{red}{green}\right) \quad (1)$$

$$\log Chlaartificial = 2.919 - 2.011*\left(\frac{\ln red}{\ln swir1}\right) + 1.449*\left(\frac{red}{green}\right) - 1.441*\left(\frac{\ln red}{\ln blue}\right) \quad (2)$$

$$\log Chlageneral = 3.599 - 0.63*\left(\frac{blue}{red}\right) - 2.183*\left(\frac{\ln red}{\ln swir2}\right) \quad (3)$$

The methodological framework outlined here can serve as a valuable resource for the ongoing monitoring and assessment of Chl-*a* concentrations in both natural and artificial lakes across Greece. By employing this framework, stakeholders can effectively manage and sustainably utilize water resources. This approach facilitates informed decision-making processes regarding WQ management, ensuring the preservation and protection of valuable aquatic ecosystems in Greece.

Table 2 Rotated component matrix (Rotation Method: Varimax with Kaiser Normalization, Rotation converged in 5 iterations; Markogianni et al. 2020)

Item	Component								
	PCA score 1	PCA score 2	PCA score 3	Communalities (1)	1[a]	Communalities (2)	2[a]	Communalities (3)	3[a]
Sampling's season	0.25	0.75	0.26	0.06	6	0.56	56	0.07	7
Chl-*a* (μg/l)	− 0.31	0.73	− 0.23	0.096	9.6	0.53	53	0.053	5.3
Natural/artificial	0.9	-0.03	0.06	0.81	81	0.0009	0.09	0.004	0.4
Köppen/climate type	− 0.08	0.02	0.95	0.006	0.6	0.0004	0.04	0.9	90
WFD typology	0.89	0.01	− 0.15	0.79	79	0.0001	0.01	0.023	2.3

[a] Percentages of the variables total communalities explained by each component

3 Modelling of Greek Lakes Water Quality Using Earth Observation in the Framework of the Water Framework Directive (WFD)

The primary goal of this approach is to develop robust WQ assessment models, focusing on parameters such as SDD and TP, within the framework of the National Lake Network Monitoring of Greece under the Water Framework Directive (WFD). By integrating in-situ measurements of SDD and TP concentrations with Landsat data, the trophic status of monitored lakes in Greece can be effectively evaluated.

Based on exploratory statistical analysis, it was observed that artificial lakes consistently exhibited higher SDD values compared to natural lakes across all sampling years from 2013 to 2018. Conversely, the highest TP concentrations were consistently observed in natural lakes, indicating the accumulation of TP loadings and a rising trend over the years. Additionally, Markogianni et al. (2020) noted significantly higher Chl-*a* concentrations in natural lakes compared to reservoirs. These findings align with similar studies, further highlighting the differences in WQ parameters' behavior between natural and artificial lake systems.

Indeed, studies such as that of Søballe et al. (1992) have shown that Chl-*a* concentrations typically exhibit lower levels in reservoirs compared to natural lakes. This phenomenon can be attributed to factors such as higher inorganic turbidity and elevated flushing rates (resulting in low hydraulic residence times) in reservoirs, which limit the development of phytoplankton biomass. Consequently, the higher SDD values observed in artificial lakes signify clearer water conditions. This clarity is often attributed to the increased presence of non-algal turbidities in artificial lakes when compared to natural ones, as discussed by Canfield and Bachmann (1981).

The observation of TP values in artificial lakes is significant, particularly concerning the understanding of the nutrient distribution's dynamics. Artificial lakes, due to their man-made nature and often regulated water flow, tend to experience nutrient loss, especially phosphorus, through settling in a downstream direction. This phenomenon underscores the importance of considering the location of sampling stations in WQ monitoring efforts.

One of the major distinctions between artificial and natural lakes is the presence of a trophic gradient in artificial lakes. As documented by Søballe et al. (1992), artificial lakes often exhibit a gradient in trophic status, with eutrophic conditions typically found in the upper reaches of the reservoir transitioning to oligotrophic conditions closer to the dam (Virginia 2007). This gradient underscores the complex interplay of factors such as nutrient input, sedimentation, and hydraulic dynamics within artificial lake ecosystems.

Then, a correlation analysis was conducted among in-situ co-orbital WQ data, including Chl-*a*, SDDs, and TP, and provided valuable comprehension into their interrelationships, thus enhancing the accuracy of the WQ models. The correlation matrix revealed several significant relationships, including a high positive correlation between TP and Chl-*a* (0.85), indicating that higher TP concentrations correspond to increased algal growth.

Conversely, high negative correlations were observed between SDD and both TP (− 0.84) and Chl-*a* (− 0.83). These findings suggest that as TP concentrations rise, leading to elevated algal biomass (Chl-*a* concentration; Virginia 2007), water transparency decreases, as indicated by reduced SDD. This relationship is consistent with the understanding that increased nutrient levels, particularly TP, directly stimulate algal growth while simultaneously reducing water clarity.

An additional explanation for the decrease in SDD with increasing TP concentration was provided by Heiskary and Wilson (2005). They demonstrated that a portion of the phosphorus may become associated with suspended particles resulting from soil erosion and transported downstream by rivers. This sediment-bound phosphorus can contribute to increased turbidity in the water column, further reducing SDD measurements.

The subsequent stepwise MLR analysis applied to the available TP and SDD datasets aimed to establish optimal quantitative models irrespective of lake characteristics. This analysis was followed by additional MLR procedures focusing on distinguishing between natural and artificial lakes, with the objective of generating distinct RS-derived models tailored to each lake type. The combination of these models was then utilized to evaluate the trophic status of the lakes.

The MLR analysis conducted on Landsat data and SDDs yielded three optimal models for assessing SDD across all lakes, natural lakes, and artificial lakes.

The SDD general model (Eq. 4), which is applicable to all lakes, exhibited a correlation coefficient (R) of 0.78 and a root mean square error (RMSE) of 0.24 m. The SDD natural model (Eq. 5) demonstrated a high correlation coefficient (R) of 0.95 and a low RMSE of 0.14 m.

The SDD artificial model (Eq. 6), tailored for artificial lakes, achieved a correlation coefficient (R) of 0.62 and a RMSE of 0.1 m, providing reliable accuracy for assessing SDD in artificial lake environments.

$$SQRT(Secchi)_{general} = 1.215 - 2.479 * \left(blue + red + \frac{red}{blue}\right) + 3.394 * \left(\frac{\ln green}{\ln swir2}\right) \quad (4)$$

$$SQRT(Secchi)_{natural} = 1.172 - (1.003 * \log chl\text{-}a) - (1.031 * \log red) \quad (5)$$

$$SQRT(Secchi)_{artificial} = 3.927 - 1.365 * \left(\frac{green}{blue}\right) - 0.318 * \left(\frac{red}{swir1}\right) - 0.361 * \log chl\text{-}a \quad (6)$$

The application of the $\text{Secchi}_{\text{general}}$ model (Eq. 4) to both natural and artificial lakes provided a valuable understanding into its effectiveness across different lake types. Interestingly, the model demonstrated superior performance in natural lakes compared to reservoirs, as water transparency in artificial lakes is significantly influenced by non-algal sources of turbidity. This observation is consistent with findings by Lind (1986), who highlighted the inadequacy of using Chl-*a* to estimate SDD in waters where even moderate levels of non-algal turbidity are present. In contrast, Lorenzen (1980) proposed considering this type of turbidity when evaluating reservoirs. Despite these considerations, many scientists argue that SDD data can be calibrated for each lake or reservoir, making them suitable for WQ monitoring purposes.

The study findings revealed that MLR analyses related to TP did not produce a statistically acceptable model for the reservoirs. However, robust models were successfully developed for both the general TP assessment ($TP_{general}$; R = 0.71; RMSE = 1.41 mg/L; Eq. 7) and TP assessment in natural lakes ($TP_{natural}$; R = 0.93; RMSE = 1.43 mg/L; Eq. 8). These models exhibited reliable values of tested statistical indices, indicating their effectiveness in estimating TP concentrations in the respective lake types.

$$\text{Log } TPgeneral = -1.425 + 0.452 * \log Chla - 0.573 * \left(\frac{\ln red}{\ln swir1}\right) \quad (7)$$

$$\text{Log } TPnatural = -0.633 - (0.704 * \log Secchi) - 0.392 * \left(\frac{green}{red}\right) \quad (8)$$

The absence of a statistically acceptable model for artificial lakes may be partly attributed to the observed time lag in phytoplankton's consumption of TP in these environments. This phenomenon complicates the relationship between TP and parameters such as Chl-*a* or SDD in reservoirs. Further limnological research is needed to gain a deeper understanding of the dynamics within these lake systems (Song et al. 2012). Conversely, in natural lakes, SDD emerged as a strong predictor of TP concentrations, indicating its utility in assessing WQ in these environments.

Following the model development, trophic status classification was conducted using Carlson's Trophic State Index (TSI), as outlined by Carlson (1977). Initially, TSI was calculated for all lakes in the study, and then the classification was further refined for natural-only and artificial-only lakes. Carlson's method utilizes SDD (in meters), a logarithmic transformation (Ln) of Chl-*a* concentration (in micrograms per liter), and TP measurements (in micrograms per liter). This index provides a numerical scale for categorizing lakes into classes based on their trophic status.

The three types of TSIs, namely general, natural, and artificial, were calculated using the published satellite-derived Chl-*a* assessment models (Eqs. 1–3) and the specially designed WQ models for SDD (Eqs. 4–6) and TP (Eqs. 7–8). Significant deviations between satellite-derived TSI values and respective in-situ measurements were observed, particularly in reservoirs and shallower lakes (mean depth < 5 m; Fig. 2), highlighting substantial differences between natural and artificial lakes. Furthermore, the classification of lakes according to TSI values was based on Table 3.

The influence of bottom reflectance on water quality parameters (WQPs) is indeed less pronounced in deeper lakes, as supported by the current findings. In shallow waters, however, bottom reflectance can significantly affect the remotely sensed reflectance spectra, potentially leading to less reliable estimations of WQPs. Thus, validation of WQP estimates in shallow waters using in-situ data is crucial (Chen et al. 2007). Additionally, the higher misclassification of TSIs in artificial lakes can be

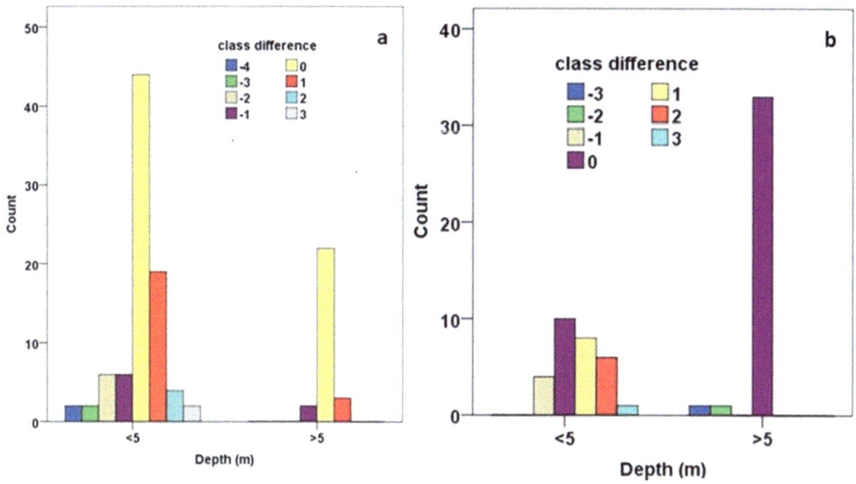

Fig. 2 Count of satellite-classified/misclassified cases belonging to **a** natural lakes (N = 112) and **b** artificial lakes (N = 64) grouped by the lake's mean depth. Numbers from − 4 up to 3 represent the class deviation between the satellite and in-situ derived TSIs while 0 indicates no differentiation. Positive and negative signs represent the direction of the deviation from oligotrophy to hyper-eutrophy and vice versa, respectively based on the corresponding in-situ TSI value (reference value; Markogianni 2023)

Table 3 Carlson's trophic state index values and classification of lakes (Carlson 1977; Prasad and Siddaraju 2012)

TSI values		Trophic status	Attributes
< 40	< 30	Oligotrophic	Transparent water
	30–40	Oligotrophic-mesotrophic	
41–50	41–48	Mesotrophic	Higher turbidity, higher algae abundance and macrophytes
	49–50	Mesotrophic-eutrophic	
51–70	51–60	Mesotrophic-eutrophic	
	61–70	Eutrophic	Usually blue-green algae blooms
> 70		Hypereutrophic	Extreme blue-green algae blooms

attributed to the greater variability of TP and SDD in reservoirs compared to natural lakes (Canfield and Bachmann 1981). Moreover, many models are developed under the assumption that phosphorus is the primary factor limiting algal growth (Kimmel et al. 1990). Nonetheless, the limitation of algal production in reservoirs can be influenced by factors beyond phosphorus, such as nitrogen availability or incident light levels (Virginia 2007). Considering that the TP concentration of artificial lakes in this study was assessed using the $TP_{general}$ model, which may not fully capture the complexities of nutrient dynamics in reservoirs, it could explain the less robust evaluation of TSI in these impoundments. Incorporating additional factors and refining the models to better account for the unique characteristics of reservoirs may improve the accuracy of TSI assessment in artificial lakes.

In summary, the study's results offer useful insights into the ongoing monitoring and prediction of WQ and trophic status in Greek lakes, facilitating their sustainable management, especially in situations where ground truth data may be scarce. A robust prediction of lake trophic status, as demonstrated in this study, is crucial for effectively monitoring eutrophication and understanding the factors driving its dynamics, particularly in the context of ongoing human activities and climate change impacts on lakes. By providing a reliable framework for continuous monitoring and assessment, this research contributes to the broader goal of preserving the health and ecological integrity of Greek lakes.

4 Atmospheric Correction Analysis of Lake WQ Models Across the WFD Lake Monitoring Network of Greece

The main goal of this approach is to assess the performance of empirically-developed WQ quantitative models for Chl-*a* (Markogianni et al. 2020), SDD, and TP (Markogianni et al. 2022) when utilizing reflectance values retrieved from Google Earth Engine (GEE) with different atmospheric correction (AC) methods compared to those used during their development.

Accurate AC is crucial for applications where small differences in surface reflectance (SR) are significant, such as the retrieval of WQ parameters (Nazeer et al. 2014; Warren et al. 2019; Pahlevan et al. 2021). Additionally, it allows for direct comparison between different image dates and sensors. AC methods can be categorized into two types: physical methods (e.g., FLAASH, ATCOR, 6S) that utilize a radiative transfer model to estimate SR, and image-based methods (e.g., DOS) that derive relevant parameters from the image itself.

Today, open-source cloud computing platforms have revolutionized the field of geospatial analysis, particularly for processing image data from satellites. Among these platforms, GEE stands out as one of the most widely used in the scientific community for satellite RS. With the continuous launch of new satellites and the capabilities offered by the GEE platform, monitoring lakes at a national scale has become more feasible and efficient than ever before.

The integration of in-situ WQ data obtained from the ongoing WFD implementation in Greece with the powerful capabilities of the GEE platform presents an exciting opportunity for lake monitoring. In our study, we aimed to assess the transferability and performance of published empirical WQ algorithms (Eqs. 1–8) when applied to Landsat (7 + ETM/8 OLI) and Sentinel 2 SR values retrieved from the GEE platform.

Specifically, GEE-Landsat and -Sentinel 2 SR data was retrieved for the year 2018 from the WFD lake sampling stations. These SR values were subjected to different AC methods, including LaSRC (for Landsat 8 OLI), LEDAPS (for Landsat 7 ETM +), and Sen2Cor (for Sentinel 2). Additionally, manually downloaded SR images were pre-processed using the AC DOS1 method.

These SR values obtained from both GEE and manually processed images were matched with the corresponding in-situ WQ data collected in 2018 within ± 7 days of satellite overpasses. This matching process yielded 192 pairs for Landsat and 210 pairs for Sentinel 2 sensor, respectively.

In hereby research effort, empirical WQ models for Chl-*a*, SDD, and TP were applied twice, employing two types of SR data: those corrected using the DOS1 method and those obtained from the GEE platform. In addition to this, the double application of WQ models was performed separately for Landsat (7 ETM+/8 OLI) and Sentinel 2 data to harmonize their inherent differences.

After applying the WQ models twice and obtaining double quantifications of each WQ element at each sampling station, we performed linear regression analyses. These analyses yielded linear equations for each sensor, representing the corrected WQ models that harmonize the AC differences between the DOS1-corrected and GEE-corrected SR data.

The linear equations obtained from the regression analyses indicated strong associations, with R^2 values ranging from 0.68 to 0.98. These equations provide a robust framework for predicting WQ parameters using SR data from both Landsat and Sentinel 2 sensors, and are presented in the Appendix (Table 6).

Based on aforementioned analysis' results, the DOS1 method was the most effective for quantifying lake WQ elements in almost all cases and for both Landsat and Sentinel 2 sensors. However, there were a few exceptions:

a. For the Chl-a_{general} and $\text{Secchi}_{\text{natural}}$ models using Landsat data, the LaSRC and LEDAPS methods were proved more effective.
b. For the Chl-a_{general} and $\text{Secchi}_{\text{general}}$ models using Sentinel 2 data, the Sen2Cor method presented better performance.

These findings, as summarized in Table 4 for the 2018 dataset, suggest that the choice of AC method can impact the performance of WQ models, and certain methods may be more suitable depending on the specific WQ parameter and sensor used.

Further studies, including those by Nazeer et al. (2014), Doña et al. (2014), El Alem et al. (2021), and Abdelal et al. (2022), provide additional support for the effectiveness of the DOS method in WQ monitoring of inland waters. Specifically, El Alem et al. (2021) conducted a comparison of image-based and physical correction models for retrieving suspended particulate matter (SPM) concentrations in lakes across the United States and Canada using Landsat imagery. Their findings suggested that image-based models, particularly the COST and DOS methods, are more suitable for retrieving SPM concentrations in inland waters compared to physical models, especially when the inputs of the physical atmospheric parameters are not well controlled.

The underlying assumption with physical methods is that they typically rely on two or more near-infrared (NIR) or shortwave infrared (SWIR) wavebands, where the marine signal is assumed to be zero in open ocean waters. However, in inland waters, this assumption may not hold true, leading to inaccuracies in the retrieved WQ parameters.

Table 4 Summarized results indicating the best performance of empirical WQ models employing (a) different AC processors (2018 dataset) and (b) GEE-derived reflectance values (2019, 2020) and exploration of the correction necessity via the application of the sensor-specific models

2018	Chl-$a_{general}$	Chl-$a_{natural}$	Chl-$a_{artificial}$	$Secchi_{general}$	$Secchi_{natural}$	$Secchi_{artificial}$	$TP_{general}$	$TP_{natural}$	$TP_{artificial}$
Landsat	LaSRC, LEDAPS	DOS1	ALL	DOS1	LaSRC, LEDAPS	DOS1	DOS1	DOS1	DOS1
Sentinel 2	Sen2Cor	DOS1	DOS1	Sen2Cor	DOS1	DOS1	ALL	DOS1	DOS1
2019	Chl-$a_{general}$	Chl-$a_{natural}$	Chl-$a_{artificial}$	$Secchi_{general}$	$Secchi_{natural}$	$Secchi_{artificial}$	$TP_{general}$	$TP_{natural}$	$TP_{artificial}$
Landsat	X		X	X	X	X			X
Sentinel 2	X	X					X	X	
Enhancement									
Landsat	No*	Yes	No	Yes	Yes	Yes*	Yes	No	Yes
Sentinel 2	No	Yes	No*	Yes	Yes	Yes*	Yes	No	No model
2020	Chl-$a_{general}$	Chl-$a_{natural}$	Chl-$a_{artificial}$	$Secchi_{general}$	$Secchi_{natural}$	$Secchi_{artificial}$	$TP_{general}$	$TP_{natural}$	$TP_{artificial}$
Landsat	X		X	X	X	X	X		X
Sentinel 2		X					X	X	
Enhancement									
Landsat	No	Yes	No	Yes*	Yes	Yes*	Yes	No	Yes
Sentinel 2	No*	No*	No*	No	Yes	No	Yes	No	No model

X symbol denotes the best performance among the sensors used; No* denotes that only reduced MAPE values were observed while Yes* denotes increased MAPE values; Markogianni (2023)

Indeed, the presence of particulate matter in Case-2 waters, such as inland water bodies, significantly influences the signal in the near-infrared (NIR) and shortwave infrared (SWIR) bands. In these waters, the reflectance over inland water bodies cannot be assumed to be negligible, as it would be in open ocean waters. Consequently, applying maritime correction methods over inland water bodies can lead to low or even negative water reflectance values in the visible bands (El Alem et al. 2021). As a result, applying these methods can cause confusion and inaccuracies in the retrieval of WQ parameters from RS data in inland water bodies.

Validation of the initial and corrected sensor-specific WQ models, involved their combination with available in-situ WQ datasets of years 2019 and 2020. This resulted in 239 matched pairs for Landsat and 242 matched pairs for Sentinel 2 SR in 2019, and 220 matched pairs for Landsat and 286 matched pairs for Sentinel 2 SR in 2020.

To assess the performance of the models, error metrics such as Mean Absolute Percentage Error (MAPE), RMSE, and Normalized Root Mean Square Error (NRMSE) were used. These metrics provide cognizance into the accuracy and reliability of the models in predicting WQ parameters based on satellite-derived data.

By comparing the performance of the initial and corrected WQ models using these error metrics for the years 2019 and 2020, it is possible to evaluate the effectiveness of the AC methods and their impact on the accuracy of the predictions. Lower error values indicate better model performance and higher accuracy in predicting WQ parameters from satellite imagery.

The analysis revealed that sensor-specific correction of WQ models was crucial for certain parameters (Table 4). Notably, the Chl-*a* models, particularly the Chl-$a_{natural}$ model, showed significant enhancement for both satellite sensors across the validation years, except for Sentinel 2 in 2020. The SDD models (general, natural, artificial) exhibited the greatest improvement compared to Chl-*a* and TP models, except for general and artificial-only models employing Sentinel 2 images in 2020. The TP models also demonstrated refined values based on in-situ datasets, except for the natural-only ones.

Furthermore, the RMSE values ranged from 11.68 μg/l (Landsat) to 14.88 μg/l (Sentinel 2) for Chl-*a*, from 2.02 m (Landsat) to 2.57 m (Sentinel 2) for SDD, and from 0.14 mg/l (Landsat) to 0.09 mg/l (Sentinel 2) for TP. These values confirm the stability and transferability of empirically developed models, even when applying differently-corrected surface reflectance values embedded in the GEE platform. Overall, the findings underscore the importance of sensor-specific correction in improving the accuracy and reliability of satellite-derived WQ models.

To our knowledge, there are a few studies demonstrating the potential of using the GEE platform for estimating Chl-*a* concentrations at large regional scales: Lin et al. (2018) utilized in-situ Chl-*a* data from 1157 lakes in 2007 and Landsat data to develop a national-scale Chl-*a* model. They employed machine learning algorithms within GEE and achieved a well-validated model with a RMSE of 34.9 μg/L. Wang et al. (2020) employed GEE to automatically generate match-up points from multisensor satellite observations and ground WQ samples. They then developed a SVM model to map Chl-*a* concentrations across 12 lakes in the tri-state region of Kentucky, Indiana, and Ohio (USA). The RMSE of the SVM model trained by Landsat 8 OLI imagery was 4.42 μg/L. Kislik et al. (2022) analyzed four spectral indices to retrieve Chl-*a* data for algal bloom identification in two highly dynamic freshwater reservoirs using Sentinel 2 MSI data in GEE. They found that the Normalized Difference Chlorophyll Index (NDCI) most accurately identified Chl-*a* across all study sites, with an RMSE of 0.02 μg/l, followed by NDVI.

Except for GEE utilization, other studies of remotely estimation of Chl-*a* concentrations have yielded comparable and even higher RMSE values than ours. For instance, Bonansea et al. (2018) achieved an RMSE of 18.47 μg/l in the Rio Tercero reservoir, Argentina's largest artificial reservoir. Similarly, Doña et al. (2014) reported an RMSE of 40 μg/l across various lakes and ponds in Spain. Additionally, Zhang et al. (2020) developed an SVM model on Landsat 8 OLI images to estimate the Chl-*a* concentrations of multiple lakes in China, while they reported an RMSE of 22.64 μg/L.

In reference to the SDD assesment, it's interesting to see the range of RMSE values reported in studies focusing on Secchi disk estimation models based on RS techniques. Zhang et al. (2021) found that the average RMSE value among recent studies (since 1993) is 1.13 m, with the highest recorded value being 1.7 m by Allan et al. (2011). This comparison highlights the variability in the performance of different Secchi disk estimation models and underscores the importance of ongoing research and refinement in this area.

Examining the different performances of TP models was also challenging to interpret, with the $TP_{general}$ model requiring correction while the model specifically developed for natural-only lakes performs efficiently without correction. The corrected $TP_{general}$ model produced RMSE values of 0.23 mg/l (Landsat) and 0.14 mg/l (Sentinel 2), while the $TP_{natural}$ model achieved RMSE values of 0.18 mg/l (Landsat) and 0.11 mg/l (Sentinel 2). Interestingly, when applying the $TP_{general}$ model to artificial lakes, the RMSE value was notably lower at 0.02 mg/l. This underscores the importance of tailoring models to specific lake types and considering the effectiveness of AC methods. During a thorough literature review, none recent study was detected utilizing GEE for the quantification of TP concentrations in lakes.

Even though the WQ models were developed based on MLRs analyses and Landsat 7 +ETM and 8 OLI images, their efficiency was indicated when were applied in GEE images, despite the pre-processing differentiation. This initially underscores the robustness and transferability of the developed WQ models and then highlights the potential of utilizing GEE for large-scale WQ monitoring and assessment. Hence, the findings suggest that GEE's public data is indeed adequate for mapping Chl-*a*, SDD, and TP concentrations across a large geographical region, specifically at a national scale like Greece. Furthermore, the combination of the spatio-temporally stable WQ models with GEE-retrieved SR data, provide scientists and Greek competent authorities with the means to leverage this vast repository of data. This allows to a greater extent for the mapping of long-term trends in WQ of lakes and the identification of underlying factors and potential pollutant threats.

5 General Conclusions and Limitations of Empirical-Based Lake WQ Monitoring

The hereby established methodological framework for modelling WQ in Greek lakes using RS technology has successfully developed WQ models and monitored the trophic status of optically diverse Greek lakes, particularly classified as Case 2 waters. Key WQ elements such as Chl-*a* and TP concentrations, as well as SDDs, were assessed using fine spatial resolution RS image data, typically ranging from 10 to 30 m. These parameters were then employed in Carlson's TSI equation to facilitate the assessment of the lakes' trophic status.

The developed WQ models successfully accounted for the spectral configuration differences among the Landsat sensors (7 ETM+ and 8 OLI). However, the assessment of WQ and trophic status in both artificial and shallow lakes (with a mean depth of less than 5 m) proved to be less successful. This outcome highlights the need for a deeper limnological research to better understand the primary limiting factors of Chl-*a* production and the sources of turbidity, both algal and non-algal, especially in reservoirs. Furthermore, it's noteworthy that no statistically acceptable model was generated for TP in artificial lakes. This could be partly attributed to the observed time lag for phytoplankton to consume TP in such lakes. Therefore, further research is essential to elucidate the dynamics of TP cycling in artificial lake systems.

The delivered WQ models have demonstrated their capability to support continuous lake monitoring and sustainable management at a national scale. By utilizing Earth observation techniques, these models offer near real-time assessments at finer spatial resolutions compared to similar large-scale applications worldwide. This advancement enables more effective lake WQ monitoring by providing valuable information on spatial variability.

The double employment of differently atmospheric corrected reflectance values (DOS1, LaSRC, LEDAPS, and Sen2Cor) in WQ models highlighted the effectiveness of DOS1 for quantifying lake WQ elements across almost all cases and sensors (Landsat/Sentinel 2). This underlines the applicability of empirically-developed WQ models to both archived and future Landsat and Sentinel 2 image data embedded in the GEE platform, streamlining the pre-processing procedure. Additionally, sensor-corrected models demonstrated improved assessment accuracy on a case-by-case basis, with SDD models (general, natural, artificial) showing the highest improvement, followed by TP and Chl-*a* models. The evaluation of WQ models for the years 2019 and 2020 confirmed the sufficiency of GEE public data for mapping Chl-*a*, SDD, TP concentrations, and trophic status at a country level (Greece).

However, there are some factors that may hindered a more accurate lake WQ assessment and are summarized as follows:

(a) The intense optical diversity characterizing the under study Greek lakes. Considering the high number (50) of the studied lakes and the fact that their optical properties may vary resembling either to both Case 1 or Case 2 waters or classes within Case 2 waters, there has been an effort to distinguish distinct OWTs among them based on statistical analysis. However, hierarchical cluster analysis incorporating in-situ WQ constituents (concentrations of Chl-a, TP and SDDs) and Landsat response (respective reflectance at bands blue, green, red, NIR and SWIRS) did not manage to yield reliable results. Classification efforts employed datasets including all, average and median values of WQ data, and median values of WQ data grouped by seasonal sampling. The inclusion of seasonal data is important as season indirectly affects the eutrophication degree of water bodies, influencing factors such as algae growth, nutrient influx from rainfall, and suspended matter re-suspension due to wind. Despite these efforts though, the optical diversity of the lakes still remains a significant challenge for accurate WQ assessment.

(b) The uncertainty regarding the accuracy of the location of sampling stations. Factors such as winds, seasonal variations, changes in water level, and ease of proximity can lead to potential transpositions even when revisiting the same sampling sites. This uncertainty is particularly relevant for artificial lakes, which often exhibit a trophic gradient, ranging from eutrophic in the upper reaches to oligotrophic near the dam. As nutrients, especially phosphorus, settle downstream in reservoirs, the WQ conditions may vary significantly across different parts of the lake. To address this challenge, it is proposed that water samples from reservoirs be collected from at least two to three stations. The first station should be located near the point where rivers or streams drain into the lake, the second station in the middle of the lake (consistent with current practices), and the third station close to the dam. This approach would provide a more comprehensive understanding of WQ conditions across different parts of the reservoir, helping to account for spatial variations in nutrient concentrations and trophic status.

(c) The influence of non-algal sources of turbidity on water transparency in artificial lakes presents a challenge for estimating SDD using Chl-*a* concentrations. Studies by Canfield and Bachmann (1981) and Lind (1986) have highlighted the limitations of using Chl-*a* to estimate SDD in waters with significant non-algal turbidity. This limitation can affect the accuracy of the $Secchi_{artificial}$ model developed in this research, which relies on the logarithmic transformation of Chl-*a*. To address this issue, it may be necessary to explore alternative approaches for estimating SDD in artificial lakes, particularly those with high levels of non-algal turbidity. This could involve incorporating additional WQ parameters or developing separate models tailored specifically for these conditions. Additionally, further research into the sources and dynamics of non-algal turbidity in artificial lakes could provide valuable information for improving SDD estimation in these environments.

(d) Validating the implementation of the DOS1 AC method would indeed be crucial for ensuring the accuracy and reliability of the WQ assessment results. While DOS1 is a widely used method in the Earth Observation community, its effectiveness can vary depending on factors such as atmospheric conditions, sensor characteristics, and the specific characteristics of the study area. To address this concern, it would be advisable to conduct a validation study to assess the performance of the DOS1 AC method in the context of the study area and the specific satellite imagery used. This validation could involve comparing reflectance values before and after correction with ground truth measurements of WQ parameters collected at the same time as the satellite imagery. Additionally, sensitivity analyses could be performed to evaluate the robustness of the DOS1 method under different atmospheric conditions and across different types of water bodies.

(e) The dependence on favorable climatic conditions for satellite imagery acquisition is indeed a common limitation in RS-based WQ monitoring. Atmospheric effects such as cloud cover, aerosols, and haze can obstruct the satellite's view of the Earth's surface, making it challenging to acquire cloud-free and high-quality images. This fact induced the loss of a significant amount of field data, which otherwise could have contributed to the development of even stronger WQ models.

Furthermore, future pathways must be configured to advance the management of an increasingly successful WQ assessment in Greek lakes using Earth Observation applications. In summary, literature indicates that certain AC methods outperform others in assessing specific WQ parameters. As the DOS1 AC method lacks validation, a critical future step is to explore alternative AC methods to gauge their broader impact on the predictive accuracy of WQ models and identify the most effective one. To conduct a robust statistical analysis of AC method effectiveness, ground truth measured Rrs values, obtained through handheld or shipborne spectroradiometers, are necessary to compare and evaluate AC corrected pixels. With a larger dataset of optical property observations, it becomes feasible to categorize match-ups by water type, offering additional insights and potential targeting of water bodies for specific AC processing. Intercomparing AC methods will help determine the optimal AC algorithm and identify the most suitable one not only for each WQ constituent but also for each OWT.

Considering the rationale of aligning appropriate AC methods with respective OWTs, it is considered essential to classify the under studied 50 lakes into distinct OWT categories. Generally, lakes and reservoirs exhibit WQ variations not only across the study area (Greece) but also within regions and individual water bodies. The objective of this classification is to ensure that each OWT is associated with specific bio-optical conditions, thereby reflecting the prevalence of individual or grouped WQ element concentrations. More specifically, resampling OWT class spectra to match Landsat and Sentinel 2 spectra—except for WQ constituents' concentrations, as previously performed—and identifying their inter-similarities would streamline OWT classification in an efficient and accurate manner. Utilizing OWTs aids in developing WQ algorithms for optically complex waters and concurrently assists in selecting an effective AC processor for a particular region of interest.

Furthermore, future research endeavors will explore non-linear methodologies for the development of WQ models. Empirical approaches, particularly machine learning techniques, demonstrate proficiency in managing complex non-linear relationships inherent in WQ RS data. Given the non-parametric characteristics of the data employed in this study, conducting a sensitivity analysis among artificial neural networks, genetic algorithms, and support vector machines could offer valuable support for a more precise evaluation of WQ constituents. Moreover, identifying the most optimal non-linear method and comparing it with the current linear WQ models would enhance understanding of the behavior of WQ parameters overall.

Regarding satellite sensors, the majority of applications for monitoring and managing WQ in inland waters have relied on images from multispectral and medium to high spatial resolution satellites, such as Landsat and Sentinel-2-MSI. However, there is a growing interest in spaceborne hyperspectral sensors due to their ability to capture hundreds of narrow spectral bands simultaneously, which simplifies the retrieval of WQ parameters. Therefore, future efforts will involve experimenting with hyperspectral image data from sensors like PRISMA, DESIS, HISUI, and EnMAP to assess WQ of Greek lakes.

Acknowledgements The author would like to declare that hereby chapter constitutes a formed part of Markogianni Vassiliki's Ph.D. thesis (2023) entitled "Water quality assessment in Greek lakes by using remote sensing and statistical modelling". Moreover, the author would like to express her gratitude to the three (3) members of her PhD Supervisory Committee Dr. Dionissios P. Kalivas, Dr. Elias Dimitriou and Dr. George P. Petropoulos for their valuable assistance during the writing of the publications, part of the results of which are presented in this chapter.

The author wishes also to thank for the freely shared WQ data used in this study, collected in the framework of the National Water Monitoring Network for lakes. The network is supervised by the Directorate for the Protection and Management of Water Resources of the Special Secretariat for Waters of the Ministry of Environment and Energy and implemented by the Goulandris Natural History Museum, Greek Biotope/Wetland Center (EKBY) according to the Joint Ministerial Decision 140384/2011.

Appendix

See Tables 5 and 6.

Table 5 Main characteristics of the lakes included in the National Lake Monitoring Network in Greece (WFD) (Mavromati et al. 2018; Markogianni et al. 2020)

No.	National name station	Surface (km^2)	(N)atural/ (A)rtificial	Mean depth (m)	No.	National name station	Surface (km^2)	(N)atural/ (A)rtificial	Mean depth (m)
1	Lake Ladona	–	A	–	28	Lake Petron	11.91	N	3.1
2	Lake Pineiou	19.64	A	15.1	29	Lake Zazari	2.98	N	3.95
3	Lake Stymfalia	–	N	1.31	30	Lake Cheimaditida	9.82	N	1.01
4	Lake Feneou	0.47	A	10.5	31	Lake Kastorias	30.87	N	3.7
5	Lake Kremaston	68.43	A	47.2	32	Lake Sfikias	3.96	A	23.2
6	Lake Kastrakiou	25.58	A	33.2	33	Lake Asomaton	2.46	A	20.8
7	Lake Stratou	7.02	A	9.6	34	Lake Polyfytou	63.49	A	22.4
8	Lake Tavropou	21.46	A	15.0	35	Lake Lesser Prespa A	–	N	3.95
9	Lake Lysimacheia	10.87	N	3.5	36	Lake Lesser Prespa B		N	–
10	Lake Ozeros	10.57	N	3.8	37	Lake Great Prespa A	–	N	17
11	Lake Trichonida	93.53	N	29.6	38	Lake Great Prespa B		N	–
12	Lake Amvrakia	13.14	N	23.4	39	Lake Doirani 1	33.25	N	4.6
13	Lake Voulkaria	7.38	N	0.96	40	Lake Doirani 2		N	–
14	Lake Saltini	–	N	–	41	Lake Pikrolimni	6.30	N	1.2
15	Lake Mornou	17.50	A	38.5	42	Lake Koroneia	–	N	3.8

(continued)

Table 5 (continued)

No.	National name station	Surface (km^2)	(N)atural/ (A)rtificial	Mean depth (m)	No.	National name station	Surface (km^2)	(N)atural/ (A)rtificial	Mean depth (m)
16	Lake Evinou	2.68	A	31.5	43	Lake Volvi	70.36	N	12.3
17	Lake Pigon Aoou	11.44	A	20.8	44	Lake Kerkini	–	A	2.19
18	Lake Pournariou	19.28	A	29.8	45	Lake Leukogeion	0.83	A	4.05
19	Lake Pamvotida	21.82	N	5.3	46	Lake Ismarida	–	N	0.9
20	Lake Pournariou II	0.56	A	11.7	47	Lake Platanovrysis	2.99	A	26.4
21	Lake Marathona	2.17	A	15.8	48	Lake Thisavrou	13.43	A	38.4
22	Lake Dystos	–	N	–	49	Lake Gratinis	0.80	A	14.2
23	Lake Yliki	19.96	N	20.1	50	Lake N. Adrianis	–	A	–
24	Lake Paralimni	9.96	N	2.99	51	Lake Kourna	–	N	15
25	Lake Karlas	–	A	0.9	52	Lake Bramianou	–	A	10.1
26	Lake Smokovou	–	A	–	53	Lake Faneromenis	0.33	A	9.98
27	Lake Vegoritida	47.67	N	26.52					

Table 6 Regression analysis basic statistics between the resulted values after the two-fold employment of SR (DOS1- and LaSRC; LEDAPS; Sen2Cor-corrected reflectance in GEE) in WQ models (developed by Markogianni et al. 2020, 2022) concerning datasets of year 2018

No.	WQ element	Equation (corrected WQ models)	N	R	R^2	Std. error of the estimate	Durbin-Watson	Sensor
1a	Chl-$a_{general}$ (DOS1)	LogChl-$a_{general}$ = 0.221 + 0.61 * (logChl-$a_{general}$_GEE)	115	0.88	0.78	0.19	1.99	Landsat
1b	Chl-$a_{general}$ (GEE reflect.)							
2a	Chl-$a_{natural}$ (DOS1)	LogChl-$a_{natural}$ = − 0.109 + (0.747 * logchl-$a_{natural}$_GEE)	26	0.87	0.76	0.23	2.03	
2b	Chl-$a_{natural}$ (GEE reflect.)							
3a	Chl-$a_{artificial}$ (DOS1)	LogChl-$a_{artificial}$ = 0.279 + (0.209 * logchl-$a_{artificial}$_GEE)	33	0.39	0.15	0.28	1.68	
3b	Chl-$a_{artificial}$ (GEE reflect.)							
4a	$Secchi_{general}$ (DOS1)	SQRTSecchi$_{general}$ = 0.671 + (0.737 * SQRTSecchi$_{general}$_GEE)	95	0.91	0.83	0.25	1.99	Landsat
4b	$Secchi_{general}$ (GEE reflect.)							
5a	$Secchi_{natural}$ (DOS1)	SQRTSecchi$_{natural}$ = 0.079 + (0.875 * SQRTSecchi$_{natural}$_GEE)	26	0.97	0.94	0.14	2.32	
5b	$Secchi_{natural}$ (GEE reflect.)							
6a	$Secchi_{artificial}$ (DOS1)	SQRTSecchi$_{artificial}$ = 1.391 + (0.475 * SQRTSecchi$_{artificial}$_GEE)	33	0.88	0.78	0.15	1.72	
6b	$Secchi_{artificial}$ (GEE reflect.)							
7a	$TP_{general}$ (DOS1)	LogTP$_{general}$ = − 0.127 + (0.925 * LOGTP$_{general}$_GEE)	28	0.99	0.98	0.03	2.25	
7b	$TP_{general}$ (GEE reflect.)							
8a	$TP_{natural}$ (DOS1)	LogTP$_{natural}$ = − 0.177 + (0.796 * LOGTP$_{natural}$_GEE)	40	0.96	0.93	0.08	1.83	
8b	$TP_{natural}$ (GEE reflect.)							
9a	Application of $TP_{general}$ model on artificial lakes (DOS1)	LogTP$_{artificial}$ = − 0.143 + (0.905 * LOGTP$_{artificial}$_GEE)	11	0.99	0.98	0.04	1.89	

(continued)

Table 6 (continued)

No.	WQ element	Equation (corrected WQ models)	N	R	R^2	Std. error of the estimate	Durbin-Watson	Sensor
9b	Application of $TP_{general}$ model on artificial lakes (GEE reflect.)							
10a	Chl-$a_{general}$ (DOS1)	LogChl-$a_{\_general}$ = 0.338 + (0.329 * logchl-$a_{general}$_GEE)	64	0.6	0.36	0.43	2.18	Sentinel 2
10b	Chl-$a_{general}$ (GEE reflect.)							
11a	Chl-$a_{natural}$ (DOS1)	LogChl-$a_{natural}$ = 0.125 + (0.549 * logchl-$a_{natural}$_GEE)	23	0.82	0.68	0.28	0.74	
11b	Chl-$a_{natural}$ (GEE reflect.)							
12a	Chl-$a_{artificial}$ (DOS1)	LogChl-$a_{artificial}$ = 0.06 + (0.299*logchl-$a_{artificial}$_GEE)	33	0.38	0.15	0.62	1.9	
12b	Chl-$a_{artificial}$ (GEE reflect.)							
13a	$Secchi_{general}$ (DOS1)	$SQRTSecchi_{general}$ = 1.646 + (0.291 * $SQRTSecchi_{general\_}$ GEE)	103	0.42	0.17	0.65	1.86	
13b	$Secchi_{general}$ (GEE reflect.)							
14a	$Secchi_{natural}$ (DOS1)	$SQRTSecchi_{natural}$ = 0.233 + (0.841 * $SQRTSecchi_{natural\_}$ GEE)	24	0.97	0.95	0.13	1.44	
14b	$Secchi_{natural}$ (GEE reflect.)							
15a	$Secchi_{artificial}$ (DOS1)	$SQRTSecchi_{artificial}$ = 1.490 + (0.399 * $SQRTSecchi_{\_artificial\_}$ GEE)	36	0.79	0.62	0.24	1.89	
15b	$Secchi_{artificial}$ (GEE reflect.)							
16a	$TP_{general}$ (DOS1)	$LogTP_{general}$ = − 0.086 + (0.963 * $LOGTP_{general_}$GEE)	27	0.99	0.98	0.04	1.84	Sentinel 2
16b	$TP_{general}$ (GEE reflect.)							
17a	$TP_{natural}$ (DOS1)	$LogTP_{natural}$ = − 0.178 + (0.742 * $LOGTP_{natural_}$GEE)	42	0.93	0.87	0.11	2.37	
17b	$TP_{natural}$ (GEE reflect.)							

(continued)

Table 6 (continued)

No.	WQ element	Equation (corrected WQ models)	N	R	R^2	Std. error of the estimate	Durbin-Watson	Sensor
18a	Application of $TP_{general}$ model on artificial lakes (2 RECORDS) (DOS1)	No linear regression analysis between TP values (accrued from the application of general model to artificial lakes) derived both from GEE and manual analysis (2018 dataset) due to few records						
18b	Application of TP general model on artificial lakes (2 RECORDS) (GEE Reflect.)							

Different reflectance products are referred as DOS1 and as GEE reflect. for LaSRC; LEDAPS and Sen2Cor correction methods, Markogianni (2023)

References

Abdelal Q, Assaf, NM, Al-Rawabdeh A, Arabasi S, Nathir AR (2022) Assessment of Sentinel-2 and Landsat-8 OLI for small-scale inland water quality modeling and monitoring based on handheld hyperspectral ground truthing. J Sens 2022, Article ID 4643924, 19 p. https://doi.org/10.1155/2022/4643924

Allan MG, Hamilton DP, Hicks BJ, Brabyn L (2011) Landsat remote sensing of chlorophyll a concentrations in central north island lakes of New Zealand. Int J Remote Sens 32:2037–2055

Austin RW, Petzold TJ (1981) Water colour measurements. In: Gower J (ed) Oceanography from space. Plenum, New York, NY, USA, pp 239–256

Bonansea M, Rodriguez C, Pinotti L (2018) Assessing the potential of integrating Landsat sensors for estimating chlorophyll—a concentration in a reservoir. Hydrol Res 49(5):1608–1617

Bonansea M, Rodriguez MC, Pinotti L, Ferrero S (2015) Using multi-temporal Landsat imagery and linear mixed models for assessing water quality parameters in Río Tercero reservoir (Argentina). Remote Sens Environ 15:28–41

Busse LB, Simpson JC, Cooper SD (2006) Relationships among nutrients, algae, and land use in urbanized southern California streams. Can J Fish Aquat Sci 63:2621–2638

Canfield D Jr, Bachmann R (1981) Prediction of total phosphorus concentrations, chlorophyll a, and secchi depths in natural and artificial lakes. Can J Fish Aquat Sci 38(4):414–423. https://doi.org/10.1139/f81-058

Carlson RE (1977) A trophic state index for lakes. Limnol Oceanogr 22:361–369

Chen J, Quan WT (2012) Using Landsat/TM imagery to estimate nitrogen and phosphorus concentration in Taihu Lake, China. IEEE J-STARS 5(1):273–280. https://doi.org/10.1109/JSTARS.2011.2174339

Chen Q, Zhang Y, Hallikainen M (2007) Water quality monitoring using remote sensing in support of the EU water framework directive (WFD): a case study in the Gulf of Finland. Environ Monit Assess 124:157–166

Dewidar K, Khedr A (2001) Water quality assessment with simultaneous Landsat-5 TM at Manzala Lagoon, Egypt. Hydrobiologia 457(1–3):49–58

Doña C, Sánchez MJ, Caselles V, Domínguez JA, Camacho A (2014) Empirical relationships for monitoring water quality of lakes and reservoirs through multispectral images. IEEE J Sel Top Appl Earth Obs Remote Sens 7(5)

Duan W, He B, Takara K, Luo P, Nover D, Sahu N, Yamashiki Y (2013) Spatiotemporal evaluation of water quality incidents in Japan between 1996 and 2007. Chemosphere 93:946–953

EEA (2018) EuropeanWaters—assessment of status and pressures. EEA Report No 7/2018. https://doi.org/10.2800/303664

El Alem A, Lhissou R, Chokmani K, Oubennaceur K (2021) Remote retrieval of suspended particulate matter in inland waters: image-based or physical atmospheric correction models? Water 13:2149. https://doi.org/10.3390/w13162149

Fatima R (2018) Studies on physical, chemical and bacteriological characteristics on quality of spring water in Hajigak iron ore mine, Bamyan Province, Central Afghanistan. Open J Geol 8:313–332

Fuller LM, Aichele SS, Minnerick RJ (2002) Predicting water quality by relating Secchi-disk transparency and chlorophyll a measurements to satellite imagery for Michigan Inland Lakes, August 2002, 2004, U.S. Geological Survey Scientific Investigations Report 2004-5086, 25 p

Gholizadeh MH, Melesse AM, Reddi L (2016) A comprehensive review on water quality parameters estimation using remote sensing techniques. Sensors 16:1298. https://doi.org/10.3390/s16081298

Gons JH, Auer TM (2004) Some notes on water color in Keweenaw Bay (Lake Superior). J Great Lakes Res 30(Supplement 1):481–489. ISSN 0380-1330. https://doi.org/10.1016/S0380-1330(04)70408-6

Gons JH, Auer TM, Effler WS (2008) MERIS satellite chlorophyll mapping of oligotrophic and eutrophic waters in the Laurentian Great Lakes. Remote Sens Environ 112(11):4098–4106. ISSN 0034-4257. https://doi.org/10.1016/j.rse.2007.06.029

Gordon HR, Morel A (1983) Remote assessment of ocean color for interpretation of satellite visible imagery. A review. In: Barber RT, Mooers NK, Bowman MJ, Zeitzschel B (eds) Lecture notes on coastal and estuarine studies. Springer, New York, NY, USA, p 114

Gunn JM, Snucins E, Yan ND, Arts MT (2001) Use of water clarity to monitor the effects of climate change and other stressors on oligotrophic lakes. Environ Monit Assess 67:69–88

Healey FP, Hendzel LL (1979) Indicators of phosphorus and nitrogen deficiency in five algae in culture. Can J Fish Aquat Res 36:1364–1369

Heiskary S, Wilson B (2005) Minnesota lake water quality: developing nutrient criteria, 3rd ed. Minnesota Pollution Control Agency, St. Paul, MN, USA

IOCCG (2000) Remote sensing of ocean colour in coastal, and other optically-complex, waters. Reports of the International Ocean-Colour Coordinating Group, No. 3 (ed: Sathyendranath S). IOCCG, Dartmouth, NH, Canada

Isenstein EM, Park M-H (2014) Assessment of nutrient distributions in Lake Champlain using satellite remote sensing. J Environ Sci 26:1831–1836. https://doi.org/10.1016/j.jes.2014.06.019

Japitana M, Burce M (2019) A satellite-based remote sensing technique for surface water quality estimation. Eng Technol Appl Sci Res 9:3965–3970

Jiang D, Matsushita B, Setiawan F, Vundo A (2019) An improved algorithm for estimating the Secchi disk depth from remote sensing data based on the new underwater visibility theory. ISPRS J Photogramm Remote Sens 152:13–23

Kimmel BL, Lind OT, Paulson LJ (1990) Reservoir primary production. In: Thorton KW, Kimmel BL, Payne FE (eds) Reservoir limnology. Ecological perspectives. Wiley, New York

Kislik C, Dronova I, Grantham ET, Kelly M (2022) Mapping algal bloom dynamics in small reservoirs using Sentinel-2 imagery in Google Earth Engine. Ecol Indicators 140:109041. ISSN 1470-160X. https://doi.org/10.1016/j.ecolind.2022.109041

Knoll LB, Sharma S, Denfeld BA, Flaim G, Hori Y, Magnuson JJ et al (2019) Consequences of lake and river ice loss on cultural ecosystem services. Limnol Oceanogr Lett 4(5):119–131. https://doi.org/10.1002/lol2.10116

Kutser T, Arst H, Miller T, Käärmann L, Milius A (1995) Telespectrometrical estimation of water transparency, chlorophyll-a and total phosphorus concentration of Lake Peipsi. Int J Remote Sens 16:3069–3085

Li J et al (2016) MODIS observations of water color of the largest 10 lakes in China between 2000 and 2012. Int J Digit Earth 9(8):788–805

Lin S, Novitski LN, Qi J, Stevenson RJ (2018) Landsat TM/ETM+ and machine-learning algorithms for limnological studies and algal bloom management of inland lakes. J Appl Remote Sens 12:1

Lind OT (1986) The effect of non-algal turbidity on the relationship of Secchi depth to chlorophyll a. Hydrobiologia 140:27–35. https://doi.org/10.1007/BF00006726

Lisi PJ, Hein CL (2019) Eutrophication drives divergent water clarity responses to decadal variation in lake level. Limnol Oceanogr 64(S1):S49–S59

Lorenzen MW (1980) The use of chlorophyll-secchi disk relationships. Limnol Oceanogr 25:371–372

Markogianni VV (2023) Water quality assessment in Greek lakes by using remote sensing and statistical modelling. Ph.D. thesis, Agricultural University of Athens. http://hdl.handle.net/10329/7833

Markogianni V, Kalivas D, Petropoulos GP, Dimitriou E (2020) Estimating chlorophyll-*a* of inland water bodies in Greece based on Landsat data. Remote Sens 12:2087. https://doi.org/10.3390/rs12132087

Markogianni V, Kalivas D, Petropoulos GP, Dimitriou E (2022) Modelling of Greek lakes water quality using earth observation in the framework of the water framework directive (WFD). Remote Sens 14:739. https://doi.org/10.3390/rs14030739

Mavromati E, Kagalou I, Kemitzoglou D, Apostolakis A, Seferlis M, Tsiaoussi V (2018) Relationships among land use patterns, hydromorphological features and physicochemical parameters of surface waters: WFD lake monitoring in Greece. Environ Process 5(Suppl. S1):139–151

McCullough IM (2012) Remote estimation of regional lake clarity with Landsat TM and MODIS satellite imagery. Electronic Theses and Dissertations, pp 1744

Meyer MF, Labou SG, Cramer AN et al (2020) The global lake area, climate, and population dataset. Sci Data 7:174. https://doi.org/10.1038/s41597-020-0517-4

Nazeer M, Nichol JE, Yung Y-K (2014) Evaluation of atmospheric correction models and Landsat surface reflectance product in an urban coastal environment. Int J Remote Sens 35(16):6271–6291. https://doi.org/10.1080/01431161.2014.951742

Neil C, Spyrakos E, Hunter PD, Tyler AN (2019) A global approach for chlorophyll-a retrieval across optically complex inland waters based on optical water types. Remote Sens Environ 229:159–178. ISSN 0034-4257. https://doi.org/10.1016/j.rse.2019.04.027

Nikolaidis NP, Phillips G, Poikane S, Várbíro G, Bouraoui F, Malago A, Lilli MA (2022) River and lake nutrient targets thatsupport ecological status: European scale gap analysis and strategiesfor the implementation of the Water Framework Directive. Sci Total Environ 813(March):151898. https://doi.org/10.1016/j.scitotenv.2021.151898

Nima P, Mangin A, Balasubramanian SV, Smith B, Alikas K, Arai K, Barbosa C, Bélanger S, Binding C, Bresciani M, Giardino C, Gurlin D, Fan Y, Harmel T, Hunter P, Ishikaza J, Kratzer S, Lehmann MK, Ligi M, Ma R, Martin-Lauzer F-R, Olmanson L, Oppelt N, Pan Y, Peters S, Reynaud N, de Carvalho LAS, Simis S, Spyrakos E, Steinmetz F, Stelzer K, Sterckx S, Tormos T, Tyler A, Vanhellemont Q, Warren M (2021) ACIX-Aqua: a global assessment of atmospheric correction methods for Landsat-8 and Sentinel-2 over lakes, rivers, and coastal waters. Remote Sens Environ 258:112366. https://doi.org/10.1016/j.rse.2021.112366

Odermatt D, Gitelson A, Vittorio EV, Schaepman M (2012) Review of constituent retrieval in optically deep and complex waters from satellite imagery. Remote Sens Environ 118:116–126.

Peel MC, Finlayson BL, McMahon TA (2007) Updated world map of the Köppen-Geiger climate classification. Hydrol Earth Syst Sci 11:1633–1644. https://doi.org/10.5194/hess-11-1633-2007

Pérez-Fuentetaja A, Dillon PJ, Yan ND et al (1999) Significance of dissolved organic carbon in the prediction of thermocline depth in small Canadian shield lakes. Aquat Ecol 33:127–133. https://doi.org/10.1023/A:1009998118504

Pizani F, Maillard P (2022) The determination of water quality parameters by remote sensing technologies: 2000–2020. Universidade Federal de Minas Gerais, [S. l.], 1–30. https://doi.org/10.13140/RG.2.2.32203.87849

Prasad AD, Siddaraju (2012) Carlson's Trophic State Index for the assessment of trophic status of two lakes in Mandya district. Adv Appl Sci Res 3

Randolph K, Wilson J, Tedesco L, Li L, Pascual DL, Soyeux E (2008) Hyperspectral remote sensing of cyanobacteria in turbid productive water using optically active pigments, chlorophyll-a and phycocyanin. Remote Sens Environ 112:4009–4019

Ruiz-Verdu A, Simis SGH, de Hoyos C, Gons HJ, Pepa- Martvnez R (2008) An evaluation of algorithms for the remote sensing of cyanobacterial biomass. Remote Sens Environ 112:3996–4008

Sagan V, Peterson TK, Maimaitijiang M, Sidike P, Sloan J, Greeling AB, Samar M, Adams C (2020) Monitoring inland water quality using remote sensing: potential and limitations of spectral indices, bio-optical simulations, machine learning, and cloud computing. Earth-Sci Rev 205:103187

Salmaso N, Tolotti M (2021) Phytoplankton and anthropogenic changes in pelagic environments. Hydrobiologia 848:251–284. https://doi.org/10.1007/s10750-020-04323-w

Schindler DW, Curtis PJ, Bayley SE et al (1997) Climate-induced changes in the dissolved organic carbon budgets of boreal lakes. Biogeochemistry 36:9–28. https://doi.org/10.1023/A:1005792014547

Søballe DM, Kimmel BL, Kennedy RH, Gaugush RF (1992) Reservoirs. In: Hackney CT, Adams SM, Martin WH (eds) Biodiversity of the Southeastern United States: aquatic communities 1992. Wiley, New York, pp 421–474

Song K, Li L, Li S et al (2012) Hyperspectral remote sensing of total phosphorus (TP) in three central Indiana water supply reservoirs. Water Air Soil Pollut 223:1481–1502. https://doi.org/10.1007/s11270-011-0959-6

Spyrakos E, O'donnell R, Hunter PD, Miller C, Scott M, Simis SG, Neil C, Barbosa CC, Binding CE, Bradt S et al (2018) Optical types of inland and coastal waters. Limnol Oceanogr 63:846–870

Strobel CJ, Paul J, Hughes M, Buffum HW, Brown BS, Summers K (2000) Using information on spatial variability of small estuaries in designing large-scale estuarine monitoring programs. Environ Monit Assess 63:223–236

Topp S, Pavelsky T, Jensen D, Simard M, Ross M (2020) Research trends in the use of remote sensing for inland water quality science: moving towards multidisciplinary applications. Water 12:169

Virginia Water Resources Research Institute and State University Blacksburg. Nutrients in lakes and reservoirs-Aliterature review for use in nutrient criteria development. VWRRC Special report SR34-2007. Virginia (2007). Downloaded from: http://www.vwrrc.vt.edu. Accessed on 10 Oct 2020

Wang YP, Xia H, Fu J, Sheng GY (2004) Water quality change in reservoirs of Shenzhen, China: detection using LANDSAT/TM data. Sci Total Environ 328:195–206

Wang L, Xu M, Liu Y, Liu H, Beck R, Reif M, Emery E, Young J, Wu Q (2020) Mapping freshwater chlorophyll-*a* concentrations at a regional scale integrating multi-sensor satellite observations with Google Earth Engine. Remote Sens 12:3278. https://doi.org/10.3390/rs12203278

Warren MA, Simis SGH, Martinez-Vicente V, Poser K, Bresciani M, Alikas K et al (2019) Assessment of atmospheric correction algorithms for the Sentinel-2A MultiSpectral Imager over coastal and inland waters. Remote Sens Environ 225:267–289. https://doi.org/10.1016/j.rse.2019.03.018

Watanabe FSY, Miyoshi GT, Rodrigues TWP, Bernardo NMR, Rotta LHS, Alcântara E, Imai NN (2020) Inland water's trophic status classification based on machine learning and remote sensing data. Remote Sens Appl Soc Environ 19:100326. ISSN 2352-9385. https://doi.org/10.1016/j.rsase.2020.100326

Whyte A, Feredinos K, Petropoulos GP (2018) A new synergistic approach for monitoring wetlands using Sentinels-1 and 2 data with object-based machine learning algorithms. Environ Model Softw 104:40–57

Williamson CE, Morris DP, Pace ML, Olson OG (1999) Dissolved organic carbon and nutrients as regulators of lake ecosystems: resurrection of a more integrated paradigm. Limnol Oceanogr 44:795–803

Wu CF, Wu JP, Qi JG, Zhang LS, Huang HQ, Lou LP et al (2010) Empirical estimation of total phosphorus concentration in the mainstream of the Qiantang River in China using Landsat TM data. Int J Remote Sens 31(9):2309–2324

Yang Z, Reiter M, Munyei N (2017) Estimation of chlorophyll-a concentrations in diverse water bodies using ratio-based NIR/Red indices. Remote Sens Appl Soc Environ 6:52–58

Zhang T, Huang M, Wang Z (2020) Estimation of chlorophyll-*a* concentration of lakes based on SVM algorithm and Landsat 8 OLI images. Environ Sci Pollut Res 27(13):14977–14990. https://doi.org/10.1007/s11356-020-07706-7

Zhang Y, Zhang Y, Shi K, Zhou Y, Li N (2021) Remote sensing estimation of water clarity for various lakes in China. Water Res 192:116844. https://doi.org/10.1016/j.watres.2021.116844

Utilization of the Self-organizing Map for Limnological Modelling

Ekaterini Hadjisolomou

Abstract Artificial Neural Networks (ANNs) are an advanced modelling technique applied in numerous scientific fields, including limnology. This study explores the application of the Kohonen Self-organizing Map (SOM), which is an unsupervised ANN, utilized to explore the interactions of water quality parameters and how the associated environmental data samples are grouped into clusters with different environmental properties. Therefore, the SOMs can be used for assessing the overall status of lakes. Regarding eutrophication phenomenon, which is a major issue for Mediterranean lakes, the SOMs have been proved to be an excellent modelling tool for examining the trophic status of lakes, superior to many other modelling techniques (e.g., Cluster Analysis). This is because the SOMs can provide insight into complex limnological data, but also to analyze multidimensional environmental data and simplify them into visual information with the use of the SOM's Component Planes. Taking SOM-based modelling studies of water quality as examples and their practical implementation, this study provides a guideline on how to develop a SOM model and to analyze the outputs. Additionally, the role of SOM models as management tools for lakes is pointed out and gives directions for future limnological research.

Keywords Lakes · Water quality parameters · Self-organizing Map · Data clustering · Parameters interactions

1 Introduction

Limnological models are conceptualized as a mathematical representation of the lake's ecosystem. They aim to mimic a phenomenon taking place into the lake by integrating the associated environmental variables, such as physical, biological and chemical (Vasistha and Ganguly 2020). Furthermore, the inclusion of several biotic and abiotic parameters, which act as inputs into a water quality model enables us

E. Hadjisolomou (✉)
Department of Electrical Engineering, Computer Engineering and Informatics, Cyprus University of Technology, Limassol, Cyprus
e-mail: e.hadjisolomou@cut.ac.cy

K. Stefanidis et al. (eds.), *Lakes of the Eastern Mediterranean and Balkans*,
https://doi.org/10.1007/978-3-032-02168-7_7

to examine the impact of these parameters on the simulated phenomenon/parameter (Hadjisolomou et al., 2023). For example, Mellios et al. (2015) managed to investigate the interrelations between the nutrients cycling and the trophic state of a Greek shallow lake by utilizing limnological modelling. While, in another water quality modelling study of Hadjisolomou et al. (2016) the synergistic effect of the input environmental parameters on the algal production mechanism in Pamvotida Lake (Greece) was calculated by applying a data-driven model, specifically an Artificial Neural Network (ANN).

ANNs is a subgroup of Machine Learning (ML) models, and they have been used widely for water quality modelling purposes during the last decades (Karul et al. 2000). Initially, - at the beginning of the 90s-, only a small number of ANN modelling studies related to environmental and ecological aspects were reported in the literature (Lek and Guégan 1999); including the utilization of ANNs aiming to simulate the algal production and the prediction of harmful algal bloom (HAB) events (Teles et al. 2006). Some of the first ANN applications, which simulated algal production and eutrophication phenomenon are the studies of Karul et al. (1999), Maier and Dandy (1996), Recknagel et al. (1997) and Scardi (1996).

The ANN's mathematical ability to model complex and non-linear relationships is ideal for modelling eutrophication underlying mechanisms (Hadjisolomou et al. 2017), for that reason ANNs are characterized as a useful management tool for lakes eutrophication control (Karul et al. 1999). As stated by Chen and Zhang (2009) the ANNs can be applied successfully to investigate the cause-effect relationships and the associated impact of the environmental parameters on a complex limnological system, when other conventional methods might fail.

A special category of ANNs applied for water quality modelling tasks is the Kohonen Self-organizing Maps (SOMs). Mainly, the SOM models are used for clustering/pattering environmental data and for exploratory data analysis (data mining) of the environmental parameters. As stated by Park et al. (2003), usually the conventional multivariate analyses have been used for these classification problems; however, SOMs have been proved more efficient for ecological patterning, because very often environmental data sets contain many different variables obtained from various sampling areas, so the interactions in the data set are nonlinear and complex.

Several modelling studies based on SOM models have been utilized for water quality modelling tasks. For example, a SOM was applied by Lu and Lo (2002), where a trophic state classifier was constructed based on a SOM model. This SOM diagnosed the water quality of the Fei-Tsui Reservoir (Taiwan) during the monitoring period 1987–1995 and compared the simulated SOM results with those of the Carlson Index. In another application given by Choi et al. (2014) the SOM model was applied and hydrochemical parameters (e.g., pH, Cd, Hg, Ca, Na, K, NO_3) of bedrock groundwater of 299 sampling stations in South Korea were clustered into two main groups and by that way the water quality and the anthropogenic contamination was evaluated. In the study of Hadjisolomou et al. (2018) the SOM model was applied into the transboundary Prespa Lakes (regarding the Greek part) aiming to explore the interrelationships among several water quality parameters (like nutrients, water temperature, chlorophyll-a) which are associated with the eutrophication. The developed SOM model acted as a eutrophication control management tool and helped to understand the underlying mechanisms associated with trophic productivity.

Therefore, water quality models, like SOMs, can become valuable management tools and help us to propose solutions in the form of management scenarios; addressing complicated environmental issues, primary associated with anthropogenic activities and also the climate change effect (Hadjisolomou et al., 2023). As stated by Moriasi et al. (2012) water quality models are mainly used to assess the impacts of climate change, land use and crop management practices on the quality and quantity of water resources. SOM models have been applied successfully for the assessment of anthropogenic activities and to examine the impact of land use on the water environment. For instance, Kim et al. (2021) described the application of a modified SOM model, called the Geo-SOM, combined with linear mixed-effect models for examining the eutrophication-related parameters from 54 monitoring stations along Nakdong River basin (South Korea). Specifically, the authors managed to examine the major stressors related to land use and to extract thresholds for phosphorus and Biological Oxygen Demand parameters for eutrophication management strategies. In another water quality modelling study, Pei et al. (2022) based on a SOM and redundancy analysis modelled the water quality of the Yellow River (China) from several stations, which were representative of the land use for each subarea. The above authors resulted in useful conclusions and revealed that the most polluted station was associated with agricultural non-point sources of ammonia nitrogen pollution, but also that the land used as cropland had the most negative impact on water quality compared to the other types of land use practices.

Therefore, the contribution of SOM models in the field of limnological modelling is significant, since by utilizing SOMs serious environmental issues, like eutrophication, can be evaluated based on monitored environmental parameters and at the same time several other factors, like the land usage nearby the sampling stations can be examined by the model. Hence, the SOM models can become valuable management tools for evaluating the environmental pressure linked to human activities or even environmental changes, like the water temperature increase because of global warming. It is also important to keep in mind that very few SOM modelling studies exist for lakes in the Mediterranean region, many of which are negatively impacted by the eutrophication phenomenon.

2 Properties of the SOM

The Kohonen's SOM is a special category of ANNs, and more details of its origin can be found in the studies of Kohonen (1982, 1998, 2001). SOM is an unsupervised learning algorithm of ANNs, meaning that no human intervention is required during the learning process of the ANN (Gelao et al. 2024). The term unsupervised is attributed to the SOM, since no supervision is required during its learning process and the SOM's ability to learn and organize information without being given the associated output values for the corresponding input data (Al-Mudhaf et al. 2010).

SOM's results visualization and its dimension reduction ability are very important modelling properties, which are emphasized in many water quality modelling studies (An et al. 2016; Jahan et al. 2013; Kangur et al. 2007). An advantage of SOM is its efficiency to visualize large amounts of data through a component plane (CP) technique (Cho et al. 2009). In their study Jahan et al. (2013) stated that one of the SOM's most significant characteristics is that the results obtained from the SOM can be visualized from its topology map, meaning that the SOM can interpret information from multiple variables based on its visualization abilities (Juntunen et al. 2013). Also, from the SOM's map visual inspection, simultaneous observation of both spatial and temporal changes in water quality is enabled (Astel et al. 2007).

Based on its dimension reduction ability, the SOM projects multidimensional data into a two-dimensional grid, while associates non-linear complex relationships between the variables (Peeters et al. 2007). For the convenience of visualization, the artificial neurons (nodes) of the SOM are arranged in a limited number of dimensions, preferably two-dimensional grids of neurons (Kohonen 1982). Each neuron is associated with $w = [w_1, \ldots, w_n]$ that is a n-dimensional weight vector (or prototype vector), where parameter n equals to the dimension of the input vectors (see Fig. 1), and the neurons are connected to adjacent neurons by a neighborhood relation based on which the topology of the map is dictated (Vesanto and Alhoniemi 2000).

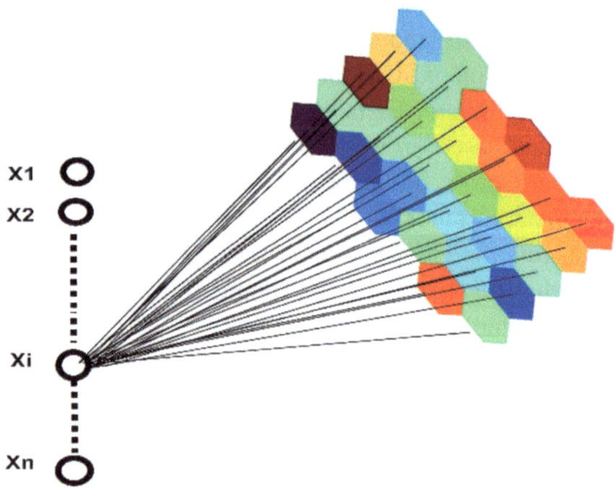

Fig. 1 Data dimension reduction from a n-dimensional space (where n the number of input variables) to a 2-dimensional space (XY) with the utilization of SOM. The measured data are rearranged to a 2D map as illustrated, where the synaptic weights connect the x-th input variable with the nodes of the SOM

3 The SOM Algorithm

The SOM algorithm follows several steps based on an iterative procedure. The learning procedure of the SOM algorithm is analytically explained in the studies of Kangas and Kohonen (1996) and Kohonen (2001). Initially, the connection weights are initialized with small random values and then the lowest value of the Euclidean distance between the input and weight vectors is calculated (Choi et al. 2014). Then the Best Matching Unit (BMU) is calculated (Zhang et al. 2008), whereas BMU is defined as the output neuron that wins the "competition", meaning the output neuron that has the most similar weight vector to the input sample (Greco et al. 2024). Finally, the stimulated neurons adjust their weights by minimizing the squared Euclidean distance between the input and weight vectors (Nayak et al. 2024). This means that the BMU and its neighboring (adjacent) neurons are changing their weights for each iteration, to reduce the distance between the input vector and the weights (Kohonen 2001).

The learning procedure of the SOM algorithm (Hadjisolomou 2019) is synopsized into the following steps (An et al. 2016; Chrobak et al. 2021; Hadjisolomou et al. 2018):

1. From the given data set, which consists of n columns of variables and m rows of measured samples, a $m \times n$ matrix is created. For avoiding bias in the classification, because of different unit's magnitude of the different variables, the variables are normalized before presenting them to the SOM.
2. Initialize the weight vector with random values.
3. An input vector is presented to the network and then the distances between the given input vector and the i-weight vector are calculated based on a distance metric. The Euclidean distance (symbolized as D_i) is the most frequently used metric. The mathematical equation for the Euclidean distance is given as below:

$$D_i = \sqrt{\left(p_{ij} - w_{ij}\right)^2}; \quad i = 1, 2, \ldots, S \tag{1}$$

where S is the number of output neurons, p_{ij} represents the j element of the input vector, and w_{ij} symbolizes the j element of the i-weight vector.

4. The smallest distance is chosen and the BMU is calculated by the following equation:

$$\|x - m_c\| = min(\|x - m_i\|) \tag{2}$$

where $\| \ \|$ symbolizes the distance measure, x the input vector, m the weight vector, and c the subscription of the weight vector for the winning neuron.

5. The weights of the BMU and the unit within its neighborhood ratio $N(t)$ are updated in order the new reference vectors to be closer to the input. During that process the neighborhood ratio function $h_{ci}(t)$ and the learning ratio are decreasing along with the number of iterations of the model, forcing by that way

the network to converge. The weight vector is updated based on the following relationship:

$$w_{ij}(t+1) = w_{ij}(t) + a(t)h_{ci}(t)\big(p_{ij} - w_{ij}(t)\big) \tag{3}$$

where $w_{ij}(t+1)$ is the weight vector at the learning step $t+1$.

6. The process goes on based on an interactive way and steps 3–5 are repeated, until a predefined number of iterations is done and then the process starts again from step 3.

After training the SOM, its quality must be assessed. The SOM quality is usually measured with the use of quantization error (QE) and the topographic error (TE) (Zhuang and Fu 2024). The QE indicates if the neurons of the trained SOM adapt well to the input variables and the TE represents how well the trained SOM keeps the topography of the analyzed data. The QE is defined as the mean distance between each data point (Kohonen 2001) and the associated BMU and samples with high QE values are considered possible outliers (Gelao et al. 2024). The TE gives the percentage of the data vectors for which the first BMU and the second BMU are not neighboring units (Kohonen 2001). The mathematical formulas, which describe the QE and the TE are the following (Tsai et al. 2017):

$$E_{QE} = \frac{1}{m}\sum_{p=1}^{m}\left\|X_p - M_{c(p)}\right\| \tag{4}$$

where E_{QE} is the quantization error; m is the number of input vectors of the SOM; X_p the data vectors; $M_{c(p)}$ the winner vectors of X_p.

$$E_{TE} = \frac{1}{m}\sum_{p=1}^{m}u\left(X_p\right) \tag{5}$$

where m is the number of input vectors of the SOM; X_p is a data vector. The u function equals to $u(X_p) = 1$ if the first and second BMUs of X_p are not adjacent to each other; otherwise, the $u(X_p) = 0$.

There are no specific rules for choosing SOM's parameters like global topology type (sheet, cylinder and toroid); local topology type (hexagonal and rectangular); neighbor function, map dimension (Al-Mudhaf et al. 2010). However, a very commonly used rule for finding the SOM map dimension is the one proposed by Vesanto and Alhoniemi (2000) and has the following formula:

$$M \approx 5\sqrt{n} \tag{6}$$

Another way for finding the optimum map size of the SOM is to find for which neurons number the quantization error (QE) and the topographic error (TE) are minimized (Wang et al. 2014), by examining several SOM with different sizes. The

TE and QE are decreasing gradually as the map size of the trained SOM is increasing and the optimum size is based on local minimum values for the parameters TE and QE (Céréghino and Park 2009). This procedure is described in the study of Ruiz-Varona et al. (2022), where the authors selected the SOM's best size (a SOM with 20 × 32 neurons) based on the QE and TE values evaluation. The authors observed that when the number of neurons was increasing (with respect to the chosen size), then the QE was decreasing slowly.

The input data before being presented into the SOM network must be pre-processed, because different magnitude of the water quality variables may result in unequal importance for the weight update procedure, and by that way the SOM model performance is improved (Çinar and Merdun 2009). Furthermore, all the input variables are treated by the SOM in the same way (Garća and González 2004). In their study Zhang et al. (2009) emphasized that the SOM modelling results might be affected by an inappropriate pre-processing method, and they suggested a way to resolve this problem by applying the Shapiro–Wilk normality test to examine if the input variables were following normal distributions.

Several data pre-processing (transformation) methods are proposed for the SOM's input parameters normalization (Çinar and Merdun 2009; Garća and González 2004; Peeters et al. 2007; Zhang et al. 2009), such as the logarithmic normalization, the zero-mean (z-score) normalization, the Min–Max normalization.

4 The U-Matrix and the Component Planes

The SOM is visualized with the usage of the U-matrix and the component planes (CPs), and by that way the clustering structure of the SOM map is presented (Dong et al. 2012). The unified distance matrix (U-matrix) or otherwise known as "Interneuron distance matrix" (Garća and González 2004) is very important. The U-matrix of the trained SOM is visualizing the distances between the SOM map neurons (Vesanto and Alhoniemi 2000). The U-matrix is graphically presented as in Fig. 2. Usually, the distances between the neighboring neurons of the SOM are visualized by using gray shading scales between each neuron, however this is not a norm and other color shading scales can be used.

The darker the shade the largest is the Euclidean distance between the neurons of the SOM map, therefore a light shade between the neurons of the SOM map indicates a small Euclidean distance (Vesanto 1999). The darkest shades on the U-matrix (in grayscale) represent the biggest distances between neighboring data and therefore the borders between the clusters are defined (Gorzałczany and Rudziński 2020; Recknagel et al. 2006), and by that way it is possible to identify the clusters of the input environmental data as defined by the U-matrix visualization.

The CPs are used to visualize the value of the SOM's input variables in each map unit (Vesanto and Alhoniemi 2000). The CPs are considered as a good tool to evaluate the interrelationships between the several water quality parameters (Hadjisolomou et al. 2023; Tsai et al. 2017). For example, in the water quality modelling study of

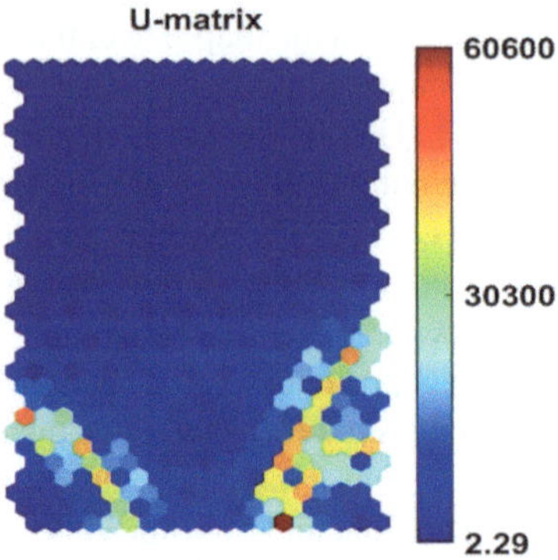

Fig. 2 U-matrix of a trained SOM map, where the distances between the neurons are visualized by the colorbar. The red color shades indicate long distance between the neurons, while the blue color shades indicate shorter distance

Hadjisolomou et al. (2018) the interrelationships between the SOM input variables can be observed by inspecting the CPs of the associated limnological parameters measured in the Greek part of the Prespa Lakes (see Fig. 3).

Through the CPs the exploratory potential among the SOM's input variables is provided and therefore the CPs visualization are making the SOM neural network to be characterized as a useful tool in exploratory data analysis and clustering of multivariate data sets (Peeters et al. 2007), while the visualization of inter-specific association among the CPs is supporting the SOMs role as a useful tool in exploratory data analysis (Giraudel and Lek 2001). The CPs (Vesanto and Alhoniemi 2000) interpretation is based on degrading color shadings. Each CP is associating the value range of every input variable of the trained SOM with each map unit (Vesanto 1999).

As previously stated, the SOM algorithm is used for exploring purposes, to discern any reasonable relationships among the environmental variables, without prior knowledge of the water quality data set characteristics. For that reason, the term "correlation hunting" is imported by Vesanto (1999) and as stated by the author the CPs can be used for correlation hunting among the SOM's input variables. The possibility of having even local correlations might exist, when two CPs resemble each other only in some regions (Barreto and Pérez-Uribe 2007). So, the term correlation is not limited only to just linear correlation but also includes nonlinear correlation and local (partial) correlation between the SOM's input variables (Vesanto and Ahola 1999).

The SOM can identify the corresponding correlations and similarities between the water quality variables (Dong et al. 2012). Particularly, when there is a strong positive correlation between two water quality variables, then the associated CPs are very identical; and when there is strong negative correlation between two water quality variables, then the associated CPs are identical but with reverse degrading of the color shading (Zhang et al. 2015). For example, in Fig. 3 by examining the CPs of the Electric Conductivity (EC) parameter and the Sechi Disc Depth (SD)

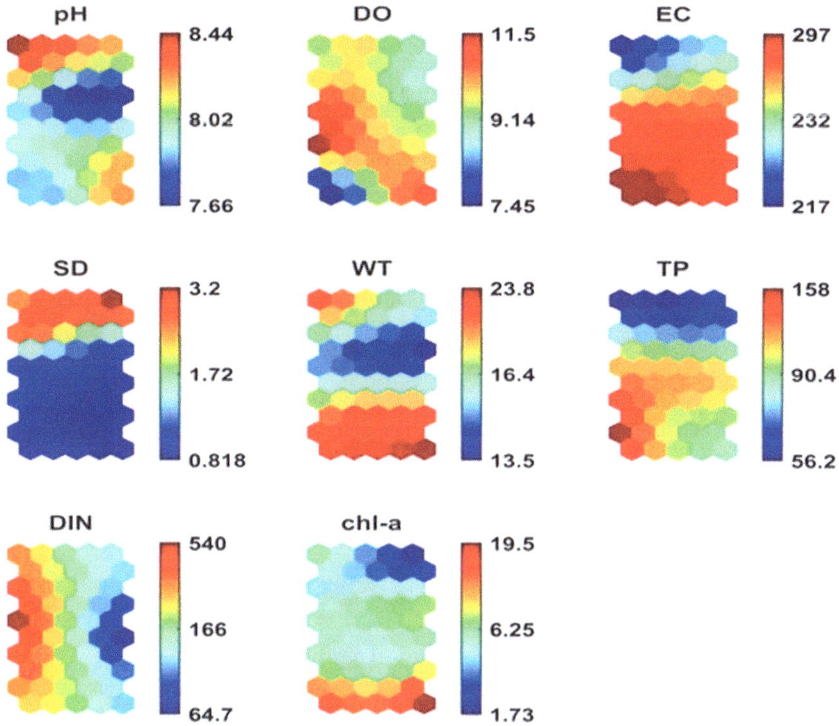

Fig. 3 Visualization of the component planes (CPs) associated with the trained SOM that examines the descriptive characteristics of the water quality in the Prespa Lakes based on several measured limnological parameters (taken from Hadjisolomou et al. 2018)

parameter, it is observed that there is strong negative correlation between these two water quality parameters, since the CPs are very similar in a reverse way, meaning that the dark scale coloring (red) areas are following an identical reverse degradation towards the light scale coloring (blue) areas.

Spherical implementation of the SOM results requires some visualization and clustering tools to be utilized, in order to have substantial information on the input data distribution and interrelationships among the SOM's input variables (Bieroza et al. 2009). Besides the U-matrix and the CPs, the sample distribution map, where the associated SOM was created based on a demo limnological data set (Fig. 4) and the hit histogram (Fig. 5) (taken from Hadjisolomou et al. 2018) are also used for implementing the results of a trained SOM.

With the use of the sample distribution map for each node the most frequent BMU with an assigned data set index (e.g. sampling site number) is associated (Bieroza et al. 2009; Zhang et al. 2015), meaning that with the help of the sample distribution map the sampling sites (or every other possible assigned index) are distributed and clustered analogously. By that way the sampling sites can be related to the CPs

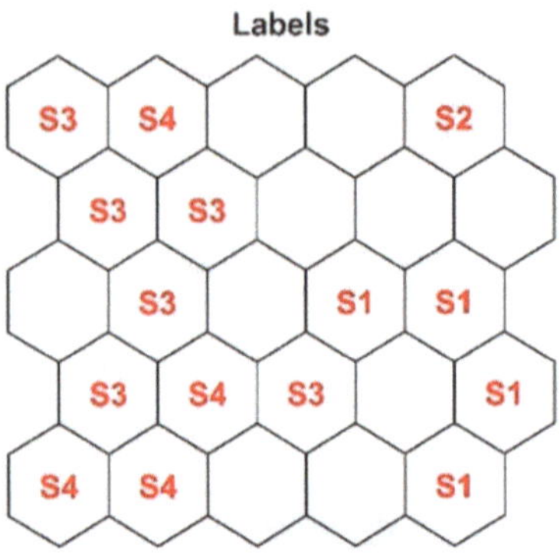

Fig. 4 A sample distribution map of a demo limnological data set. The four sampling stations (S1, S2, S3, S4) are classified into the sample distribution map. The labels within the nodes represent the sampling stations

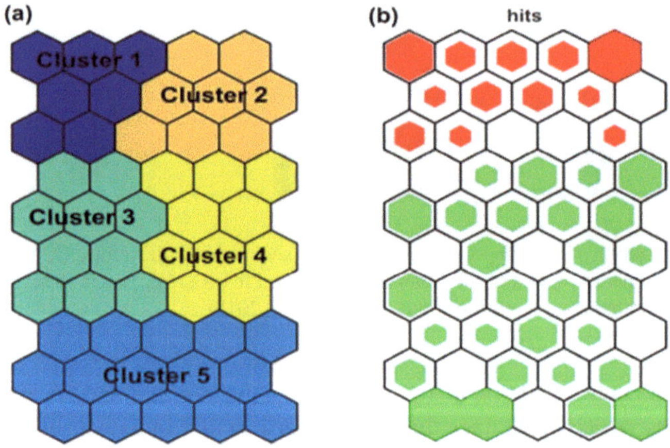

Fig. 5 **a** Clustering of the SOM based on the *K*-means algorithm. **b** Hits histogram of SOM analysis representing the density of SOM hits. The green represents data from Lake Mikri Prespa, the red represents data from Lake Megali Prespa, and the empty nodes are associated with the absence of data samples (taken from Hadjisolomou et al. 2018)

maps, while associations are enabled regarding the environmental conditions, which correspond for each sampling site.

The hit histogram of the trained SOM is displaying the samples distribution on the SOM map. For example, in the limnological modelling study of Hadjisolomou et al. (2018), the hit histogram and the clustering of the created SOM are utilized to evaluate the environmental data samples measured from the Prespa Lakes (Lakes Mikri Prespa and Megali Prespa). Figure 5a shows the clustering of the SOM based on the *K*-means algorithm, where the data samples are clustered into five clusters. The trained SOM displays the number of sample hits for each cell (node). Figure 5b corresponds to the hit histogram of the trained SOM and displays the density distribution of the sample hits for each cell (node), where the more sample hits for each cell the more intense (having larger surface) is the color inside it.

In conclusion, it can be said that this visualization of the hit histogram of SOM analysis displays the density of the SOM hits based on a distinction of the data sample. According to Bieroza et al. (2009) the greater the spread of water quality samples on the hit histogram, the greater the variation of the water quality properties that are observed. This is related to the fact that the water quality samples assigned to distant neurons of the SOM map are different, while the samples assigned in neighboring neurons are similar (Płóciennik et al. 2015).

It must be noted that when referring to the sample distribution map, each cell/neuron of the SOM map gives the name of an assigned index (e.g., the name of a sampling site into the lake) corresponding to the most frequent best matching sample, while multiple data samples might correspond to this winning neuron (Zhang et al. 2015). Regarding the hit histogram, the density of data samples that are associated with the winning neuron is illustrated, where the colored area in the hexagon cell is proportional to the number of training data samples placed in that cell during SOM's training phase (Moosavi et al. 2015).

5 The Cluster Structuring Index

The influence of each input variable on the organization of the SOM map is measured with the use of an index, called the Cluster Structuring Index (CSI or SI). SI was initially created to assess which one of the measured diatom species has the strongest influence on the SOM map's organization (Park et al. 2006). As stated by Park et al. (2005), the idea behind the SI derived from the observation that a SOM's cluster is an assembly of neighboring units, and therefore when a variable is specified into a certain cluster and has smaller difference between the weights of SOM units. The SI for the ith variable can be calculated based on the formula given below (Park et al. 2005, 2006):

$$SI = \sum_{j=1}^{S} \sum_{k=1}^{J-1} \frac{\left|w_{ij} - w_{ik}\right|}{\left\|r_j - r_k\right\|} \tag{7}$$

where the terms w_{ij} and w_{ik} symbolize the connection weights for the ith input variable of the j and k neurons respectively; $\left\|r_j - r_k\right\|$ is the topological distance between the units j and k; and S is the total number of SOM's output units.

According to Tison et al. (2005), the SI has high values for variables found in high abundance of frequency in samples gathered in the same area of the SOM map, while the SI has low values for variables observed in many data samples of different clusters. It is observed that the existence of a strong gradient for an input variable generates a high SI value, while the existence of a weak gradient for an input variable generates a low SI value (Tsai et al. 2017).

The SI is calculated for the needs of several water quality modelling studies utilizing a SOM neural network methodology For example, in the study of Tsai et al. (2017) a SOM neural network was created to explore the multi-relationships between the fish species and the water quality parameters, where eight water quality parameters were measured (Dissolved Oxygen; total phosphorus; Water Temperature; pH; electric conductivity; suspended solids; chemical oxygen demand; ammonia). In that study, SOM managed to produce meaningful patterns between the 25 fish species and the water quality parameters, corresponding to 276 heterogeneous data measured across Taiwan. The authors computed the SI for the water quality parameters, where it was found that the Dissolved Oxygen parameter had the biggest influence of the monitored water quality parameters on the organization of the SOM map.

A computational problem that might appear during the modelling procedure with a SOM, is the existence of too many input variables. This may result in difficulties regarding the evaluation of the variables interactions based on the CPs visualization, because of the large number of input variables. In their study Tison et al. (2005), they addressed this problem by adopting the SI to reduce the number of input variables. Specifically, it was decided to keep only the variables having SI over 90 and by that way the number of input variables was reduced by three, while no significant quantities of ecological information were lost. The authors, in order to exam if ecological information was preserved for the new reduced data set, calculate the regression determination coefficient (R^2) between SI values of the new reduced data set and the initial data set. In the study of Park et al. (2006), a data set of diatom community data, which were sampled from 836 sites with 941 species throughout the French hydrosystem, was examined by applying the SOM model. The authors addressed the problem of too many input variables by also utilizing the SI.

6 Clustering of the SOM

The clustering of the trained SOM prototypes is usually based on the U-matrix evaluation (Joo and Jeong 2005). However, two other clustering techniques are also used: the K-means algorithm clustering method and the hierarchical clustering method, which is based on a dendrogram that shows the dissimilarity of the SOM's cells (Gevrey et al. 2004). An analytical presentation of these three clustering methods of the SOM's prototypes is found in the studies of Gonçalves et al. (2008) and Vesanto and Alhoniemi (2000).

When the clustering is done based on the U-matrix's observation, the clusters are detected after visual inspection of the U-matrix. The darker shades of the colorbar are associated with large Euclidean distances for the neighboring prototypes, and by that way the borders between the formed clusters are indicated. The application of the K-means algorithm for clustering the SOM's prototypes is widely used in SOM water quality modelling by many researchers (Garća and González 2004; Juntunen et al. 2013; Kim and An 2015; Li et al. 2017) and the *K*-means clustering algorithm is applied on the nodes of the trained SOM. It must be noted that, when the number of clusters is increased, then the algorithm is sensitive to outliers because the number of samples in each cluster is decreasing (Vesanto and Alhoniemi 2000). The Davies–Bouldin index calculates the optimal number of the *K* clusters for the prototypes of the trained SOM (Farsadnia et al. 2014). The Davies–Bouldin index (where more information is given in the study of Davies and Bouldin (1979)) associates its minimum value with the optimal number of clusters of the SOM (Çinar and Merdun 2009).

Finally, when SOM's prototypes clustering is done with the use of the hierarchical clustering method, then a dendrogram is involved and the usage of the Ward linkage method based on the Euclidean distance (Chon 2011). Several water quality modelling studies based on SOMs are applying the hierarchical clustering method (Choi et al. 2014; Kwon et al. 2012; Lek-Ang et al. 2007; Oh et al. 2007).

The clustering of the SOM prototypes based on the different clustering methods can give useful results and reveal several mechanisms related to the functioning of the limnological system (Hadjisolomou et al. 2018). However, the clusters examination shouldn't be limited only to examine their statistical properties (e.g. mean, medium, maximum, minimum), but also additional aspects of the clustering should be taken into consideration, e.g., if the formed cluster has sub-clusters (Vesanto and Alhoniemi 2000).

7 Discussion

Lakes are governed by complicated environmental mechanisms and the limnological data are usually non-linear and complex. Therefore, the selection of the proper modelling method is important when working with limnological data sets. Usually, ANN models produce better results than classical modeling techniques. Besides ANNs ability not to be affected by the non-linearity, ANNs are capable to perform well in limited data situations (Hadjisolomou et al. 2021; Jain and Kumar 2006). For example, the process-based techniques for water quality modelling require a large amount of monitored data samples and a large number of environmental variables (Karul et al. 2000; Park et al. 2003), while data-driven modelling techniques need fewer input parameters and data points than the process-based techniques (Mohammed et al. 2022; Palani et al. 2008).

The SOM models, besides being more reliable as modelling techniques, also possess some modelling advantages when compared to other modelling techniques (e.g., statistical methods). For example, the phenomenon of dendrogram overcrowding is observed, when dendrograms grow so much and are overloaded, that brunches become overcrowded. This drawback of dendrograms is increasing analogously with sample size and might cause a problem of dendrogram. In a relevant modelling study of Sanchez-Martos et al. (2002), Cluster Analysis and SOM are used to model the groundwater quality based on data set of 168 observations and 9 variables. During the Cluster Analysis application on the raw (initial) data set the phenomenon of dendrogram overcrowding is intensively observed, where it is difficult to identify the groups, and which samples have similar water quality. According to the authors, this problem is attributed to the big number of sampling points in the data set. However, the SOM model, managed to represent each sampling point clearly on the activation map, while the resulting dendrogram was easily implemented. Synopsizing these findings, it is concluded that the SOM method is producing results, which can be more easily implemented with the use of a dendrogram, comparing with the hierarchical clustering methodologies. According to Astel et al. (2007) both Cluster Analysis algorithms and SOM methodologies have the advantage of visualization ability, but the SOM has three additional advantages: (i) SOM visualization enables us to present similarity between positive correlated variables but is also able to present similarity between negative correlated variables; (ii) SOM visualization can detect outliers, corresponding to data samples that don't belong to a well-organized group; (iii) The projection of variables similarity contains semi-quantitative information about the distribution of a given parameter in the sampling locations space.

Water modelling studies that are applying both PCA and SOM methods, are usually producing similar results (Samecka-Cymerman et al. 2007). However, it is observed that the SOM ability to classify environmental samples is more advanced than the PCA method ability. In the study of Choi et al. (2014) the SOM and PCA methods were used to classify groundwater monitoring data in South Korea. Because the groundwater samples were widely dispersed, it was noticed that the PCA results were not satisfactory to classifying the samples and that the variables loading values did not produce any meaningful information. At the same time, SOM managed to relate the groundwater samples with the variables and to classify the groundwater samples into different groups.

In contrast with the traditional statistical methodologies that are not able to present well the information of big data sets, the SOM is suitable for visualization of high-dimensional data; because the SOM converts complex statistical relationships between data sets into simple geometric relationships on a low-dimensional display (Várbíró et al. 2007). According to Lee and Scholz (2006) the outstanding performance of the SOM model comparing with other modelling techniques is related to the SOM's potential of clustering and classifying data, but also relies to the SOM's ability to detect outliers, which are displaying in particular parts of the map without affecting the remaining parts and each outlier takes its place in one unit of the map.

Finally, some additional modelling advantages of the SOM comparing with other modelling methodologies are the following: it's ability to handle missing data; can deal with nonlinearities of the system; the SOM model can be developed from data without requiring a mechanistic knowledge of the system; the SOM is easily and quickly updated; can handle irregular data/noise reduction; SOM gives an advanced visualization method of the results (Çinar and Merdun 2009; Vesanto and Alhoniemi 2000). Regarding SOM's advanced visualization abilities derived from the CPs, the SOM enables the visualization of parameters interactions, while at the same time the sample units and the variables concentration can be displayed in the same figure (Giraudel and Lek 2001).

Therefore, the SOM results are more specific and enable direct assessment of the variables interactions because of SOM's advanced visualization abilities. Considering SOM's other modelling merits (e.g., not being affected by the nonlinearity or data heterogeneity), then the SOM models are ideal for limnological applications. In the case of data scarcity, an issue often observed in limnology, it is recommended the creation of fused databases (e.g., satellite data and in situ measurements) based on which several more specialized SOMs can be developed, like seasonal SOM models; since it is observed that the seasonality effect is significantly impacting Mediterranean lakes. In conclusion, the utilization of SOM models for limnological applications is good modelling practice and useful information is extracted, which can serve as a guideline for lakes management and restoration.

Competing Interests The author has no conflicts of interest to declare that are relevant to the content of this chapter.

References

Al-Mudhaf HF, Astel AM, Selim MI, Abu-Shady ASI (2010) Self-organizing map approach in assessment spatiotemporal variations of trihalomethanes in desalinated drinking water in Kuwait. Desalination 252(1–3):97–105. https://doi.org/10.1016/J.DESAL.2009.10.017

An Y, Zou Z, Li R (2016) Descriptive characteristics of surface water quality in Hong Kong by a self-organising map. Int J Environ Res Public Health 13(1):115. https://doi.org/10.3390/IJERPH13010115

Astel A, Tsakovski S, Barbieri P, Simeonov V (2007) Comparison of self-organizing maps classification approach with cluster and principal components analysis for large environmental data sets. https://doi.org/10.1016/j.watres.2007.06.030

Barreto SMA, Pérez-Uribe A (2007) Improving the correlation hunting in a large quantity of SOM component planes. In: Lecture notes in computer science (including subseries lecture notes in artificial intelligence and lecture notes in bioinformatics), LNCS, vol 4669, part 2, pp 379–388. https://doi.org/10.1007/978-3-540-74695-9_39

Bieroza M, Baker A, Bridgeman J (2009) Exploratory analysis of excitation-emission matrix fluorescence spectra with self-organizing maps as a basis for determination of organic matter removal efficiency at water treatment works. J Geophys Res Biogeosci 114(G4):01–07. https://doi.org/10.1029/2009JG000940

Céréghino R, Park YS (2009) Review of the self-organizing map (SOM) approach in water resources: commentary. Environ Model Softw 24(8):945–947. https://doi.org/10.1016/J.ENVSOFT.2009.01.008

Chen LH, Zhang XY (2009) Application of artificial neural networks to classify water quality of the Yellow River. Adv Soft Comput 54:15–23. https://doi.org/10.1007/978-3-540-88914-4_3

Cho HS, Choi KH, Lee SD, Park YS (2009) Characterizing habitat preference of Eurasian river otter (*Lutra lutra*) in streams using a self-organizing map. Limnology 10(3):203–213. https://doi.org/10.1007/S10201-009-0275-7/METRICS

Choi BY, Yun ST, Kim KH, Kim JW, Kim HM, Koh YK (2014) Hydrogeochemical interpretation of South Korean groundwater monitoring data using self-organizing maps. J Geochem Explor 137:73–84. https://doi.org/10.1016/J.GEXPLO.2013.12.001

Chon TS (2011) Self-organizing maps applied to ecological sciences. Eco Inform 6(1):50–61. https://doi.org/10.1016/J.ECOINF.2010.11.002

Chrobak G, Kowalczyk T, Fischer TB, Chrobak K, Szewrański S, Kazak JK (2021) Combining indicators for better decisions—algorithms vs experts on lakes ecological status assessment. Ecol Ind 132:108318. https://doi.org/10.1016/J.ECOLIND.2021.108318

Çinar Ö, Merdun H (2009) Application of an unsupervised artificial neural network technique to multivariant surface water quality data. Ecol Res 24(1):163–173. https://doi.org/10.1007/S11284-008-0495-Z

Davies DL, Bouldin DW (1979) A cluster separation measure. IEEE Trans Pattern Anal Mach Intell PAMI 1(2):224–227. https://doi.org/10.1109/TPAMI.1979.4766909

Dong Y, Scholz M, Harrington R (2012) Statistical modeling of contaminants removal in mature integrated constructed wetland sediments. J Environ Eng 138(10):1009–1017. https://doi.org/10.1061/(ASCE)EE.1943-7870.0000572

Farsadnia F, Rostami Kamrood M, Moghaddam Nia A, Modarres R, Bray MT, Han D, Sadatinejad J (2014) Identification of homogeneous regions for regionalization of watersheds by two-level self-organizing feature maps. J Hydrol 509:387–397. https://doi.org/10.1016/J.JHYDROL.2013.11.050

Garća HL, González IM (2004) Self-organizing map and clustering for wastewater treatment monitoring. Eng Appl Artif Intell 17(3):215–225. https://doi.org/10.1016/J.ENGAPPAI.2004.03.004

Gelao V, Fornasaro S, Briguglio SC, Mattiussi M, De Martin S, Astel AM, Barbieri P, Licen S (2024) Self-organizing maps: an AI tool for identifying unexpected source signatures in non-target screening analysis of urban wastewater by HPLC-HRMS. Toxics 12(2):113. https://doi.org/10.3390/TOXICS12020113/S1

Gevrey M, Rimet F, Park YS, Giraudel JL, Ector L, Lek S (2004) Water quality assessment using diatom assemblages and advanced modelling techniques. Freshw Biol 49(2):208–220. https://doi.org/10.1046/J.1365-2426.2003.01174.X

Giraudel JL, Lek S (2001) A comparison of self-organizing map algorithm and some conventional statistical methods for ecological community ordination. Ecol Model 146(1–3):329–339. https://doi.org/10.1016/S0304-3800(01)00324-6

Gonçalves ML, Netto MLA, Costa JAF, Zullo Junior J (2008) An unsupervised method of classifying remotely sensed images using Kohonen self-organizing maps and agglomerative hierarchical clustering methods. Int J Remote Sens 29(11):3171–3207. https://doi.org/10.1080/01431160701442146

Gorzałczany MB, Rudziński F (2020) Evolution of SOMs' structure and learning algorithm: from visualization of high-dimensional data to clustering of complex data. Algorithms 13(5):109. https://doi.org/10.3390/A13050109

Greco M, Giarnetti S, Giovenale E, Taschin A, Leccese F, Doria A, Senni L (2024) THz data analysis and self-organizing map (SOM) for the quality assessment of hazelnuts. Appl Sci 14(4):1555. https://doi.org/10.3390/APP14041555

Hadjisolomou E (2019) Modelling of environmental parameters in lake environments using neural networks. Ph.D. thesis, University of Patras. https://doi.org/10.12681/EADD/45209

Hadjisolomou E, Stefanidis K, Papatheodorou G, Papastergiadou E (2016) Assessing the contribution of the environmental parameters to eutrophication with the use of the "PaD" and "PaD2" methods in a hypereutrophic lake. Int J Environ Res Public Health 13(8). https://doi.org/10.3390/ijerph13080764

Hadjisolomou E, Stefanidis K, Papatheodorou G, Papastergiadou E (2017) Evaluating the contributing environmental parameters associated with eutrophication in a shallow lake by applying artificial neural networks techniques. Fresenius Environ Bull 26(5)

Hadjisolomou E, Stefanidis K, Papatheodorou G, Papastergiadou E (2018) Assessment of the eutrophication-related environmental parameters in two Mediterranean lakes by integrating statistical techniques and self-organizing maps. Int J Environ Res Public Health 15(3). https://doi.org/10.3390/ijerph15030547

Hadjisolomou E, Stefanidis K, Herodotou H, Michaelides M, Papatheodorou G, Papastergiadou E (2021) Modelling freshwater eutrophication with limited limnological data using artificial neural networks. Water (Switzerland) 13(11). https://doi.org/10.3390/w13111590
Hadjisolomou E, Rousou M, Antoniadis K, Vasiliades L, Kyriakides I, Herodotou H, Michaelides M (2023) Data-driven models for evaluating coastal eutrophication: a case study for Cyprus. Water 15(23):4097. https://doi.org/10.3390/W15234097
Jahan R, Choi HC, Park YS, Park YC, Seo JH, Choi JK (2013) Implementation of self-organizing maps (SOM) to analyses of environmental parameters and phytoplankton biomass in a macrotidal estuary and artificial lake. J Mar Biol Assoc UK 93(1):1–12. https://doi.org/10.1017/S0025315412000616
Jain A, Kumar A (2006) An evaluation of artificial neural network technique for the determination of infiltration model parameters. Appl Soft Comput 6(3):272–282. https://doi.org/10.1016/J.ASOC.2004.12.007
Joo GJ, Jeong KS (2005) Modelling community changes of cyanobacteria in a flow regulated river (the lower Nakdong River, S. Korea) by means of a self-organizing map (SOM). In: Modelling community structure in freshwater ecosystems, pp 273–287. https://doi.org/10.1007/3-540-26894-4_24
Juntunen P, Liukkonen M, Lehtola M, Hiltunen Y (2013) Cluster analysis by self-organizing maps: an application to the modelling of water quality in a treatment process. Appl Soft Comput 13(7):3191–3196. https://doi.org/10.1016/J.ASOC.2013.01.027
Kangas J, Kohonen T (1996) Developments and applications of the self-organizing map and related algorithms. Math Comput Simul 41(1–2):3–12. https://doi.org/10.1016/0378-4754(96)88223-1
Kangur K, Park YS, Kangur A, Kangur P, Lek S (2007) Patterning long-term changes of fish community in large shallow Lake Peipsi. Ecol Model 203(1–2):34–44. https://doi.org/10.1016/J.ECOLMODEL.2006.03.039
Karul C, Soyupak S, Yurteri C (1999) Neural network models as a management tool in lakes. Hydrobiologia 408:139–144. https://doi.org/10.1023/A:1017007313690
Karul C, Soyupak S, Çilesiz AF, Akbay N, Germen E (2000) Case studies on the use of neural networks in eutrophication modeling. Ecol Model 134(2–3):145–152. https://doi.org/10.1016/S0304-3800(00)00360-4
Kim JY, An KG (2015) Integrated ecological river health assessments, based on water chemistry, physical habitat quality and biological integrity. Water 7(11):6378–6403. https://doi.org/10.3390/W7116378
Kim HG, Hong S, Chon TS, Joo GJ (2021) Spatial patterning of chlorophyll a and water-quality measurements for determining environmental thresholds for local eutrophication in the Nakdong River basin. Environ Pollut 268:115701. https://doi.org/10.1016/J.ENVPOL.2020.115701
Kohonen T (1982) Self-organized formation of topologically correct feature maps. Biol Cybern 43(1):59–69. https://doi.org/10.1007/BF00337288/METRICS
Kohonen T (1998) The self-organizing map. Neurocomputing 21(1–3):1–6. https://doi.org/10.1016/S0925-2312(98)00030-7
Kohonen T (2001) Self-organizing maps, p 30. https://doi.org/10.1007/978-3-642-56927-2
Kwon YS, Li F, Chung N, Bae MJ, Hwang SJ, Byoen MS, Park SJ, Park YS (2012) Response of fish communities to various environmental variables across multiple spatial scales. Int J Environ Res Public Health 9(10):3629–3653. https://doi.org/10.3390/IJERPH9103629
Lee BH, Scholz M (2006) A comparative study: prediction of constructed treatment wetland performance with K-nearest neighbors and neural networks. Water Air Soil Pollut 174(1–4):279–301. https://doi.org/10.1007/S11270-006-9113-2/METRICS
Lek S, Guégan JF (1999) Artificial neural networks as a tool in ecological modelling, an introduction. Ecol Model 120(2–3):65–73. https://doi.org/10.1016/S0304-3800(99)00092-7
Lek-Ang S, Park YS, Ait-Mouloud S, Deharveng L (2007) Collembolan communities in a peat bog versus surrounding forest analyzed by using self-organizing map. Ecol Model 203(1–2):9–17. https://doi.org/10.1016/J.ECOLMODEL.2006.01.007

Li X, Sha J, Wang ZL (2017) Chlorophyll-a prediction of lakes with different water quality patterns in China based on hybrid neural networks. Water 9(7):524. https://doi.org/10.3390/W9070524

Lu RS, Lo SL (2002) Diagnosing reservoir water quality using self-organizing maps and fuzzy theory. Water Res 36(9):2265–2274. https://doi.org/10.1016/S0043-1354(01)00449-3

Maier HR, Dandy GC (1996) The use of artificial neural networks for the prediction of water quality parameters. Water Resour Res 32(4):1013–1022. https://doi.org/10.1029/96WR03529

Mellios N, Kofinas D, Laspidou C, Papadimitriou T (2015) Mathematical modeling of trophic state and nutrient flows of Lake Karla using the PCLake model. Environ Process 2(1):S85–S100. https://doi.org/10.1007/S40710-015-0098-Y/FIGURES/7

Mohammed H, Michel Tornyeviadzi H, Seidu R (2022) Emulating process-based water quality modelling in water source reservoirs using machine learning. J Hydrol 609:127675. https://doi.org/10.1016/J.JHYDROL.2022.127675

Moosavi V, Aschwanden G, Velasco E (2015) Finding candidate locations for aerosol pollution monitoring at street level using a data-driven methodology. Atmos Meas Tech 8(9):3563–3575. https://doi.org/10.5194/AMT-8-3563-2015

Moriasi DN, Wilson BN, Douglas-Mankin KR, Arnold JG, Gowda PH (2012) Hydrologic and water quality models: use, calibration, and validation. Trans ASABE 55(4):1241–1247. https://doi.org/10.13031/2013.42265

Nayak P, Trivedi V, Gupta S, Booba PB, Soloveva OV, Rozhdestvenskiy OI, Joshi A (2024) Artificial neural network-based clustering in wireless sensor networks to balance energy consumption. Cogent Eng 11(1). https://doi.org/10.1080/23311916.2024.2384652

Oh HM, Ahn CY, Lee JW, Chon TS, Choi KH, Park YS (2007) Community patterning and identification of predominant factors in algal bloom in Daechung Reservoir (Korea) using artificial neural networks. Ecol Model 203(1–2):109–118. https://doi.org/10.1016/J.ECOLMODEL.2006.04.030

Palani S, Liong SY, Tkalich P (2008) An ANN application for water quality forecasting. Mar Pollut Bull 56(9):1586–1597. https://doi.org/10.1016/J.MARPOLBUL.2008.05.021

Park YS, Verdonschot PFM, Chon TS, Lek S (2003) Patterning and predicting aquatic macroinvertebrate diversities using artificial neural network. Water Res 37(8):1749–1758. https://doi.org/10.1016/S0043-1354(02)00557-2

Park YS, Gevrey M, Lek S, Giraudel JL (2005) Evaluation of relevant species in communities: development of structuring indices for the classification of communities using a self-organizing map. In: Modelling community structure in freshwater ecosystems, pp 369–380. https://doi.org/10.1007/3-540-26894-4_31

Park YS, Tison J, Lek S, Giraudel JL, Coste M, Delmas F (2006) Application of a self-organizing map to select representative species in multivariate analysis: a case study determining diatom distribution patterns across France. Eco Inform 1(3):247–257. https://doi.org/10.1016/J.ECOINF.2006.03.005

Peeters L, Bação F, Lobo V, Dassargues A (2007) Exploratory data analysis and clustering of multivariate spatial hydrogeological data by means of GEO3DSOM, a variant of Kohonen's self-organizing map. Hydrol Earth Syst Sci 11(4):1309–1321. https://doi.org/10.5194/HESS-11-1309-2007

Pei L, Wang C, Zuo Y, Liu X, Chi Y, Wang Y, Wang W, Liang P, Wang C, Zuo Y, Liu X, Chi Y (2022) Impacts of land use on surface water quality using self-organizing map in middle region of the Yellow River basin, China. Int J Environ Res Public Health 19(17):10946. https://doi.org/10.3390/IJERPH191710946

Płóciennik M, Kruk A, Michczyńska DJ, Birks JB (2015) Kohonen artificial neural networks and the IndVal index as supplementary tools for the quantitative analysis of palaeoecological data. Geochronometria 42(1):189–201. https://doi.org/10.1515/GEOCHR-2015-0021

Recknagel F, French M, Harkonen P, Yabunaka KI (1997) Artificial neural network approach for modelling and prediction of algal blooms. Ecol Model 96(1–3):11–28. https://doi.org/10.1016/S0304-3800(96)00049-X

Recknagel F, Talib A, van der Molen D (2006) Phytoplankton community dynamics of two adjacent Dutch lakes in response to seasons and eutrophication control unravelled by non-supervised artificial neural networks. Eco Inform 1(3):277–285. https://doi.org/10.1016/J.ECOINF.2006.01.001
Ruiz-Varona A, Lacasta J, Nogueras-Iso J (2022) Geo-information self-organizing maps to evaluate multidimensional trajectories of shrinkage in Spain. https://doi.org/10.3390/ijgi11020077
Samecka-Cymerman A, Stankiewicz A, Kolon K, Kempers AJ (2007) Self-organizing feature map (neural networks) as a tool in classification of the relations between chemical composition of aquatic bryophytes and types of streambeds in the Tatra national park in Poland. Chemosphere 67(5):954–960. https://doi.org/10.1016/J.CHEMOSPHERE.2006.11.001
Sanchez-Martos F, Aguilera PA, Garrido-Frenich A, Torres JA, Pulido-Bosch A (2002) Assessment of groundwater quality by means of self-organizing maps: application in a semiarid area. Environ Manage 30(5):716–726. https://doi.org/10.1007/S00267-002-2746-Z/METRICS
Scardi M (1996) Artificial neural networks as empirical models for estimating phytoplankton production. Mar Ecol Prog Ser 139(1–3):289–299. https://doi.org/10.3354/MEPS139289
Teles LO, Vasconcelos V, Pereira E, Saker M, Vasconcelos V (2006) Time series forecasting of cyanobacteria blooms in the Crestuma Reservoir (Douro River, Portugal) using artificial neural networks. Environ Manage 38(2):227–237. https://doi.org/10.1007/S00267-005-0074-9/FIGURES/4
Tison J, Park YS, Coste M, Wasson JG, Ector L, Rimet F, Delmas F (2005) Typology of diatom communities and the influence of hydro-ecoregions: a study on the French hydrosystem scale. Water Res 39(14):3177–3188. https://doi.org/10.1016/J.WATRES.2005.05.029
Tsai WP, Huang SP, Cheng ST, Shao KT, Chang FJ (2017) A data-mining framework for exploring the multi-relation between fish species and water quality through self-organizing map. Sci Total Environ 579:474–483. https://doi.org/10.1016/J.SCITOTENV.2016.11.071
Várbíró G, Ács É, Borics G, Érces K, Fehér G, Grigorszky I, Japport T, Kocsis G, Krasznai E, Nagy K, Nagy-László Z, Pilinszky Z, Kiss KT (2007) Use of self-organizing maps (SOM) for characterization of riverine phytoplankton associations in Hungary. Large Rivers 17(3–4):383–394. https://doi.org/10.1127/LR/17/2007/383
Vasistha P, Ganguly R (2020) Water quality assessment of natural lakes and its importance: an overview. Mater Today Proc 32:544–552. https://doi.org/10.1016/J.MATPR.2020.02.092
Vesanto J (1999) SOM-based data visualization methods. Intell Data Anal 3(2):111–126. https://doi.org/10.1016/S1088-467X(99)00013-X
Vesanto J, Ahola J (1999) Hunting for correlations in data using the self-organizing map. ICSC Academic Press, pp 279–285. https://research.aalto.fi/en/publications/hunting-for-correlations-in-data-using-the-self-organizing-map
Vesanto J, Alhoniemi E (2000) Clustering of the self-organizing map. IEEE Trans Neural Netw 11(3):586–600. https://doi.org/10.1109/72.846731
Wang B, Li H, Sun D (2014) Social-ecological patterns of soil heavy metals based on a self-organizing map (SOM): a case study in Beijing, China. Int J Environ Res Public Health 11(4):3618–3638. https://doi.org/10.3390/IJERPH110403618
Zhang L, Scholz M, Mustafa A, Harrington R (2008) Assessment of the nutrient removal performance in integrated constructed wetlands with the self-organizing map. Water Res 42(13):3519–3527. https://doi.org/10.1016/J.WATRES.2008.04.027
Zhang L, Scholz M, Mustafa A, Harrington R (2009) Application of the self-organizing map as a prediction tool for an integrated constructed wetland agroecosystem treating agricultural runoff. Bioresour Technol 100(2):559–565. https://doi.org/10.1016/J.BIORTECH.2008.06.042
Zhang Y, Liang X, Wang Z, Xu L (2015) A novel approach combining self-organizing map and parallel factor analysis for monitoring water quality of watersheds under non-point source pollution. Sci Rep 5(1):1–12. https://doi.org/10.1038/srep16079
Zhuang Y, Fu R (2024) Quantifying the contribution of atmospheric circulation to precipitation variability and changes in the US Great Plains and southwest using self-organizing map-analogue. Atmos Chem Phys 24(3):1641–1657. https://doi.org/10.5194/ACP-24-1641-2024

Hydromorphological Analysis for the Assessment and Evaluation of the Environmentally Minimum Lake Level

Charalampos Doulgeris

Abstract The determination of the environmentally minimum water level in lakes is crucial, as it provides water managers with a tool for the protection of lake ecosystems, although, it could also start a conflict with water uses. The assessment of minimum level is quite challenging and depends on biotic and abiotic factors, related to biological communities and hydromorphological features of the lake ecosystem. Herein, a hydromorphological method is presented for assessing the minimum lake level using the water level-volume relationship and the annual inflow of surface water into the lake. In addition, a hydrological evaluation of the assessed minimum level is applied to ensure that the minimum level could be safely used by water managers over a range of wet or dry hydrological years. The hydromorphological method is demonstrated in three lake ecosystems of Northern Greece, where the minimum level has also been estimated with other methods, based on a morphological analysis and on water level requirements of macrophytes species. The hydromorphological method is straightforward and provides estimations of minimum lake level that satisfy the key requirements of species.

Keywords Ecological lake level · Lake hydrology · Ecosystem approach · IWRM

1 Introduction

Freshwater ecosystems must be protected and managed properly, as, despite the significant technological progress, societies depend strongly on the capacity of ecosystems to produce goods and services for social and economic development. To this direction, integrated policies and strategies should be developed and implemented to resolve the complex and interrelated conservation and management problems of aquatic ecosystems. Integrated Water Resources Management (IWRM) can

C. Doulgeris (✉)
Soil & Water Resources Institute (SWRI), Hellenic Agricultural Organisation (ELGO-DIMITRA), Sindos, Greece
e-mail: ch.doulgeris@elgo.gr

K. Stefanidis et al. (eds.), *Lakes of the Eastern Mediterranean and Balkans*,
https://doi.org/10.1007/978-3-032-02168-7_8

be described as the coordinated development and management of water, land, and related resources to maximize the resulting economic and social welfare in an equitable manner without compromising the sustainability of vital ecosystems (Roy et al. 2011). IWRM is one form of the ecosystem approach as a strategy for the integrated management of land, water, and living resources that promotes conservation and sustainable use in an equitable way (Millennium Ecosystem Assessment 2005). Although ecosystem approach to IWRM gains ground (Roy et al. 2011; Doulgeris et al. 2012), its implementation is challenging, as the ecosystem approach should involve all relevant sectors of society and scientific disciplines (Secretariat of the Convention on Biological Diversity 2004). Furthermore, the ecosystem approach in catchments is based on the notion that water, biodiversity, and environmental protection require establishing interdisciplinary, inter-sectoral and inter-institutional initiatives (Al-Jayyousi and Bergkamp 2008).

The serious threats and risks that face many freshwater lakes worldwide, due to irrational water use and lack of effective management, jeopardise the provision of invaluable ecosystem services that are essential for human wellbeing. The environmentally minimum water level in lakes, or in other words, minimum lake level or ecological lake level, can be defined as the water level below which no further water withdrawal at its catchment occurs to further decrease lake's level. The determination of the environmentally minimum water level in lakes is critical, as it provides water managers and competent authorities with a decisive tool for the sustainable management and protection of their ecosystems. The assessment of lakes' environmentally minimum water level depends on biotic and abiotic factors of the lake ecosystem, related to biological communities and hydromorphological features. However, these factors might not be entirely known, and in addition, their relationship to lake level fluctuations is complex and not fully understood. Consequently, the assessment of environmentally minimum lake level is quite challenging, and thus, in many case studies expert judgement is adopted without avoiding biased estimations of the minimum level.

The methods for the environmentally minimum water level in lakes are categorized into morphological methods (Li et al. 2007; Shang 2013; Shang and Shang 2018), methods that analyse historical data of water level (Xu et al. 2004; Leeper and Ellison 2017; Wen et al. 2020), methods that combine hydrological and morphological features (Doulgeris et al. 2020a), habitat analysis methods and species-environment models (Abbaspour and Nazaridoust 2007; Cui et al. 2009; Mjelde et al. 2013; Yang et al. 2020; Yi et al. 2020). Shang (2013) proposed the lake surface area method that defines minimum lake level from the relationship of lake's water level-area-volume considering protecting ecosystems and meeting water demands. Doulgeris et al. (2020a) estimated minimum lake level by applying a hydromorphological analysis that combines the lake's hydrology and morphology, considering the water inflows from the lake's hydrological catchment and the lake's water level-volume relationship. Wang et al. (2020) applied six different methods to estimate the minimum water level in Lake Poyang (China) based on historical lake level data, analysis of morphological characteristics of the lake's bottom and water level requirements of fish fauna. Bao et al. (2019) analysed the hydrological regime of wetland systems of Momoge (China) and proposed suitable water depths to favour the growing of aquatic plants, benthic, phytoplankton and fish fauna. Petriki et al. (2020) assess the optimal water level fluctuation in four Mediterranean lakes to ensure the protection and sustainability of their biological communities by applying an analysis of hydromorphological features and expert judgment based on the ecological requirements of macrophyte and fish species. Yi et al. (2020) simulated the environmental conditions in shallow Lake Baiyangdian (China) using a habitat mathematical model of common reed (*Phragmites australis*), relating water level fluctuation and reed growth to estimate the lake's minimum ecological water level. Doulgeris et al. (2024) followed a multidisciplinary approach by combining morphological, hydrological and biological (macrophytes, benthic macroinvertebrates and fish) elements to determine the minimum water level in Lake Volvi (Greece).

The objective of this chapter is to present a hydromorphological analysis to assess and evaluate the environmentally minimum water level of lake ecosystems. In the following, the applied methodologies are presented, and three case studies of Mediterranean lakes are examined. The minimum lake level calculated by the hydromorphological method is compared with other methods based on a morphological analysis and on water level requirements of species, such as aquatic macrophytes.

2 Methodology

2.1 Hydrological Conditions in Lake Ecosystems

The study of hydrological conditions in lake ecosystems is quite crucial, not only for water resources management assessments, but also for environmental impact assessments related to the biological communities and ecosystem services of lake ecosystems. In addition, it is important to emphasize that these assessments are interlinked and mutually reinforcing. Understanding the hydrology of lake ecosystems is fundamental for making informed decisions and promoting the sustainable management of water resources and the preservation of important ecosystems. Some key points highlighting the above argument are the following:

Water Resources Management: Assessing the hydrological conditions in lake ecosystems provides valuable information for managing water resources effectively. Lakes are often a vital source of freshwater for various human activities, including drinking water supply, irrigation, and industrial use. Knowledge of the water balance, flow patterns, and water quality in lakes is essential for sustainable water resource planning and allocation.

Biological Communities: Hydrological conditions have a significant influence on the biological communities within lake ecosystems. Water level fluctuations, temperature variations, and nutrient dynamics are key factors that shape the habitat and influence the distribution and behaviour of aquatic species. Studying hydrology helps ecologists and conservationists understand and protect these communities and the biodiversity they support. For example, variations in water levels can influence the availability of suitable breeding and foraging areas for aquatic species.

Environmental Impact Assessments: When development projects, such as dams, urban expansion, or industrial activities, are proposed near or within lake ecosystems, it is essential to assess their potential environmental impact. This includes evaluating how changes in hydrological conditions might affect the lake and its surrounding environment. Environmental impact assessments help ensure that development projects are carried out sustainably, minimizing harm to the ecosystem and the services it provides.

Water Quality: Hydrological conditions can impact the water quality of lakes. Factors such as rainfall, runoff, and sedimentation patterns influence the concentration of pollutants, nutrients, and other substances in the water. Understanding these conditions is crucial for managing and improving water quality, which is not only important for aquatic life but also for human health and recreational activities.

Ecosystem Services: Lake ecosystems provide various ecosystem services, including water purification, recreational opportunities, and support for biodiversity. Monitoring and assessment of hydrological conditions helps ensure the continued provision of these services. For instance, understanding the flow of nutrients through the ecosystem can help maintain water quality and support the food web.

Climate Change Resilience: As climate change alters precipitation patterns and temperatures, lake ecosystems may be particularly vulnerable. Studying hydrological conditions allows scientists and policymakers to assess how these changes will impact lakes and develop strategies to enhance their resilience.

In short, the study of hydrological conditions in lake ecosystems is essential for a holistic understanding of these complex environments. It not only supports the sustainable management of water resources but also plays a vital role in safeguarding the ecological integrity and ecosystem services of lakes. In other words, it helps to ensure the long-term health and sustainability of these ecosystems, the communities that depend on them, and the services they provide to both the environment and society.

Hydrologic analysis and modelling in a lake catchment is usually developed to assess the water inflows and outflows of the lake (Fig. 1), taking also into account the pressures imposed in the catchment by the water users, i.e. irrigation, industry et cetera. More specifically, a rainfall-runoff model could be used to assess the water inflow from the catchment to the lake, using meteorological data on a monthly or daily time step. The rainfall-runoff model could be combined with a water balance model of the lake for the simulation of its water level. The water balance of a lake is simulated based on a simple algebraic equation that takes into account the water inflows and outflows:

$$V(i+1) = V(i) + Q_{in} - Q_{out} + P - E + G_{in} - G_{out} - Q_{users} \quad (1)$$

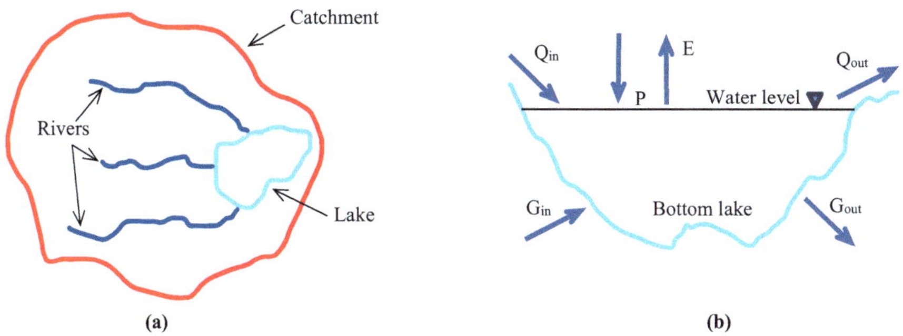

Fig. 1 Key hydrological elements in a lake catchment (**a**), water inflows-outflows in a lake (**b**)

where V is the volume of water stored in the lake in time i and i + 1, Q_{in} is the surface inflow into the lake from the catchment, Q_{out} is the surface outflow, P is precipitation and E is evaporation in/from lake's surface, G_{in}/G_{out} are the groundwater inflow/outflow, and Q_{users} is the water volume pumped by water users (if any). The terms in Eq. 1 are expressed in water volume units, e.g. m^3.

Equation 1 is usually applied to a monthly time step, but shorter intervals can also be used. At each time step in Eq. 1, the lake's water level can be easily found from the stored water volume using the water level–volume curve. Also, the precipitation and evaporation terms are calculated more accurately, if the lake surface area is updated at each time step, using the water level–surface area curve.

2.2 *Climate Change Impact*

Climate change is having a significant impact on lakes, particularly freshwater lakes, and one of the most pronounced and immediate effects is the modification of hydrological regime. This impact is mainly related to the decrease in rainfall and increase in temperature, as well as the combined effect of these changes on reducing lake catchment runoff. In addition, climate change may lead to shifts in precipitation patterns, which can result in more intense and erratic rainfall. This can lead to increased runoff into lakes, potentially causing water level fluctuations, erosion, and changes in water quality. Prolonged droughts, on the other hand, can lead to decreased water levels, which can have various ecological and socioeconomic consequences. Regarding changing temperature regimes, the rising temperatures may influence the thermal structure of lakes. Warmer surface waters can affect the mixing of the water column, which, in turn, can impact nutrient cycling and the distribution of aquatic species. In some cases, this can lead to the expansion of invasive species and alterations in the composition of biological communities. Also, higher temperatures can lead to increased evaporation from the lake surface, potentially causing water levels to drop.

Changes in temperature and precipitation patterns can influence nutrient loading into lakes. Increased runoff from agricultural areas due to heavy rainfall, for example, can lead to higher nutrient concentrations in lakes, contributing to eutrophication and harmful algal blooms. In addition, modifications to hydrological regimes can impact the overall water quality of lakes. Variations in temperature and flow can affect the distribution of pollutants and nutrients in the water, which has consequences for both aquatic life and the usability of the water for various purposes. In a nutshell, climate change is causing significant modifications to the hydrological regimes of lakes, and these changes have far-reaching ecological, social, and economic implications.

To assess the changes and the impacts caused by climate change in a lake ecosystem, a combination of climatic, hydrological and biodiversity models may be applied. The combined modelling framework, in conjunction with lake ecosystem monitoring, would enable the formulation of a series of measures for the mitigation of the impacts due to climate change. Several researchers have discussed the potential impacts of climate change on lakes (among others Woolway et al. 2020). Mediterranean lakes, in particular, face on-going climate impacts and specific threats, and consequently the risk of the loss of ecosystems services is imminent. Dimitriou and Mousoulis (2010) studied the impact of climate change at Trichonis Lake (Greece) for two climate scenarios and found that the lake water levels will show a decrease from 12 to 24.2 cm, and an increase of total nitrogen concentrations up to 10%. Fatoric et al. (2014) explored the climate change effects in Mediterranean protected coastal wetlands, Aiguamolls de l'Empordà (Spain) and Kotychi-Strofylia (Greece), and concluded that about two-thirds of the interviewed stakeholders in both areas perceived their coastal wetlands as unsustainable. Doulgeris et al. (2016) studied the impacts of climate change on the hydrology of Lake Cheimaditida and Lake Kerkini, Northern Greece, and concluded that their surface area will undergo a decrease from 20 to 37% and from 5 to 14%, respectively. Doulgeris et al. (2020b) studied nine climate scenarios and revealed contradictory water balances and associated water level fluctuations in Lake Vegoritida (Greece) for the projected period 2021–2050. Specifically, three scenarios show that the lake level will remain at its current level or higher while the other six scenarios show that a significant decline of lake level will occur.

Addressing and mitigating the impacts of climate change on lakes is essential for preserving these vital freshwater ecosystems and the services they provide to society. This includes adopting adaptive strategies, sustainable water management practices, and global efforts to reduce greenhouse gas emissions.

2.3 Minimum Lake Level

2.3.1 Morphological Method

The morphology of a lake ecosystem, i.e. the lake's shape and volume, is usually expressed by the hypsographic curve, which describes the relationship between water level and surface area, S(H), and the water level–volume curve, V(H). Lake surface area typically increases with its volume, but the increase rate described by the slope of surface area–volume relationship, S(V), may increase or decrease with the volume, depending on lake morphology. Based on the above, Shang (2013) proposed a morphological method for the assessment of minimum lake level considering a critical value of the lake volume storage. The critical volume storage corresponds to the inflection point of the surface area–volume curve and can be used for the determination of the minimum water level in lakes, similarly to the wetted perimeter method for assessing the minimum environmental flow in rivers (Gippel and Stewardson 1998).

Lake biodiversity increases with its surface area (Browne 1981), and thus, lake surface area could be an index for the protection of lake ecosystems. On the other hand, the water volume stored in a lake could be taken as an index of the available water to meet the water requirements of the economic activities in the catchment. The morphological method proposed by Shang (2013) attempts to consider both the protection of lake ecosystems and the fulfilment of water requirements by calculating the minimum water level using the following multi-objective optimization model:

$$\max\ H_1 = s(v) = S(V)/V_{max} \tag{2}$$

$$\min\ H_2 = v = V/V_{max} \tag{3}$$

where H is the water level, S and V are the lake surface area and volume, respectively. and s and v are the dimensionless surface area and volume, respectively. The first objective, Eq. 2, represents the maximisation of lake surface area, considering that the biodiversity of a lake is favoured as the lake surface area increases. The second objective, Eq. 3, represents the minimisation of lake storage to meet the water requirements to the greatest possible extent. The transformation of the multi-objective optimisation problem into a single objective problem is based on the ideal point method considering two non-negative weighted factors, w_1 and w_2, and is written as:

$$\min\{w_1 v + w_2[1 - s(v)]\} \tag{4}$$

Weighted factor w_1 expresses the minimization of lake volume to meet water requirements and w_2 expresses the maximization of lake surface area to provide more habitats to lake ecosystem. If the weighted factors have the same value, i.e., $w_1 = w_2 = 50\%$, the two objectives of the optimisation model are expressed equally. Doulgeris et al. (2017) applied the method in Mediterranean lakes and showed that a rational estimation of the minimum level is provided only when the adoption of an "environmental scenario", where $w_1 = 30\%$ and $w_2 = 70\%$, is followed. According to the "environmental scenario", the protection of the biodiversity is favoured against the water requirements of economic activities. Nevertheless, it must be mentioned that a key limitation of this method is that it depends exclusively on lake's morphological features and does not consider the lake's water potential, and particularly the water inflow into the lake from its catchment area.

2.3.2 Hydromorphological Method

The hydrological regime in a lake's catchment fundamentally determines its water level fluctuation. In many lake ecosystems, the key water balance parameter is the surface inflow into the lake from its catchment. Thus, the annual water volume that inflows to a lake from its catchment area can be considered as a critical value of the lake volume storage (V_{cr}) that can be used for the assessment of minimum water level in lakes. Based on this concept, Doulgeris et al. (2020a) proposed the hydromorphological method, in accordance with which, the minimum lake level (H_{min}) can be calculated by using the maximum volume of lake (V_{max}), the critical volume of lake (V_{cr}), and the water level–volume curve of the lake, expressed as $V = f(H)$. More specifically, the next two steps are followed to calculate the minimum lake level:

Algorithm 1 Procedure for assessing the minimum water level in lake ecosystems using the hydromorphological method

Calculate minimum volume by subtracting critical volume from maximum volume:

$$V_{min} = V_{max} - V_{cr}$$

Calculate minimum level from the water level–volume curve:

$$H_{min} = f^{-1}(V_{min})$$

Even though the above procedure is simple and straightforward, the estimation of critical volume (V_{cr}) might be under further consideration. As mentioned before, the lake's critical volume could be estimated by the annual water volume that inflows to the lake from its catchment; however, the annual inflow varies significantly from year to year depending on the hydrological and meteorological conditions in the lake's catchment that differ from wet to dry years. To obtain a representative value, a long time series (preferably at least for twenty years) of annual inflows to the lake from its catchment should be available. Considering that usually this kind of data is unavailable or incomplete, a simulated time series of annual inflows should be created. This could be done by studying the rainfall-runoff relationship of lake's catchment using a rainfall-runoff model. For most of the cases, it will be sufficient to apply the rainfall-runoff model using monthly meteorological data. In addition, observations of catchment runoff should be available to calibrate the rainfall-runoff model. Having the dataset of annual inflows to the lake from its catchment, either by observed or simulated catchment runoff, the critical volume (V_{cr}) could be estimated as the corresponding annual inflow with a probability of exceedance of a certain percentage, such as 50%, or in other words, as the median value of the annual inflow timeseries.

2.3.3 Biological Requirements of Species

A completely different approach to the above-mentioned methods is to estimate the minimum lake level based on the requirements of the lake's flora and fauna species on seasonal water level fluctuations by applying habitat analysis methods and species-environment models. Even thought, this approach sounds more appropriate from an ecological point of view, however, it could be quite challenging to be applied in lake ecosystems as different species might have different and contradictory requirements on water level fluctuation. In addition, these methods require extensive observation and long-term datasets for species that usually are not available. Thus, in many cases, scientists follow a semi-empirical approach based on their knowledge and previous experience from working in specific lake ecosystems. This approach is usually applied to species of macrophytes, fish fauna, avian fauna and benthic macroinvertebrates. Initially, it is examined what the species are and their population or density, accordingly; then, the seasonal water level requirements of species, and especially the most sensitive ones to changes in water level, are determined; and finally, based on previous experience and expert judgment, the optimal water level regime and the minimum level is estimated. This process may involve a degree of biased assessment of minimum lake level driven by the expert's belief to promote environmental protection and ecosystems preservation over meeting the current or future water needs of economic activities, such as irrigated agriculture. Nevertheless, the analysis of species' requirements is quite important as it highlights critical periods of the year, during which the water level fluctuation must be as small as possible to safeguard the ecological integrity and ecosystem services of lakes.

2.3.4 Evaluation of Assessed Minimum Level

The environmentally minimum lake level, regardless of the methodology applied to its estimation, should also be hydrologically evaluated. The hydrological evaluation is required as the proposed minimum lake level does not vary from year to year, considering that otherwise it could be impractical to be applied by the water managers of a lake ecosystem. However, water level fluctuation is directly related to the water volume that inflows to the lake from its hydrological catchment, which varies significantly between a wet and a dry hydrological year. Thus, it is necessary to check if water inflow is sufficient to cause water level rising from the minimum level to the maximum (overflow) level to ensure renewal of lake's water and conserve water balance for ecosystems.

In this context, the lake level at the beginning of each hydrological year (1st of October) is assumed to be at the lake's environmentally minimum level. Then, based on the surface inflow (Q_{in}) of each year and for a period of at least twenty years, it is checked how many years the water level is rising to the maximum level. It is anticipated that for at least half of the years, the water level should rise to its maximum level. Furthermore, for the years that the lake level will rise to the maximum level, the excess of water volume could be considered as "overflow"; for the other years, that the lake level will not rise to the maximum level, the water volume that is missing to approach the maximum level could be considered as "deficit". Particularly, the calculation steps are (Doulgeris et al. 2020a):

Algorithm 2 Procedure for hydrological evaluation of the proposed minimum water level in lake ecosystems

Calculation of minimum volume from minimum level and the water level–volume curve:

$$V_{min} = f(H_{min})$$

Calculation of water volume to rise the lake's level from minimum to maximum level:

$$V_r = V_{max} - V_{min}$$

For each Year
Calculation of the "overflow" or "deficit" lake's water volume:

If $Q_{in} > V_r$ *Then* overflow $= Q_{in} - V_r$
If $Q_{in} < V_r$ *Then* deficit $= Q_{in} - V_r$

End Year

The above simple procedure can also be used to compare methods that assess the environmentally minimum level, regarding their effectiveness in the renewal of lake's water.

3 Case Studies

3.1 Introduction

In this section, the hydromorphological method is applied to assess the environmentally minimum level in three lake ecosystems of Northern Greece, namely Cheimaditida, Vegoritida and Volvi (Fig. 2). In addition, the minimum level assessed by the hydromorphological method is hydrologically evaluated and compared with the minimum level assessed by the morphological method and the analysis of water level requirements of macrophyte species. The morphology of the three lakes is sufficiently described by the water level-surface area-stored volume curves (Fig. 3). The quality of level-area-volume data is crucial for estimating the minimum lakes' level using the hydromorphological method, as well as to evaluate hydrologically the proposed minimum level. The hydromorphological features of the three lakes are given in Table 1.

Fig. 2 Case studies of lake ecosystems for the assessment of the environmentally minimum water level with the hydromorphological method

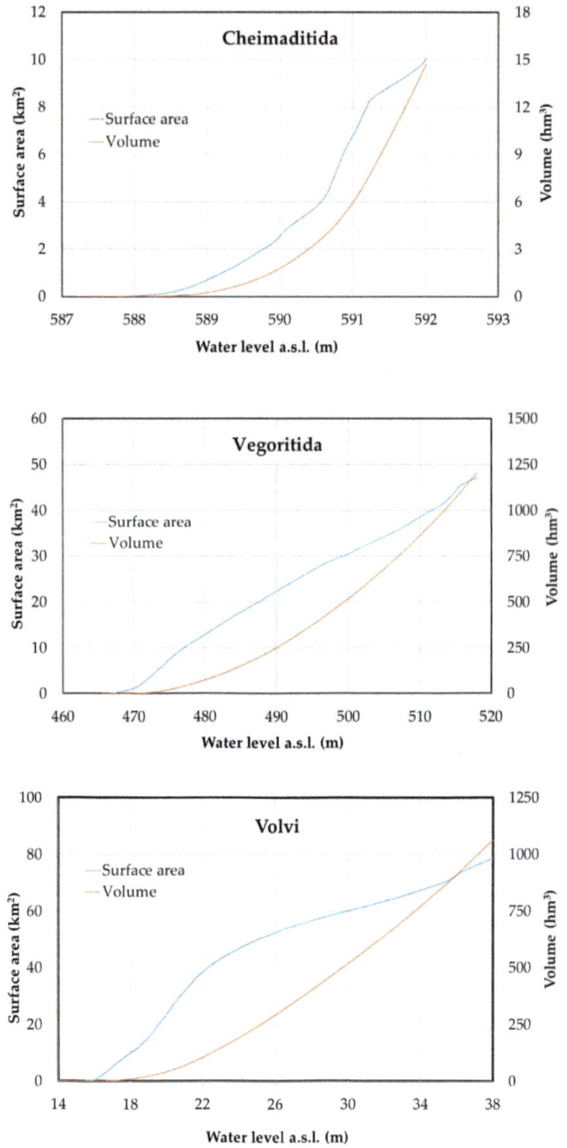

Fig. 3 Water level–surface area–volume curves in Lakes Cheimaditida, Vegoritida and Volvi (*source data* Greek Biotope/Wetland Centre—National Monitoring Network)

3.2 *Lake Cheimaditida*

Lake Cheimaditida is located in the water district of Western Macedonia, Northern Greece, and is part of the Natura 2000 network. It covers an area of 10.1 km^2 and holds a water volume of 14.7 hm^3 at its maximum water level (592 m a.s.l.). Its average water depth is 1.5 m and its maximum depth is approximately 4.5 m. Its hydrological catchment covers an area of 228 km^2 and includes also Lake Zazari

Table 1 Hydromorphological features in Lakes Cheimaditida, Vegoritida and Volvi

	Cheimaditida	Vegoritida	Volvi
Maximum water level a.s.l. (m)	592	518	36
Maximum surface area (km^2)	10.1	47.2	72.5
Maximum stored volume (hm^3)	14.7	1206	911
Average depth (m)	1.5	26	12.5
Maximum depth (m)	4.5	52	21

and River Sklithros. Lake Zazari is located 2 km northwest of Lake Cheimaditida and overflows through a weir and a joining canal into the Lake Cheimaditida. River Sklithros constitutes the main source of surface water for lakes Zazari and Cheimaditida. The economy in the catchment is mainly based on farming and livestock while the income from fishing is low. Irrigation water is abstracted from the lakes and groundwater.

The water level in Lake Cheimaditida fluctuates between 591 and 592 m (a.s.l.) according to the data provided by the Greek Biotope/Wetland Centre for the period 2012–2022. The minimum water level is usually observed during October and the maximum during April. The water balance and rainfall-runoff relationship were simulated in Lakes Zazari and Cheimaditida for the period 1980–2015 using the MIKE BASIN model and the NAM rainfall-runoff model (Doulgeris et al. 2020b). Based on this, the inflow of surface water into Lake Cheimaditida was estimated for a period of 35 hydrological years and is given in Fig. 4.

In Table 2 is given the environmentally minimum water level in Lake Cheimaditida using the hydromorphological method. Initially, the critical water volume is

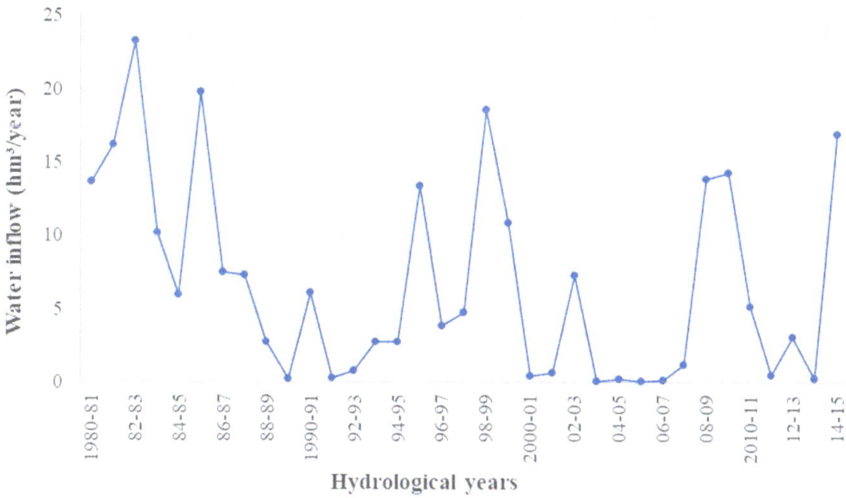

Fig. 4 Inflow of surface water into Lake Cheimaditida

Table 2 Minimum level in Lake Cheimaditida using the hydromorphological method

Critical volume, hm^3 (from Fig. 4)	4.77
Minimum volume, hm^3 ($V_{max} - V_{cr}$)	9.93
Minimum level, m a.s.l. (from Fig. 3)	591.5
Water level drop, cm	50
Decrease in surface area, %	12.3
Decrease in volume, %	32.4

found as the median value of the annual inflow timeseries. Then, the minimum water level is calculated from minimum volume and the water level-volume curve. At the minimum level of 571.5 m, which means that the water level could drop up to 50 cm in a hydrological year, the surface area decreases 12.3% and the water volume 32.4%, compared to the corresponding maximum values. Figure 5 shows the hydrological evaluation of minimum level in Lake Cheimaditida following the procedure described in paragraph 2.3.4. We notice that for 20 out of 35 years, the lake overflows downstream as the water level approaches its maximum level, favouring the lake's water renewal. Using the morphological method, Doulgeris et al. (2020a) estimated the minimum level to 591.3 m, i.e. 20 cm lower than the hydromorphological method, while Zervas (2016) and Petriki et al. (2020) applied an analysis of water level requirements of macrophyte species and proposed the minimum level to 591.5 m during February–May and 591 m from June to January.

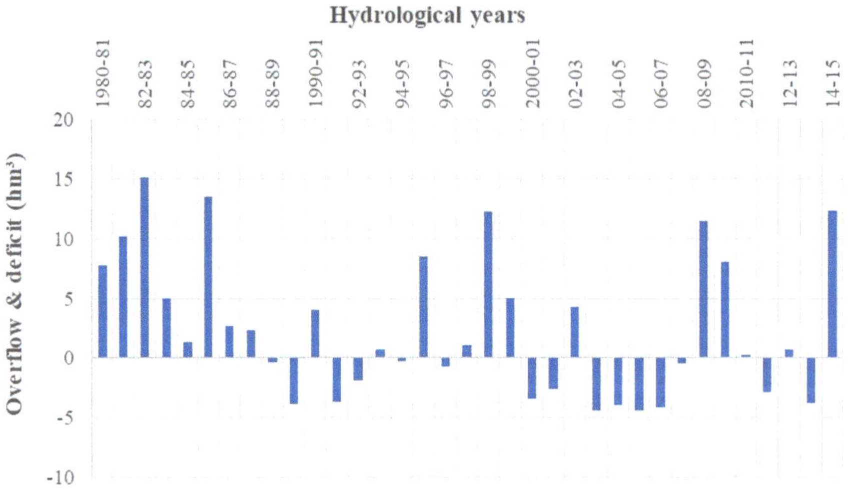

Fig. 5 Hydrological evaluation of minimum level in Lake Cheimaditida

3.3 *Lake Vegoritida*

Lake Vegoritida is in the water district of Western Macedonia in Northern Greece. Lake Vegoritida is part of Natura 2000 network and is included in a Site of Community Importance area (SCI), according to the Council Directive 92/43/EEC. Its hydrological catchment covers an area of 2145 km^2 with an average annual precipitation of 579 mm and an average temperature of 11.4 °C, for the historical period 1980–2015. Key water inflows in Lake Vegoritida are the excess of surface water from Lake Petron flowing to Vegoritida through a tunnel, and the water flow of River Pentavryso. A process that also influences the water balance of the lake is associated with the existence of karst sinkholes (katavothres) as a characteristic of its limestone bed (Antonopoulos and Gianniou 2003). The economic activities in the catchment that exert pressures in the lake ecosystem are mainly associated nowadays with agriculture. However, the water level in Lake Vegoritida has undergone great changes in the past decades caused by severe water abstraction by the Public Power Corporation directly from the lake and its catchment (Doulgeris et al. 2020b), affecting not only the conditions of natural environment, but also the social and economic ones. Nevertheless, the rising of water level triggered a discussion for the lake's maximum water level that becomes a source of conflict among stakeholders (Doulgeris and Argyroudi 2019).

The water level in Lake Vegoritida fluctuates between 514 and 519 m (a.s.l.) according to the data provided by the Greek Biotope/Wetland Centre for the period 2012–2022. However, the last three years, the water level varies from 516.5 to 518 m. The lake covers an area of 47.2 km^2 and holds a water volume of 1206 hm^3 at its maximum water level (518 m a.s.l.), while its average water depth is 26 m and its maximum depth is 52 m. The water balance and rainfall-runoff relationship were simulated in Lake Vegoritida, as well as in Lakes Petron, Cheimaditida and Zazari that are encompassed in the hydrological catchment of Lake Vegoritida, for the period 1980–2015 using the MIKE BASIN model and the NAM rainfall-runoff model (Doulgeris et al. 2020b). Based on the above analysis, the inflow of surface water into Lake Vegoritida was estimated for a period of 35 hydrological years and is given in Fig. 6.

In Table 3 is given the environmentally minimum water level in Lake Vegoritida using the hydromorphological method. The critical water volume is found as the median value of the annual inflow timeseries and the minimum water level is calculated from minimum volume and the water level-volume curve. At the minimum level of 517.6 m, i.e. the water level drops up to 40 cm in a hydrological year, the surface area and the volume decrease 0.6% and 1.6% respectively, compared to the maximum ones. Figure 7 shows the hydrological evaluation of minimum level in Lake Vegoritida (see also paragraph 2.3.4), based on which the water level approaches the maximum level for 17 out of 35 years. Using the morphological method, Doulgeris et al. (2020a) estimated the minimum level to 516.2 m, i.e. 1.4 m lower than the hydromorphological method, while Zervas (2016) and Petriki et al. (2020) applied

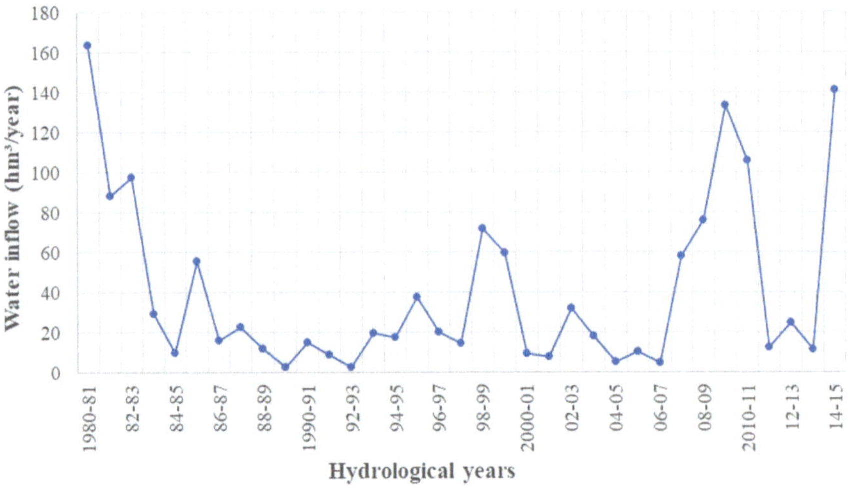

Fig. 6 Inflow of surface water into Lake Vegoritida

Table 3 Minimum level in Lake Vegoritida using the hydromorphological method

Critical volume, hm^3 (from Fig. 6)	19.5
Minimum volume, hm^3 ($V_{max} - V_{cr}$)	1186.5
Minimum level, m a.s.l. (from Fig. 3)	517.6
Water level drop, cm	40
Decrease in surface area, %	0.6
Decrease in volume, %	1.6

an analysis of water level requirements of macrophyte species and proposed the minimum level to 517.5 m during February-May and 517 m from June to January.

3.4 *Lake Volvi*

Lake Volvi is in the Water District of Central Macedonia, Northern Greece, at a distance of around 50 km east of the city of Thessaloniki. It has a surface area of 72.5 km^2 and holds a water volume of 911 hm^3 at the water level of 36 m a.s.l. Its average water depth is 12.5 m and its maximum depth is 21 m. Lake Volvi with the neighboring Lake Koronia are protected under the Ramsar Convention as wetlands of international importance. Their hydrological catchment drains an area of 2061 km^2, while the sub-catchment area of Lake Volvi is 1293 km^2. Lake Volvi outflows to River Richios that ends up outflowing to Strymonikos Gulf. The lake's catchment is sparsely populated, with 31,059 inhabitants distributed across 30 settlements. The

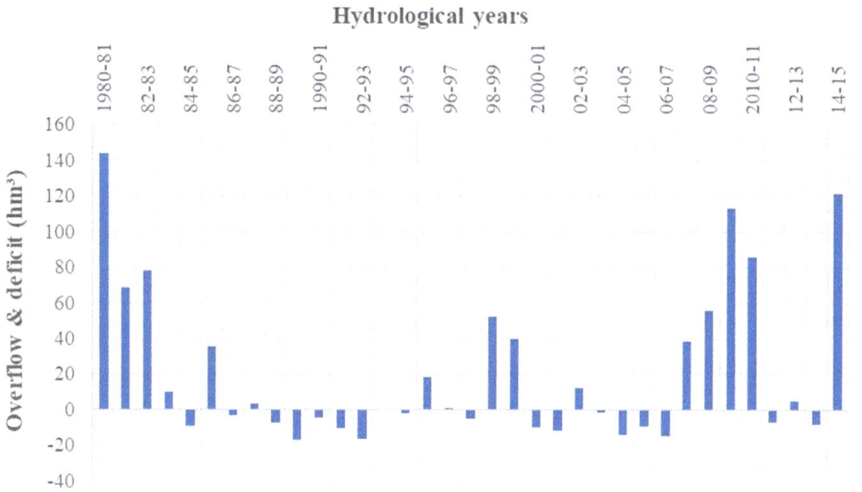

Fig. 7 Hydrological evaluation of minimum level in Lake Vegoritida

climate is characterized by mild winters and dry, hot summers, with an average annual temperature of 13.3 °C. The average annual precipitation is 519 mm.

Morphological, hydrological and meteorological data for Lake Volvi were available from the Greek Biotope/Wetland Centre and Malamataris (2019). The water level in Lake Volvi fluctuates between 34.2 m and 37.5 m (a.s.l.), based on the data for the periods 1985–1994 and 2012–2022. The minimum water level is usually observed during September–November, with an average value of 35.3 m, while the maximum level is observed in March–April, with an average of 36 m. The lake's water balance and rainfall-runoff modelling were studied for the period 2015–2019 using the MIKE HYDRO BASIN model and the NAM rainfall-runoff model (Doulgeris et al. 2024). Based on the above analysis, the annual volume of surface water inflow to the lake from its hydrological catchment, is estimated to 7.5% of the volume of annual rainfall received by the lake's catchment. The inflow of surface water into Lake Volvi (Fig. 8) was estimated for a period of 30 hydrological years (1970/1971–1999/2000), considering that a long timeseries of a sufficient number of meteorological stations were available for this period.

In Table 4 is given the minimum water level in Lake Volvi using the hydromorphological method. For Lake Volvi, the critical water volume is calculated either as the median value of the annual inflow timeseries, which corresponds to a 50% probability of exceedance of annual inflow, or as a 75% probability of exceedance of annual inflow. For the two values of critical volume, two values of the minimum level are calculated, 35.3 m (for 50%) and 35.4 m (for 75%), and as expected, the minimum level is higher (10 cm) for the 75% probability of exceedance. Even though the two values of minimum level are not quite different, the hydrological evaluation of minimum level (Fig. 9), reveals that for the 50% probability of exceedance, the water level approaches the maximum level only for 15 out of 35 years, while for the

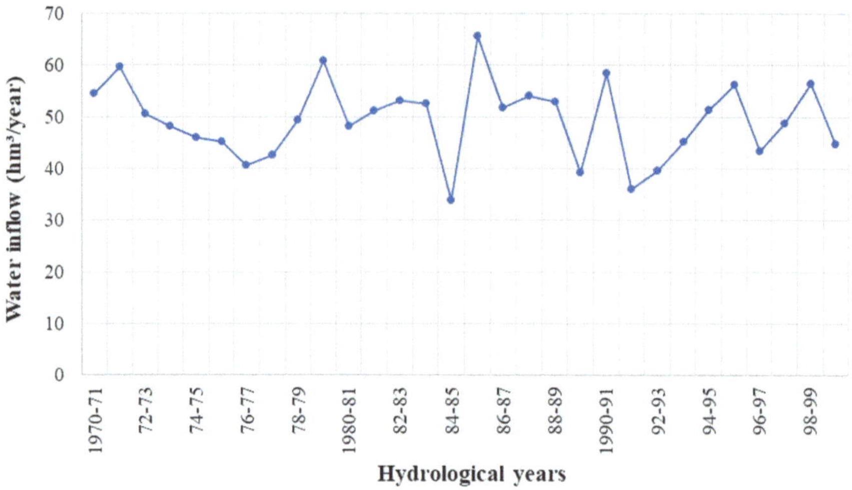

Fig. 8 Inflow of surface water into Lake Volvi

75% probability of exceedance, the number of years increased to 22 out of 35 years, which significantly increases the renewal rate of lake's water.

Using the morphological method, the minimum level was estimated to 32.5 m, i.e. almost two meters lower than both estimations of the hydromorphological method, while applying an analysis of water level requirements of macrophyte species, the minimum level was estimated to 35.8 m during February-May and 35.3 m from June to January, similarly to the estimations of the hydromorphological method (Doulgeris et al. 2024).

Table 4 Minimum level in Lake Volvi using the hydromorphological method

	*50% (median value)	*75%
Critical volume, hm^3 (from Fig. 8)	50	44.7
Minimum volume, hm^3 ($V_{max} - V_{cr}$)	861	866.3
Minimum level, m a.s.l. (from Fig. 3)	35.3	35.4
Water level drop, cm	70	60
Surface area decrease, %	3.2	2.9
Stored volume decrease, %	5.5	4.9

* Probability of exceedance of annual inflow

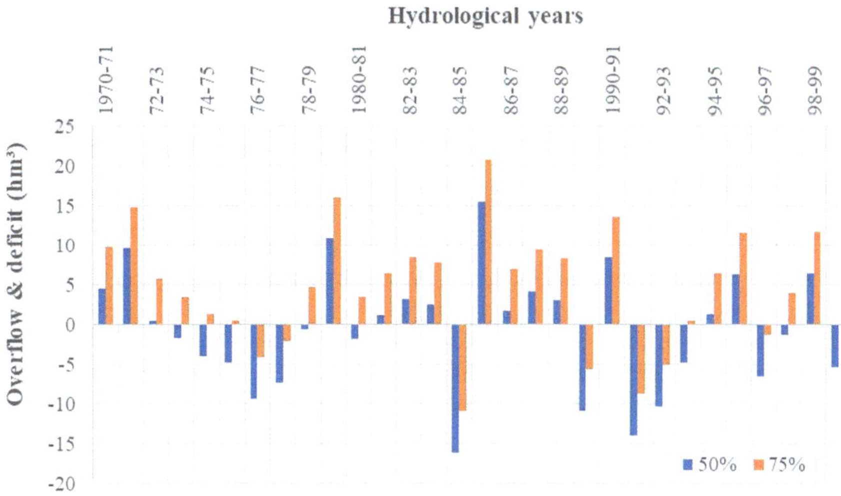

Fig. 9 Hydrological evaluation of minimum level in Lake Volvi

4 Discussion

The demonstration of the hydromorphological method in Lakes Cheimaditida, Vegoritida and Volvi showed that the environmentally minimum level should be at 50, 40 and 60 cm, respectively, below the maximum (overflow) water level. The hydromorphological method is straightforward and requires data that usually are available or can be estimated. Specifically, it requires the water level-volume relationship and the annual inflow of surface water into the lake. To obtain a representative value of the annual surface inflow might be quite challenging as it will require a well-established monitoring network that usually is not available. To overcome this, a time series of annual inflows could be simulated by using hydrological models, which are well-developed tools with a long tradition of application in water resources management.

The hydrological evaluation of proposed minimum level should always be applied to ensure that the minimum level could be safely used by water managers over a range of wet or dry hydrological years. The hydrological evaluation verified that the water potential of hydrological catchments in Lakes Cheimaditida, Vegoritida and Volvi is sufficient to raise the lake level from the minimum level, proposed by the hydromorphological method, to the maximum level. Particularly for Lake Volvi, the hydrological evaluation indicated that a slightly higher minimum level will ensure a significant increase in the renewal rate of the lake's water.

The fluctuation of water level in lakes could be quite different from year to year. Some of the measures potentially applied by water managers, to prevent that water level will be below the proposed minimum level, are related to the systematic control of water withdrawals, e.g. for irrigation, either directly from the lake or from its catchment, the monthly lake level monitoring, and the regular flow measurement of rivers in the lake's catchment.

Freshwater bodies are a source of water that supports the economic activities of society, such as irrigated agriculture, industry, etc. However, there is an ethical commitment, which is adopted by legislation and international treaties, that modern societies should promote sustainability and maintain aquatic ecosystems. In rivers, the concept of Environmental Flow Requirement (in terms of minimum flow in the river) has been widely used to describe the water regime necessary for the lotic ecosystems' maintenance. Similarly, in lakes, the Environmentally Minimum Water Level could be used to determine the minimum lake level for the lentic ecosystems' maintenance. In the context of water resources management, the minimum flow or level is usually a conflict negotiation with water uses, which is a case-specific compromise that decision-makers should resolve. To this direction, the hydromorphological method can be a valuable tool to assess simply and reliably the Environmentally Minimum Water Level in lake ecosystems.

References

Abbaspour M, Nazaridoust A (2007) Determination of environmental water requirements of Lake Urmia, Iran: an ecological approach. Int J Environ Stud 64:161–169. https://doi.org/10.1080/00207230701238416

Al-Jayyousi O, Bergkamp G (2008) Water management in the Jordan river basin: towards an ecosystem approach. In: Varis O, Biswas AK, Tortajada C (eds) Management of transboundary rivers and lakes. Springer, Heidelberg, pp 105–121

Antonopoulos V, Gianniou S (2003) Simulation of water temperature and dissolved oxygen distribution in Lake Vegoritis, Greece. Ecol Model 160:39–53. https://doi.org/10.1016/S0304-3800(02)00286-7

Bao K, Liu J, Meng B, Sun B (2019) The effects of hydrological conditions on eco-exergy of food webs in Momoge National Nature Reserve, China. Wetlands 39:601–617. https://doi.org/10.1007/s13157-018-1111-9

Browne RA (1981) Lakes as islands: Biogeographic distribution, turnover rates, and species composition in the lakes of central New York. J Biogeogr 8:75–83. https://doi.org/10.2307/2844594

Cui B, Tang N, Zhao X, Bai JA (2009) Management-oriented valuation method to determine ecological water requirement for wetlands in the Yellow River Delta of China. J Nat Conserv 17:129–141. https://doi.org/10.1016/j.jnc.2009.01.003

Dimitriou E, Moussoulis E (2010) Hydrological and nitrogen distributed catchment modeling to assess the impact of future climate change at Trichonis Lake, western Greece. Hydrogeol J 18:441–454. https://doi.org/10.1007/s10040-009-0535-y

Doulgeris C, Argyroudi A (2019) Contribution to decision-making on establishing the maximum water level for Lake Vegoritida. Lakes Reserv: Res Manag 24:24–36. https://doi.org/10.1111/lre.12249

Doulgeris C, Georgiou P, Apostolakis A, Papadimos D, Zervas D, Petriki O, Bobori D, Papamichail D, Antonopoulos V, Farcas C, Stålnacke P (2017) Assessment of the environmentally minimum lake level based on morphological features. Eur Water 58:197–202. https://www.ewra.net/ew/pdf/EW_2017_58_29.pdf

Doulgeris C, Papadimos D, Kapsomenakis J (2016) Impacts of climate change on the hydrology of two Natura 2000 sites in Northern Greece. Reg Environ Change 16:1941–1950. https://doi.org/10.1007/s10113-015-0857-3

Doulgeris C, Georgiou P, Papadimos D, Papamichail D (2012) Ecosystem Approach to Water Resources Management using the MIKE 11 modeling system in the Strymonas River and Lake Kerkini. J Environ Manage 94(1):132–143. https://doi.org/10.1016/j.jenvman.2011.06.023

Doulgeris C, Koukouli P, Georgiou P, Dalampakis P, Karpouzos D (2020a) Assessment of minimum water level in lakes and reservoirs based on their morphological and hydrological features. Hydrology 7(4):1–14, 83. https://doi.org/10.3390/hydrology7040083

Doulgeris C, Pisinaras V, Argyroudi A, Tziritis E, Hatzigiannakis E, Panagopoulos A (2020b) Impacts on water level fluctuation in Lake Vegoritida: insights for the historical and projected climate period. Desalin Water Treat 194:358–368. https://doi.org/10.5004/dwt.2020.25756

Doulgeris C, Ntislidou C, Petriki O, Zervas D, Nikolaidou R, Bobori D (2024) Assessment of environmentally minimum water level in a Mediterranean lake using morphological, hydrological and biological factors (in press)

Fatorić S, Morén-Alegret R, Kasimis C (2014) Exploring climate change effects in Euro-Mediterranean protected coastal wetlands: the cases of Aiguamolls de l'Empordà, Spain and Kotychi-Strofylia, Greece. Int J Sust Dev World 21(4):346–360. https://doi.org/10.1080/13504509.2014.888377

Gippel CJ, Stewardson MJ (1998) Use of wetted perimeter in defining minimum environmental flows. Regul Rivers: Res Mgmt 14:53–67. https://doi.org/10.1002/(SICI)1099-1646(199801/02)14:1%3c53::AID-RRR476%3e3.0.CO;2-Z

Leeper D, Ellison D (2017) Minimum levels and proposed guidance for Lake Hancock in Polk County, Florida. Southwest Florida Water Management District: Brooksville, Florida

Li X, Song Y, Li Y, Xing X, Zhang S (2007) Calculation methods of lowest ecological water level of lake. Arid Land Geogr 30:526–530

Malamataris D (2019) Integrated water resources management and sustainable development strategy in a changing climate: the case study of the Mygdonia water basin. Ph.D. Thesis, Aristotle University of Thessaloniki

Millennium Ecosystem Assessment (2005) Ecosystems and human well-being: synthesis. Available via http://www.millenniumassessment.org/documents/document.356.aspx.pdf. Accessed 7 Nov 2009

Mjelde M, Hellsten S, Ecke FA (2013) Water level drawdown index for aquatic macrophytes in Nordic lakes. Hydrobiologia 704:141–151. https://doi.org/10.1007/s10750-012-1323-6

Petriki O, Zervas D, Doulgeris C, Bobori D (2020) Assessing the ecological water level: the case of four mediterranean lakes. Water 12(11):1–22, 2977. https://doi.org/10.3390/w12112977

Roy D, Barr J, Venema HD (2011) Ecosystem approaches in integrated water resources management (IWRM): a review of Transboundary River Basins. UNEP-DHI, IISD, p 86

Secretariat of the Convention on Biological Diversity (2004) The ecosystem approach. Secretariat of the Convention on Biological Diversity, CBD Guidelines, Montreal, p 50

Shang S (2013) Lake surface area method to define minimum ecological lake level from level-area-storage curves. J Arid Land 5:133–142. https://doi.org/10.1007/s40333-013-0153-3

Shang S, Shang S (2018) Simplified lake surface area method for the minimum ecological water level of lakes and wetlands. Water 10:1056. https://doi.org/10.3390/w10081056

Wang L, Li ZQ, Liu C, Wang S, Li B, Liu X, Han Z (2020) Research on ecological water level of open lakes: a case study of Poyang Lake. In: 5th International conference on smart grid and electrical automation (ICSGEA), pp 606–610. https://doi.org/10.1109/ICSGEA51094.2020.00137

Wen D, Jin C, Shixiang G, Gang C, Jinming C, Mi Z (2020) Hydrological variability of water level of Dianchi Lake and its application. IOP Conf Ser: Mater Sci Eng 780:062048. https://doi.org/10.1088/1757-899X/780/6/062048

Woolway RI, Kraemer BM, Lenters JD et al (2020) Global lake responses to climate change. Nat Rev Earth Environ 1:388–403. https://doi.org/10.1038/s43017-020-0067-5

Xu Z, Chen M, Dong Z (2004) Researches on the calculation methods of the lowest ecological water level of lake. Acta Ecol Sin 24:2324–2328

Yang W, Xu M, Li R, Zhang L, Deng Q (2020) Estimating the ecological water levels of shallow lakes: a case study in Tangxun Lake, China. Sci Rep 10:5637. https://doi.org/10.1038/s41598-020-62454-5

Yi Y, Xie H, Yang Y, Zhou Y, Yang Z (2020) Suitable habitat mathematical model of common reed (*Phragmites australis*) in shallow lakes with coupling cellular automaton and modified logistic function. Ecol Model 419:108938. https://doi.org/10.1016/j.ecolmodel.2020.108938

Zervas D (2016) Analysis of aquatic macrophytes requirements on minimum seasonal water level and water quality in Lakes Zazari, Cheimaditida, Petron and Vegoritida (D.A1.2). Greek Biotope/Wetland Centre (EKBY), Thermi, Greece, p 63 + 2 maps

Ecosystem Services of Mediterranean Lakes: Management Implications

Ifigenia Kagalou and Dionissis Latinopoulos

Abstract Freshwaters and especially lake ecosystems, when in good status, provide multiple benefits and services to societies, making them essential for assisting in reaching several Sustainable Development Goals (SDGs). However, the ability of these ecosystems to provide services is not unlimited but is dependent on the lake-type, exploitation extent and its full functioning. Mediterranean lakes are the most threatened ecosystems and of particular importance, regarding their values and services such as flood protection, biodiversity hosting, resources provision, water quality improvement, aquifer or groundwater storage and recharge. At the same time multiple pressures, usually synergistic, threaten their integrity and biodiversity, imposing new challenges to lake ecosystem management and restoration. Thus, lake managers are required to address and mitigate the impact of multiple pressures and to disentangle such effects by selecting eco-friendly practices, without affecting the lakes' current uses. It is important that managers should provide adaptive management plans to respond before important environmental thresholds threaten human livelihoods, and ecosystem sustainability. The adoption of Ecosystem Services (ES) Approach in water management offers an added value by providing a framework for conceptualizing the link between the environment per se and the ways people value and 'use' it. This chapter aims to advance our knowledge on how to integrate the ES concept into lake management and governance, how these numerous ES are related to basic limnological functions, and how these are linked to biodiversity and water security.

Keywords Ecosystem services · Sustainable development goals · Water framework directive · Lake management

---

I. Kagalou (✉) · D. Latinopoulos
Department of Civil Engineering, School of Engineering, Democritus University of Thrace, Xanthi, Greece
e-mail: ikagkalo@civil.duth.gr

D. Latinopoulos
e-mail: dlatinop@civil.duth.gr

K. Stefanidis et al. (eds.), *Lakes of the Eastern Mediterranean and Balkans*,
https://doi.org/10.1007/978-3-032-02168-7_9

1 The Importance of Freshwater Ecosystems

The immense importance of freshwater systems extends beyond climate regulation, nutrient cycling, carbon sequestration, biodiversity protection, resource provision and recreation among others (i.e. Ebner et al. 2022; Santos et al. 2018; Grizzetti et al. 2016). However, freshwater ecosystems are among the most vulnerable systems, especially to biodiversity loss (Dudgeon et al. 2006) that have been severely impacted by anthropogenic activities (including a widespread loss of water bodies) in their basins, a tendency that peaked after the 1940s (Haase et al. 2023).

Freshwater systems face several threats working in a synergistic manner effect manifest in temporal and spatial scales while this situation is expected to intensify in the future (Eros et al. 2023). Diffuse pollution from agriculture is the most common pressure affecting surface and groundwaters. Urban areas are primary sources of micropollutants, hotspots for non-native species invasions, and generators of high-nutrient inputs while diffuse atmospheric pollution remains a significant pathway for contamination of surface waters by mercury and polycyclic aromatic hydrocarbons (PAHs). In contrast, croplands contribute to fine sediment, pesticides, nutrient-laden runoff, and play a significant role in salinization. Threats posed—mostly—by anthropogenic activities can initiate sometimes irreversible ecosystem feedbacks, surpassing self-purification abilities or even tipping points (Woolway et al. 2022). Eutrophication is such a paradigm, a complex term linked to heightened productivity, simplification of biotic communities, and a diminished capacity for organisms to adapt to the introduced nutrient load (Grizzetti and Poikane 2024). Despite a plethora of environmental directives, regulations, and action taken at regional and global levels, such as the Water Framework Directive (WFD, 2000/60/EC), the Birds Directive and the Habitats Directive (EU 1992) in Europe, or the global Strategic Plan for Biodiversity (EU, Biodiversity Strategy 2030) a weak progress has been reported (EEA 2024) while anthropogenic risks have not yet been tackled satisfactorily, biodiversity continues to decline reflecting at the "poor ecological status" (EEA 2024) and " poor conservation status (Haase et al. 2023).

Freshwaters are strongly connected with their neighboring terrestrial ecosystems relating to matter and energy (Klimaszyk and Goldyn 2020). Freshwater communities located downstream of important anthropogenic pressures like pollution or hydromorphological alterations (cities, dams, agricultural areas) have less chances for actual recovery (Haase et al. 2023). Another inhibiting factor is climate change, and more importantly the warming rates that negatively affect community structure indicators (Haase et al. 2023).Climate change amplifies pollution effects and vice versa water scarcity can also be aggravated by water pollution where it is well known that floods have negative impacts as a result of pollution from surface run-off or sewage overflow (Bisselink et al. 2020).

Acknowledging the risks to Earth system stability and human well being freshwater has been identified as one of the planetary boundaries (Gleeson et al. 2020), and further safe and just Earth (water) system Boundaries (ESVs) (Rockström et al. 2023) were defined primarily based on two fundamental safeguarding mechanisms:

(a) the flow alteration in rivers which affects the life on terrestrial and marine systems (Liu et al. 2021) and (b) the drawdown of the groundwater since groundwater aquifers sustain groundwater dependent ecosystems and further terrestrial vegetation (Kath et al. 2018). The challenge for achieving ecological sustainability of freshwaters is its recognition as "global common" associated with the delivering of crucial Ecosystem Services.

2 Lakes Functioning—Mechanisms—Climate Feedbacks

Lake types may differ in their internal mechanisms and services provision, meaning reservoirs have different tolerance in external perturbations [as mass and energy receptors (May et al. 2022)], residence time and sedimentation rates than shallow lakes, but their biological elements always act as proxies for understanding ecosystem functioning. Functioning though, may be impeded by lake aging (human enhanced- or physical). Series of mechanisms govern lake communities and ultimately lake functioning. Those elements may include a wide range of both biotic and abiotic parameters that can be assessed in different time and spatial scales, i.e. for shallow lakes, alternative stable states (plankton/ macrophyte) govern different functionings and hence provide different ecosystem services (Janssen et al. 2021).

Algae and cyanobacteria are the cornerstone of lake food webs, acknowledging also the importance of microbial loop in nutrients and material circulation. It is long known that fast (shock) changes in the environmental gradient can cause important alterations in their community composition, and hence intervene with primary production/respiration rates (Reynolds 1984). This change affects (in different time scales) ecosystem functioning and services provision (Carpenter et al. 1998). Long-term procedures also, like eutrophication, offer inputs that often exceed the capacity of the ecosystem to be "balanced", forming thus a volatile environment (Grizzetti and Poikane 2024).

So, *environmental variability* plays a major role in ecosystem functioning. The amount of inputs (nutrients, carbon, energy) and the processing (heterotrophic and autotrophic) ability of lake systems define the balance through food web structure, community composition and functioning governing ultimately the systems' productivity (Hessen et al. 2023). A vicious cycle is then formed where all (biotic and abiotic elements) are affected by the drivers and in turn affect the processes. For example, the accumulation of Phosphorus in EU lakes during the past decades from untreated wastewaters and agricultural practices (both animal breeding and crops) and the documented loading increase, have assisted to pinpoint thresholds beyond which shift to plankton turbid dominated state in the long term (Hessen et al. 2023). To get back to the previous (long-term state, not phase shift) nutrient charging has to be minimized to consume the entire load in the long term, and this usually causes a hysteresis due to internal processes. To shed light on some basic functional linkages of biota with their environment, eutrophication and nutrient surplus result in excessive algal and microbial growth, resulting finally in lowering Oxygen concentrations up to anoxic conditions. This in turn makes the environment hostile to sensitive aquatic species fostering alterations in community compositions (Spears et al. 2022). Primary production, with indicators as plankton or Chl-a, correlates to nutrients, mostly P, and precipitation (flushing) during growing season (Spears et al. 2022). Though, intense enduring rainfalls can increase P leaching from the basin and especially from agricultural lands (Jeppesen et al. 2011) that will ultimately assist the internal loading.

Hydrological cycle is also crucial for lake functioning as it affects physical processes and bio-geo-chemical cycling. Meteorological variables interact with the waterbody and the basin influencing changes in the physical and chemical environment (Stockwell et al. 2020). Drought (agricultural, social or meteorological) influences lake performance by altering a basic parameter which is lake residence time (salinization, communication with aquifer) and hence salinity due to pollutants augmented concentrations (Lynnyk 2023). The same applies for short duration events like wind and augmented precipitation, that instantly offer amounts of energy (waves, augmented inflows) and disturb the cycle. And this is because during such events, runoff, sedimentation, turbulence (water column mixing) are increased along with flushing incidents (Stockwell et al. 2020). To that, it should be noted that interactions of lake morphology with its watershed and certain geo-morphological attributes (i.e. slopes, basin area, cover type and hydrological connectivity among others) are detrimental on how lakes respond to rain events following different "patterns" (Stockwell et al. 2020). So, extreme precipitation events and resultant flash-floods (IPCC 2023) causing lakes to experience increased loading of allochtonous organic material altering biochemical processes (Weyhenmeyer et al. 2016).

It is a key to quantify the "*bearing capacity*" of lake ecosystems that, if surpassed, there is impact on human and environmental health along with services provisioning (Hessen et al. 2023). Nutrients take precedence over secondary stressors in lakes, a pattern that extends to temperature stress, commonly seen as interacting synergistically with eutrophication in lakes (Moss 2011) [i.e. lake phytoplankton seems less susceptible to the effects of multiple stressors, as long as nutrients are in the growth-limiting concentration range (Birk et al. 2020)]. To couple the above, depending on the magnitude of the change, the environmental gradient and the elements affected inside the food web, the biological response may be immediate or hysteretic given the temporal dynamics of ecological responses of the affected elements (primarily phytoplankton) (Grizzetti and Poikane 2024; Stockwell et al. 2020).

Climate can be a catalyst in stressors' effect on lake ecosystems (Herring et al. 2010) making them key sentinels of climate change (Woolway et al. 2022). Geographic extent and hence climatic variability rule the lakes differently affecting primary productivity. Generally, warmer average air temperatures, change in temperature extremes and changed rainfall extremes are foreseen (Climate Change committees state and EU) with warmer and drier summers and colder and wetter winters. Lakes (natural or artificial) play a crucial role in the global climate by releasing a significant portion of terrestrial carbon as greenhouse gasses back into the atmosphere (Raymond et al. 2013). They are expected to contribute substantial feedback to climate change (Hessen et al. 2023). Biotic parameters, such as species richness, community composition, or functional diversity, capture the intricate impacts of numerous stressors on freshwater ecosystems, encompassing those linked directly or indirectly to climate change (Herring et al. 2010). Climatic "incidents" affect hydrology, water temperature, mixing diets, light conditions (turbidity, euphotic zone), nutrients and—for deep lakes—thermocline (Stockwell et al. 2020). *Warming* and prolonged growing season in lakes is beneficial for nutrient consumption and nutrient uptake in vegetations, though, this prolongation affects also agricultural meaning perhaps the need for augmentation in fertilizer application in agriculture, and hence enhanced wash-off (Spears et al. 2022; Jeppessen et al. 2009, 2011). Warming favors cyanobacteria competition (blooms), enhancing N_2O emissions from lakes due to enhanced nitrogen fixation and so, resident microbial communities favor methanogenesis over methanotrophy (Zhu et al. 2020). On the other hand, the frequency increase in *extreme events* due to climate variability, like high precipitation and strong wind may also induce changes in lake structure and function (besides warming). Even short duration incidents like storms (windstorms) alter light, nutrient and temperature condition instantly and this works as an important disturbance (Stockwell et al. 2020) within the lake's food web, usually offering a window for plankton species competition and "stable state" shift, favoring or undermining species with different functional traits. All these restructure phytoplankton communities and dynamics by misplacing them in the water column and hence disturb the state of the web (Janssen et al. 2021).

Hessen et al. (2023), reviewed 6 major drivers, related to climate change, pushing the lake systems beyond tipping points with internal feedback loops, namely; Eutrophication driven anoxia (and internal P-loading), DOM loadings, disappearance of waterbodies, N-P limitation switch, salinization and invasive species.

These changes-factors may be assessed on a local scale but follow a regional (Mediterranean) or global trend due to the massiveness of pressures or drivers i.e. global warming at which lakes, and especially shallow lakes, show high sensitivity (Meerhoff et al. 2012). Even when the assessment is performed at a broader spatial scale, we can encounter an increase in confounding factors (longer stressor gradient) (Birk et al. 2020). At the same time Mediterranean social-ecological systems are inherently complex, and oversimplifications in ecosystem assessments may arise from an inadequate representation of the relationship between historical and present human activities, biodiversity and ecosystem services spatial patterns, as well as the seasonal and long-term variability in ecosystem services supply, flow and demand (Balzan et al., 2019). The assessment of the ES will contribute to better decision making and to better adaptation plans in the Mediterranean basin social-ecological systems.

3 Ecosystem Services

Daily (1997) with his publication in *Nature* defined the Ecosystem Services as conditions and processes through which natural ecosystems and the species that make up them, sustain and fulfill human life followed by the valuation of Ecosystem Services (ES) offering the quantification of ecosystems to human well being (Costanza et al. 1997; Haines-Young and Potschin 2018). In 2005, the Millennium Ecosystem Assessment (MEA 2005) defined the "Ecosystem services" (ES) as "the ecological characteristics, functions, or processes that *directly or indirectly* contribute to human wellbeing: that is, the benefits that people derive from functioning ecosystems". This definition was a clear distinction between "functions" which are biogeophysical processes and "services" meant the contribution to the society categorized ES into provisioning, cultural, regulation and maintenance services (MAES 2005; TEEB 2010). It was later accepted that natural and human systems co-woven ecosystem services production (Jones et al. 2016).

More recently, the Intergovernmental Science-Policy Platform on Biodiversity and Ecosystem Services (IPBES[u1]) introduced Nature's Benefits to People in 2015 (Diaz et al. 2015) changed to Nature's Contributions to People (NCP), defined as 'all the positive contributions, losses or detriments, that people obtain from nature to capture both *beneficial* and *harmful* effects of nature on people's quality of life (Pascual et al. 2017). NCP is thought to be as a supra-concept to ES, as ES were seen as too narrow to capture a broad range of worldviews, knowledge systems, and stakeholders (Kadykalo et al. 2019). Despite the debate about ES-NCP, Diaz et al. (2018), acknowledge that "the notion of NCP does not appear to differ much from the original definition of ecosystem services".

Since then there is an increased knowledge on the concept of ES and how it can bridge economic growth with ecology and further how it can foster the conservation and enhancement of biodiversity into the planning processes (Savilaasko et al. 2023; Keenan et al. 2019; De Groot et al. 2010). Hence advanced mapping and assessment methods are essential for ensuring proper consideration and integration of ES into planning practices (Golenberg et al. 2017).

Fresh waters, when in good state (in terms of quality and ecological processes), provide vital services to society, including the provision of clean drinking water, food provision, recreation and tourism, pollutant processing, nutrient cycling and biodiversity sustaining the water dependent ecosystems, energy and transportation (Reynaud and Lanzanova 2017). The spiritual importance of aquatic ecosystems has been promoted by several religions while recently the role of water in our psychological health has been highlighted (White et al. 2010; Beute et al. 2020).

In Table 1, some of the main ES provided by freshwaters are presented:

Table 1 Basic ES provided by freshwater ecosystems

Provisioning	Regulating and supporting	Cultural
Water for drinking, and non-drinking (agricultural and industrial use)	Sustain of surface and groundwater ecosystems	Recreation
Fisheries and aquacultures	Flood and erosion control	Tourism
Energy and transportation	Vegetation /primary production	Cultural values, aesthetic
Medicines	Nutrient and sediment cycling	Spiritual value
Fibers-materials	Purification and Ecosystem resilience through aquatic food web	Mental health
Genetic resources	Fish and habitat and wildlife support	Scientific research and education

3.1 Linking Lake Function to Lake ES: Management Implications Under the EU WFD

Lakes provide essential Ecosystem Services (ES) that support human well-being. However, due to anthropogenic environmental degradation, lake ecosystem conditions and related ES supply are affected (Inacio et al. 2024). For example, the potential for eutrophication, where excess nutrients lead to algal overgrowth and subsequent oxygen depletion, which can be detrimental to aquatic life affecting provisioning, regulating and cultural ES.

Lentic biodiversity, which underpins the delivery of many ecosystem services, is under threat (IUCN 2015) while pollution, harmful algal blooms, and invasive alien species including infectious disease organisms constitute the main threats for their ecosystem services delivery. Lake warming can have a positive effect in species richness, especially in cooler areas but increases the risk of invasive and alien species introduction, that can happen also as effect of biotic homogenization (Hasse et al. 2023; Kortz and Magurran 2019; Seebens et al. 2017). Similarly, augmentation in precipitation rates can have a positive effect in functional richness, though, greater increases in such rates can have adverse effects (Hasse et al. 2023). Furthermore, increased climate feedback due to warming by greenhouse gasses emissions affect in the long-term two key drivers, the eutrophication and the browning, thus shifting the lakes' function towards enhanced plankton production (algal biomass), heterotrophy and benthic oxygen depletion (Hessen et al. 2023).

The idea of the interplay is basic (Grizzetti et al. 2016) where humans pose pressures on aquatic ecosystems and these in turn provide humans with services depending on the status (sustainability and resilience). Based on Stockwell et al. (2020), some outcomes can be derived for the interplay of major lake pressures (Climate, lake, watershed, conditions, food webs) and ES.

- Ecosystem Services and Human Benefit: The health of an aquatic ecosystem directly influences the services it can provide, including recreational opportunities, water quality, and climate regulation. Maintaining a balanced ecosystem ensures these services are sustained for human benefit.

- Eutrophication and Management: a potential for eutrophication, where excess nutrients lead to algal overgrowth and subsequent oxygen depletion, which can be detrimental to aquatic life. Active management of nutrient inputs is critical to avoid such outcomes.
- Climate Change Resilience: The resilience of an ecosystem to climate change can be inferred from the functional traits of its biotic community. A diverse and adaptable community can better withstand and recover from environmental stressors.
- Feedback Loops: There are multiple feedback loops where changes in one part of the system can lead to changes in another, which can then feedback into the original change. Understanding these loops is essential for effective ecosystem management.
- Integrated Approach to Conservation: the need for an integrated approach to conservation that considers physical conditions, biotic interactions, and human impacts to maintain the ecological integrity and services of lake ecosystems.

Obviously these pressures are guided by a wide suite of driving forces that are linked to demographic, socio-cultural, economic, technological and policy changes, as well as to human conflicts and epidemics (IPBES 2019).

Since the main management challenge for the European water bodies is achieving the goals of the WFD, the ES concept could play an important role in its effective implementation. The WFD pinpoints key contributors to water degradation, mandates the reduction and control of pollution, and advocates for sustainable water use, while also setting criteria for the assessment and monitoring of water quality (Klimaszyk and Goldyn 2020). The "ecological status" expresses the quality of the structure and functioning of aquatic ecosystems associated with surface waters" (European Commission 2000) and the achievement of a good ecological status might mean ensuring the ecological integrity of freshwater ecosystems and, thereby, preserving their capacity to provide ecosystem services to humans (Brauman et al. 2007). The links between WFD and the delivery of ES is already acknowledged (Giakoumis and Voulvoulis 2018; Grizzetti et al. 2016; Vlachopoulou et al. 2014). Yet, the 2012 Blueprint to Safeguard Europe's Waters provided an overview of the state of ecosystems and their capacity to supply services, showing also that the ecosystem services approach can be useful at different stages of EU Water Policy implementation as the Water Framework Directive (WFD).Two years later, the resource document "Support Policy Development for Integration of Ecosystem Service Assessments into WFD and FD Implementation" reported by European Commission (EC) provided inputs on how to use the ecosystem services approach to improve or support the implementation of the WFD and Floods Directive (FD). Furthermore, it is well documented that, among the Articles of the WFD related to ES, Article 11 (i.e. selecting cost-effective measures in the program of measures (PoMs) is very important. Such importance lies in the fact that PoMs related to various management practices can provide important ES co-benefits (Kagalou and Latinopoulos 2020). The **WFD was not made to assess**

climate change, despite its effect on qualitative water indicators and hence the link with ES. Despite all these suggestions, the embedment of the ES concept in the water (lake) management remained weak with the "single- lake thinking" dominating (Heino et al. 2021) instead of the definition of prioritization of restoration and management measures at National European level based on ES assessment. As a consequence, the Fitness Check of EU Freshwater Policy and the Blueprint document strongly pinpointed to all European Member States and relevant Water authorities the importance of actions for protecting ecosystems and delivering services in the context of sustainable water management.

3.2 *How to Assess and How to Incorporate ES into Lake Management*

Mapping and assessing ES was the focus of the EU initiative "Mapping and Assessment of Ecosystems and their Services" (MAES) (Maes et al. 2013) providing a comprehensive body of knowledge regarding mapping and assessing ES in the EU Member States (European Commission et al. 2020a; b). The past decade, a series of initiatives at EU level have highlighted the necessity of monetary terms inclusion to enhance adoption of ecosystem services in measures programming to preserve biodiversity (La Notte and Marques 2017). At a second phase, the United Nations (UN) System of Environmental-Economic Accounting (SEEA) framework has been proposed for measuring the links between the environment and the economy. The SEEA EA (2021) measures how ecosystems, through the delivery of ecosystem services (ES), contribute to human well-being and the economy and how this contribution evolves to support decision making (Fig. 1).

3.3 *Water Framework Directive and Ecosystem Services*

Several European studies have also been conducted on the mapping, assessment and evaluation of Lake ES under the WFD (i.e. Inacio et al. 2022, 2024; Kagalou and Latinopoulos 2020; Souliotis and Voulvoulis 2021). A commonly applied methodology as the DPSIR (Millennium Ecosystem Assessment, Economics of Ecosystems and Biodiversity, CICES, IPBES) to assess the ecosystem service is based on the following working hypothesis:

Driving forces reflect socio-economic demand that is triggering a change in exerted Pressures applied on ecosystems, affecting thus the Status by changes in derived Ecosystem services altering ultimately their value.

Therefore, Driving (D), Pressure (P) and Status (S) indicators should be interpreted in the assessment of ecosystem services including quantitative indicators facilitating understanding of interplay and decision making (Balzan et al. 2019; Ebner et al.

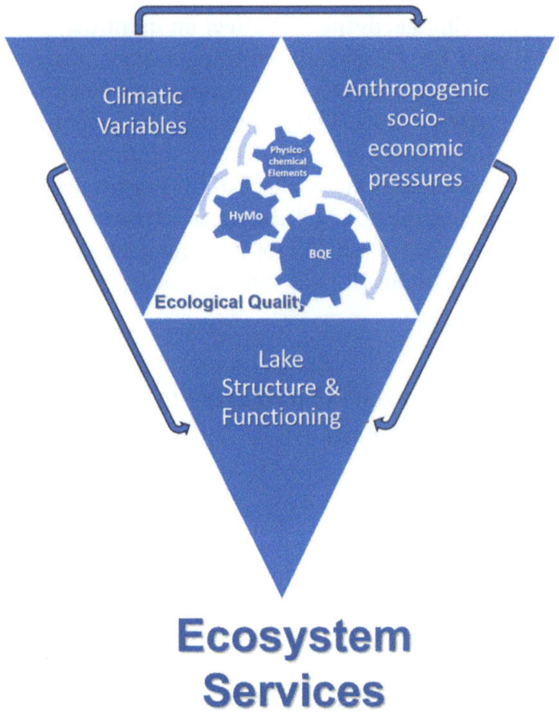

Fig. 1 Ecological quality plays a central role in the interactions among climatic factors, socio-economic characteristics, and lake structural features, shaping ecosystem services

2022). There is a wide suite of methods based either on data sources or on models applied for the mapping and assessment of the ES indicators (Vari et al. 2024.) As examples, we can refer methods applying: field data, remote sensing data, historical data series, experts opinion, statistical models, biophysical modeling, Multicriteria Decision Analysis, machine learning models, participatory models (Vari et al. 2024), Bayesian models while AI based models have also been developed (Egoh et al. 2012; Bagstad et al. 2013). Thus, evaluation on ES potential and/or ES demand is common in the literature aiming at the planning process. However the actual ES flow (i.e. how the provisional ES impacts on regulating and cultural ES in a lake and further on the dependent socio-economic system) is still not sufficiently quantified. This leads to the weak incorporation of the ES concept into the mitigation and management measures included in the Environmental planning process (as the River Basin Management Plans). In any case, ES values are closely related with Pressures, affecting State and demanding management Response and should be treated as such: "*The value of ecosystem services is the relative contribution of the ecosystem to the goal of supporting sustainable human well-being). Any decision involving trade-offs of ecosystem service implies valuation*" (Costanza et al. 2014).

3.4 Barriers and Limitations

Limitations and barriers are expected when dealing with natural processes, complex socio-ecological systems with spatial and temporal variability. Data origin is considered as a major source of uncertainty. Indicator values based on remote sensing (which is a very popular approach) data could underestimate or overestimate some parameters (i.e. chlorophyll, aquatic vegetation) thus affecting the evaluation of the ES. The lack of detailed information linking the lake water status (both in quantity and quality terms) with water uses hinders the accurate mapping and evaluation of the ES. Sometimes the applied indicators are not strictly defined for ES (potential) supply and they, often, quantify the ES demand (ES currently consumed). Caution might be appropriate in regarding the participatory methods where a deliberative dialogue with various stakeholders is not always an easy task. The Mapping and Assessment of Ecosystem Services (MAES) process in Europe proved to be a very constructive and successful way to engage stakeholders increasing their knowledge on ES values. However it seems that the most "favorable" ES are the provisioning while the regulating and maintenance are under lighted posing risks for nature conservation. Cultural ecosystem services are not yet fully integrated into operational frameworks because of the transdisciplinarity which is required to address the issue (i.e. physical, intellectual, spiritual, aesthetics interactions with nature).

4 Mediterranean Lakes and Ecosystem Services

Mediterranean natural lake systems are ancient, rich in biodiversity (and endemism) and socio-economically complex with a prolonged cultural interface whose evolution is deeply dependent on mankind (Balzan et al 2019). Climate change effects on lake biota depend on ecoregion, ecosystem type and existing stressors that affect them (Herring et al. 2010). Mediterranean environment affects lake systems in a unique way. The ecology of semi-arid to arid (especially shallow) Mediterranean lakes remains underexplored, with hydrological impacts, major ions, nutrient dynamics, and submerged macrophytes playing crucial roles in their food webs (Beklioglu et al. 2007) and hence in ES provisioning. Shallow lakes in the Mediterranean are highly sensitive to extreme climate shifts. Expanding distribution systems to regions with low precipitation, such as for irrigating semi-arid and arid areas (i.e. in the Mediterranean, Beklioglu et al. 2017), leads to exceptionally high water usage due to significant losses through evapotranspiration (Grizzetti and Poikane 2024). Further, the entire food web is different since most lakes host small omnivorous species with prolonged spawning (González-Bergonzoni et al. 2012) making weaker the top-down control (less piscivorous higher zooplankton consumption, lower phytoplankton control and cyanobacteria prevalence (Laspidou et al. 2017).

The strong connection between climate and the hydrological cycle is evident, and warming trends, particularly in the Mediterranean region, are expected to contribute to more frequent droughts. Especially, the projections foresee aridification, salinisation and desertification, in the southern parts of the Mediterranean Basin (Souissi et al. 2018). This has significant repercussions, including reductions in water levels in lakes (Wantzen et al. 2008, Beklioglu et al. 2017, 2020). Climate models for the Mediterranean foresee substantial changes in freshwater availability due to a projected decrease in precipitation (Christensen et al. 2013). Additionally, higher temperatures and increased drought occurrences are likely to elevate the demand for water resources, leading to more extensive diversions, declining water levels, and severe basin-wide water shortages (Beklioglu et al. 2020; Erol and Randhir 2012). Despite the potential reduction in external nutrient input from decreased precipitation, eutrophication may still intensify due to diminishing lake volume caused by higher evaporation and lower precipitation.

Intense water level fluctuation in Mediterranean lakes, high evaporation and water table reduction during summer, especially in shallow ones in med environment (Bucak et al. 2012).

Phytoplankton taxa are differentially adapted to a range of environmental conditions. The Mediterranean environment is characterized by a prolonged warm period, with intense solar radiation and a mild-cold period. Unfortunately the proportion of oligotrophoic systems is really low, especially when dealing with shallow lakes. All these evolutionarily shaped the warm mediterranean lakes as hosters of fast growing species, with high competition in the column, tolerant to nutrient scarcity and buoyancy ability among others. Given the P availability, N-fixing species prevail in filaments and/or colonies forms guaranteeing their prevalence, coupled with benthic dwellers and planktophagus fish, making it difficult to shift to diatom or macrophyte dominated phase. Organization in functional groups (Reynolds) can give insights in Mediterranean lakes and reservoirs, even in really particular case-studies as a multifunction artificial lake (Latinopoulos et al. 2020a; b). These can be coupled with the same nature of reservoirs where soil loss and nutrient leaching can impact lake longevity, affecting its functioning, implying that new agricultural areas should be situated away from lake water retention zones (Santos et al. 2018).

The complexity of Mediterranean ecosystems including their socio-economic aspects is such that simplifications and modeling may result in inaccurate assessment (Balzan et al. 2019), though the existing knowledge gap and the low awareness should be met.

Climate change has numerous implications for the diverse ecosystem services and benefits offered by lakes, encompassing provisioning, regulating, supporting, and cultural services. The impact of climate change on cultural ecosystem services, such as aesthetics and recreation, is evident (Ebner et al. 2022). For instance, changes like a reduction in winter ice cover can affect recreational activities like skiing, snowmobiling, and ice fishing (Knoll et al. 2019). In numerous global regions where water is a limiting factor for economic development, ensuring good water quality becomes critically significant, influencing its economic utility in areas such as water supply, recreation, and agriculture (Klimaszyk and Goldyn 2020). As stated before, it is important to pinpoint thresholds and the bearing capacity of lake systems, as it seems that, regardless trophic status, water warming can exceed our predictions and cause higher probability of harmful algal blooms occurence with adverse effects on various ecosystem services (Griffith and Gobler 2020).

In their review, Woolway et al. (2020) examined the worldwide impact of climate change on physical lake variables, highlighting reductions in winter ice cover and rises in lake surface temperatures. These alterations in lake mixing patterns (Woolway et al. 2022) and increased evaporation contribute to decreased water levels and surface area (Zhou et al. 2021), potentially affecting water quantity and quality as well as the provision of ecosystem services if not offset by increased rainfall or runoff. This worldwide effect is reflected in an increase of blooms, related with warming, with adverse effects on functioning posing obstacles in de-eutrohication measures (Ho et al. 2019). Frequency and intensity of storms are expected to increase in the coming decade for all continental areas (Lehmann et al. 2015) with projected negative effects on Mediterranean lakes (Taylor et al. 2021). Other unfavorable projections include plant mortality, reduced water quality, impaired productivity, and pest outbreaks (Balzan et al. 2019).

The aesthetic appeal of lakes may also be compromised due to an increase in algae blooms, negatively impacting tourism and lakeside property values (Woolway et al. 2022; Liu et al. 2017) despite the **Land Use and Land Cover Change (LULCC)** has favored transition to enhance touristic accessibility and infrastructures. There is an undeniable connection between landscape configuration and eco-environmental quality, underscoring the profound influence of human activities on the environment. Monitoring the effects of **LULCC** on ecosystem health and the economic value of ecosystem services (ES) is essential for informing spatial planning and restoration policies (Latinopoulos et al. 2021, Vatitsi et al. 2023). Land use and urbanization especially has damaged ES in an unprecedented level and is expected to deteriorate even more in the future (Beklioglu et al. 2017). LULCC along with climate can have a combo effect on wetland hydrological regimes (Leberger et al. 2020) since higher temperatures are expected to lead to greater water abstraction, and the anticipated increase in drought frequency will elevate irrigation demands (Hering et al. 2010). Alterations in land use might necessitate dam construction, introducing barrier effects that impact hydrological conditions. The creation of LULCC maps that can cover a historical period of time can demonstrate the changes that guided hydromorphological alterations affecting ES (Latinopoulos et al. 2020a, b).

Urban development is another factor that significantly impacts freshwater ecosystems and associated services (Wangai et al. 2019). Impervious areas assist fragmentation and increase sedimentation and so urbanization may pose detrimental effects on ecology and related ES by creating an arc of pressures around urban areas (Latinopoulos et al. 2020a, b).

Changes of water uses are a primary factor of pressures impacting habitat, water provision, and aesthetic value of lakes, while the magnitude of their impacts is mostly related to local socio-economic aspects (Schirpke and Ebner 2022). Livestock farming in proximity to the lake (Schirpke et al. 2021) and nutrient effect by adjacent agricultural areas (Heino et al. 2021) are negatively related to water related ES provision. Though, agroecosystems have the ability to foster biodiversity (Santos et al. 2022) and hence counteract the total ES balance, while at the same time

Mediterranean lakes have high nitrogen retention capacity (La Notte et al. 2015). In a Mediterranean region in southwest Europe, artificial lake construction increased carbon storage by 9.4%, but expanded agriculture reduced water services, potentially altering the positive cost–benefit analysis of lakes as climate change adaptation measures (Santos et al., 2018). The projections foresee aridification, salinisation and desertification, especially in the southern parts of the Mediterranean Basin (Souissi et al. 2018).

The significance of littoral and riparian zones for lakes is paramount, influencing hydrology, ecology, and the support of ecosystem services. In a Greek case study for a historic natural lake (Pamvotis) the long-term effect of LULC change can shed light in ES provisioning alteration. Despite 70-year changes increased the overall value of ecosystem services, it came at the cost of some provisioning and regulating services such as habitat/refugia, material and water supply, genetic resources, and waste treatment, while regulation and recreation services were enhanced (Latinopoulos et al. 2020a, b) altering trade offs. Transformations in the Mediterranean Basin, including the abandonment of traditional practices, intensified primary sector activities (agriculture, deforestation, fishing), and increased urban development, are proposed to result in biodiversity loss and reduced ecosystem services supply (Balzan et al. 2019; Liquete et al. 2016). On the other hand, based on people's preferences, elevated positive affect, and heightened perceived restorativeness were linked to both natural and constructed settings featuring water underlying the positive impact of tourism and aesthetics (White et al. 2010). Touristic interest lies also in lagoons along with intense agriculture in Meditteranean systems, making them share the same fate, despite lagoons being even more volatile systems and less studied. Combining indicators of ecosystem services with environmental sustainability ones, demonstrate that monitoring and managing both is achievable simultaneously (Mirli et al. 2022).

Generally there is a negative criticism on lake ES modeling since it is difficult to incorporate all elements and provide a solid finding, though their usefulness in acquiring basin knowledge on tradeoffs is important. An example of (most cited and applied) such a model is the Integrated Valuation of Ecosystem Services and Tradeoffs (InVEST) model (Stanford University). Xu et al. (2018), review on lake ES modeling identified gaps related to data sources, quantitative analyses robustness and linkage of ES with ecosystem functioning along with trade-offs. While lake ES modeling might be problematic, sectorial modeling like process-based hydrological modeling can assist ES modeling to assist decision making (Latinopoulos et al. 2020b; Vigerstol and Aukema 2011), and its value lies in the necessity to assess trade-offs among alternative scenarios of landscape and resource use (Brauman et al. 2007). Through computational methods someone could test different management scenarios, assess water availability and sediment regulation along with nutrient loading (as water quality proxies) to depict their effect on water-related ES. Air quality, carbon sequestration, self purification and soil creation are some of the "hydrological related services" and most of them are included in regulating ES. Hydrological (and ecohydrological) variables affecting ES are included in a qualitative, quantitative, spatial and temporal dimensions that govern water supply, supporting elements and aesthetics (Brauman et al. 2007).

5 Operation and Framework for Policy Forming

ES approach can support the inclusion of previously unrecognized services into land use and management decisions and identify opportunities for synergies, or the requirement for trade-offs between the production of private goods, common goods and public goods (Savilaakso et al. 2023; Kagalou and Latinopoulos 2020).

The perception of Brauman et al. (2007), gives a picture of a cyclic framework with components influencing each other, where policy formation (as driver of outcomes in ES management) is at the core of the framework, suggesting that policies governing ecosystem services should be informed by considerations of ecological, economic, and social values. Policy decisions should take into account the biophysical aspects of ecosystem services (production, monitoring, benefits), ecological, economic, and social considerations with continuous re-evaluation based on information and changing conditions.

Identifying the natural and anthropogenic drivers of biotic change is critical to inform effective freshwater management strategies, especially when dealing with a series of simultaneous alterations, should be based on knowledge of ecosystems' metabolic responses (Grizzetti and Poikane 2024). Historically, management of water bodies has focussed on the control of single stressors (Verdonschot et al. 2013), which were assumed to be dominant. The current bet is to deal with multiple stressors, preserving ES, safeguarding people's wellbeing within uncertainty. The use of a solid framework for integrated Ecosystem Services assessment of the costs and benefits of large scale landscape restoration have given fruitful results in the Mediterranean region (Spain) (de Groot 2022).

When it comes to restoration management based on ES, only case-specific evaluation can ensure restoration strategies effectively, avoid harm to the target lake and downstream ecosystems, remain cost-effective, enhance valuable ecosystem services delivery, reduce conflicts in public interests, and eliminate the need for repetitive interventions (Tammeorg et al. 2023). Financial feasibility and cost effectiveness of measures (Carvalho et al. 2021) is another important element for lake restoration, as it requires multidisciplinary research and close collaboration across environmental, social, political, and economic sectors (Eros et al. 2023; Tammeorg et al. 2023). Having said for the social dimensions in restoration and conservation measures, the public awareness should not be omitted, since it is a critical factor, even though geographical, gender, urbanization differences can impede these efforts (Kokkoris et al. 2023). Gradual change must be enforced in incorporating ES at the European level, with increasing public awareness as a tool (Kokkoris et al. 2023; Dick et al. 2018). Several impediments can be found to implementing ES included structural and governance deficiencies, lack of public and private leadership, complexity of language, and the lack of sustained, long-term investment (Keenan et al. 2019). Even effective restoration efforts need permanent investment (eco-socio-economic) since otherwise the ameliorated status will not be stable (Rosińska et al. 2018).

Key recommendations are: adaptive National water policy perspectives, Regional and Catchment water policy with augmented research (May et al. 2022) for deeper understanding of Interplay, functioning and mechanisms that originate from multiple stressors and drivers along with climate change effects (Hessen et al. 2023; Ebner et al. 2022; Spears et al. 2022; Stockwel et al. 2020; Hering et al. 2010) including humans as a key component (Grizzetti and Poikane 2024).

References

Bagstad KJ, Johnson GW, Voigt B, Villa F (2013) Spatial dynamics of ecosystem service flows: a comprehensive approach to quantifying actual services. Ecosyst Serv 4:117–125. https://doi.org/10.1016/j.ecoser.2012.07.012

Balzan MV, Pinheiro AM, Mascarenhas A, Morán-Ordóñez A, Ruiz-Frau A, Carvalho-Santos C, Vogiatzakis IN, Arends J, Santana-Garcon J, Roces-Díaz JV, Brotons L, Campagne CS, Roche PK, De Miguel S, Targetti S, Drakou EG, Vlami V, Baró F, Geijzendorffer IR (2019) Improving ecosystem assessments in Mediterranean social-ecological systems: a DPSIR analysis. Ecosyst People 15(1):136–155. https://doi.org/10.1080/26395916.2019.1598499

Beklioğlu M, Bucak T, Coppens J, Bezirci G, Tavşanoğlu Ü, Çakıroğlu A, Levi E, Erdoğan Ş, Filiz N, Özkan K, Özen A (2017) Restoration of eutrophic lakes with fluctuating water levels: a 20-year monitoring study of two inter-connected lakes. Water 9(2):127. https://doi.org/10.3390/w9020127

Beklioğlu M, Bucak T, Levi EE, Erdoğan Ş, Özen A, Filiz N, Bezirci G, Çakıroğlu Aİ, Tavşanoğlu ÜN, Gökçe D, Demir N, Özuluğ M, Duran M, Özkan K, Brucet S, Jeppesen E (2020) Influences of climate and nutrient enrichment on the multiple trophic levels of Turkish shallow lakes. Inland Waters 10(2):173–185. https://doi.org/10.1080/20442041.2020.1746599

Beklioglu M, Romo S, Kagalou I, Quintana X, Bécares E (2007) State of the art in the functioning of shallow Mediterranean lakes: workshop conclusions. Hydrobiologia 584(1):317–326. https://doi.org/10.1007/s10750-007-0577-x

Beute F, Davies Z, de Vries S, Glanville J, Keune H, Lammel A, Marselle M, O'Brien L, Olszewska-Guizzo A, Remmen R, Russo A, Andreucci MB (2020) Types and characteristics of urban and peri-urban blue spaces having an impact on human mental health and wellbeing. Report prepared by an EKLIPSE Expert Working Group. UK Centre for Ecology & Hydrology, Wallingford, United Kingdom

Birk S, Chapman D, Carvalho L, Spears BM, Andersen HE, Argillier C, Auer S, Baattrup-Pedersen A, Banin L, Beklioğlu M, Bondar-Kunze E, Borja A, Branco P, Bucak T, Buijse AD, Cardoso AC, Couture R-M, Cremona F, De Zwart D et al (2020) Impacts of multiple stressors on freshwater biota across spatial scales and ecosystems. Nat Ecol Evol 4(8):1060–1068. https://doi.org/10.1038/s41559-020-1216-4

Bisselink B, Bernhard J, Gelati E, Adamovic M, Guenther S, Mentaschi L, Ḋe Roo A et al (2020) Climate change and Europe's water resources. EUR, 29130, 978–9276103981

Brauman KA, Daily GC, Duarte TK, Mooney HA (2007) The nature and value of ecosystem services: an overview highlighting hydrologic services. Annu Rev Environ Resour 32(1):67–98. https://doi.org/10.1146/annurev.energy.32.031306.102758

Bucak T, Saraoğlu E, Levi EE, NiHan Tavşanoğlu Ü, İdİl ÇakiRoğlu A, Jeppesen E, BekliOğlu M (2012). The influence of water level on macrophyte growth and trophic interactions in eutrophic Mediterranean shallow lakes: a mesocosm experiment with and without fish. Freshwater Biol 57(8):1631–1642. https://doi.org/10.1111/j.1365-2427.2012.02825.x

Calvin K, Dasgupta D, Krinner G, Mukherji A, Thorne PW, Trisos C, Romero J, Aldunce P, Barrett K, Blanco G, Cheung WWL, Connors S, Denton F, Diongue-Niang A, Dodman D, Garschagen M, Geden O, Hayward B, Jones C et al (2023) IPCC, 2023: climate change 2023: synthesis report. Contribution of working groups I, II and III to the sixth assessment report of the intergovernmental panel on climate change [Core Writing Team, Lee H, Romero J (eds)]. IPCC,

Geneva, Switzerland. (First). Intergovernmental Panel on Climate Change (IPCC). https://doi.org/10.59327/IPCC/AR6-9789291691647
Carpenter SR, Caraco NF, Correll DL, Howarth RW, Sharpley AN, Smith VH (1998) Nonpoint pollution of surface waters with phosphorus and nitrogen. Ecol Appl 8(3):559–568. https://doi.org/10.1890/1051-0761(1998)008[0559:NPOSWW]2.0.CO;2
Carvalho L, Gunn IDM, Spears BM, Dobel AJ (2021) Ecological restoration. In: Ferrier R, Jenkins A (eds) Handbook of catchment management 2e. https://doi.org/10.1002/9781119531241.ch10
Christensen JH, Krishna Kumar K, Aldrian E, An S-I, Cavalcanti IFA, de Castro M, Dong W, Goswami P, Hall A, Kanyanga JK, Kitoh A, Kossin J, Lau N-C, Renwick J, Stephenson DB, Xie S-P, Zhou T (2013) Climate phenomena and their relevance for future regional climate change. In: Stocker TF, Qin D, Plattner G-K, Tignor M, Allen SK, Boschung J, Nauels A, Xia Y, Bex V, Midgley PM (eds) Climate change 2013: the physical science basis. Contribution of working group I to the fifth assessment report of the intergovernmental panel on climate change. Cambridge University Press, Cambridge, United Kingdom and New York, NY, USA
Costanza R, d'Arge R, De Groot R, Farber S, Grasso M, Hannon B, Limburg K, Naeem S, O'Neill RV, Paruelo J, Raskin RG, Sutton P, Van Den Belt M (1997) The value of the world's ecosystem services and natural capital. Nature 387(6630):253–260. https://doi.org/10.1038/387253a0
Costanza R, De Groot R, Sutton P, Van der Ploeg S, Anderson SJ, Kubiszewski I et al (2014) Changes in the global value of ecosystem services. Glob Environ Chang 26:152–158
Daily GC (1997) Introduction: what are ecosystem services? In: Daily GC (ed) Nature's services: societal dependence on natural ecosystems. Island Press, Washington DC, pp 1–10
Díaz S, Demissew S, Joly C, Lonsdale WM, Larigauderie A (2015) A Rosetta Stone for nature's benefits to people. PLoS Biol 13(1):e1002040
De Groot R, Moolenaar S, De Vente J, De Leijster V, Ramos ME, Robles AB, Schoonhoven Y, Verweij P (2022) Framework for integrated Ecosystem Services assessment of the costs and benefits of large scale landscape restoration illustrated with a case study in Mediterranean Spain. Ecosyst Serv 53:101383. https://doi.org/10.1016/j.ecoser.2021.101383
De Groot RS, Alkemade R, Braat L, Hein L, Willemen L (2010) Challenges in integrating the concept of ecosystem services and values in landscape planning, management and decision making. Ecol Complex 7(3):260–272. https://doi.org/10.1016/j.ecocom.2009.10.006
Díaz S, Pascual U, Stenseke M, Martín-López B, Watson RT, Molnár Z, Hill R, Chan KMA, Baste IA, Brauman KA, Polasky S, Church A, Lonsdale M, Larigauderie A, Leadley PW, Van Oudenhoven APE, Van Der Plaat F, Schröter M, Lavorel S et al (2018) Assessing nature's contributions to people. Science 359(6373):270–272. https://doi.org/10.1126/science.aap8826
Dick J, Turkelboom F, Woods H, Iniesta-Arandia I, Primmer E, Saarela S-R, Bezák P, Mederly P, Leone M, Verheyden W, Kelemen E, Hauck J, Andrews C, Antunes P, Aszalós R, Baró F, Barton DN, Berry P, Bugter R et al (2018) Stakeholders' perspectives on the operationalisation of the ecosystem service concept: results from 27 case studies. Ecosyst Serv 29:552–565. https://doi.org/10.1016/j.ecoser.2017.09.015
Dudgeon D, Arthington AH, Gessner MO, Kawabata Z, Knowler DJ, Lévêque C, Naiman RJ, Prieur-Richard A, Soto D, Stiassny MLJ, Sullivan CA (2006) Freshwater biodiversity: importance, threats, status and conservation challenges. Biol Rev 81(2):163–182. https://doi.org/10.1017/S1464793105006950
Ebner M, Schirpke U, Tappeiner U (2022) How do anthropogenic pressures affect the provision of ecosystem services of small mountain lakes? Anthropocene 38:100336. https://doi.org/10.1016/j.ancene.2022.100336
EEA (2024) Europe's state of water 2024: the need for improved water resilience. EEA Report 07/2024. ISBN 978-92-9480-653-6, ISSN 1977-8449. https://doi.org/10.2800/02236
Egoh B, Drakou EG, Dunbar MB, Maes J, Willemen L (2012) Indicators for mapping ecosystem services: a review. https://doi.org/10.13140/2.1.3420.2565
Erol A, Randhir TO (2012) Climatic change impacts on the ecohydrology of Mediterranean watersheds. Clim Change 114(2):319–341. https://doi.org/10.1007/s10584-012-0406-8

Erős T, Hermoso V, Langhans SD (2023) Leading the path toward sustainable freshwater management: reconciling challenges and opportunities in historical, hybrid, and novel ecosystem types. Wires Water 10(3):e1645. https://doi.org/10.1002/wat2.1645

EU 1992 (1992, July 22) Council Directive 92/43/EEC of 21 May 1992 on the conservation of natural habitats and of wild fauna and flora. OJ L 206:7–50. http://data.europa.eu/eli/dir/1992/43/oj

European Commission (2000) Directive 2000/60/EC of the European Parliament and of the Council of 23 October 2000 establishing a framework for Community action in the field of water policy. OJ L 327:1–73

European Commission, Joint Research Centre (2020a) Climate change and Europe's water resources. Publications Office. https://data.europa.eu/doi/10.2760/15553

European Commission, Joint Research Centre (2020b) Mapping and assessment of ecosystems and their services: An EU wide ecosystem assessment in support of the EU biodiversity strategy. Publications Office. https://data.europa.eu/doi/10.2760/757183

Giakoumis T, Voulvoulis N (2018) The transition of EU water policy towards the water framework directive's integrated river basin management paradigm. Environ Manage 62(5):819–831. https://doi.org/10.1007/s00267-018-1080-z

Gleeson T, Wang-Erlandsson L, Zipper SC, Porkka M, Jaramillo F, Gerten D, Fetzer I, Cornell SE, Piemontese L, Gordon LJ, Rockström J, Oki T, Sivapalan M, Wada Y, Brauman KA, Flörke M, Bierkens MFP, Lehner B, Keys P et al (2020) The water planetary boundary: interrogation and revision. One Earth 2(3):223–234. https://doi.org/10.1016/j.oneear.2020.02.009

Goldenberg R, Kalantari Z, Cvetkovic V, Mörtberg U, Deal B, Destouni G (2017) Distinction, quantification and mapping of potential and realized supply-demand of flow-dependent ecosystem services. Sci Total Environ 593–594:599–609. https://doi.org/10.1016/j.scitotenv.2017.03.130

González-Bergonzoni I, Meerhoff M, Davidson TA, Teixeira-de Mello F, Baattrup-Pedersen A, Jeppesen E (2012) Meta-analysis shows a consistent and strong latitudinal pattern in fish omnivory across ecosystems. Ecosystems 15(3):492–503. https://doi.org/10.1007/s10021-012-9524-4

Griffith AW, Gobler CJ (2020) Harmful algal blooms: a climate change co-stressor in marine and freshwater ecosystems. Harmful Algae 91:101590. https://doi.org/10.1016/j.hal.2019.03.008

Grizzetti B, Poikane S (2024) The importance of inland waters. In: Wetzel's limnology. Elsevier, pp 7–13. https://doi.org/10.1016/B978-0-12-822701-5.00002-1

Grizzetti B, Lanzanova D, Liquete C, Reynaud A, Cardoso AC (2016) Assessing water ecosystem services for water resource management. Environ Sci Policy 61:194–203. https://doi.org/10.1016/j.envsci.2016.04.008

Haase P, Bowler DE, Baker NJ, Bonada N, Domisch S, Garcia Marquez JR, Heino J, Hering D, Jähnig SC, Schmidt-Kloiber A, Stubbington R, Altermatt F, Álvarez-Cabria M, Amatulli G, Angeler DG, Archambaud-Suard G, Jorrín IA, Aspin T, Azpiroz I et al (2023) The recovery of European freshwater biodiversity has come to a halt. Nature 620(7974):582–588. https://doi.org/10.1038/s41586-023-06400-1

Haines-Young R, Potschin-Young M (2018) Revision of the common international classification for ecosystem services (CICES V5.1): a policy brief. One Ecosyst 3:e27108. https://doi.org/10.3897/oneeco.3.e27108

Heino J, Alahuhta J, Bini LM, Cai Y, Heiskanen A, Hellsten S, Kortelainen P, Kotamäki N, Tolonen KT, Vihervaara P, Vilmi A, Angeler DG (2021) Lakes in the era of global change: Moving beyond single-lake thinking in maintaining biodiversity and ecosystem services. Biol Rev 96(1):89–106. https://doi.org/10.1111/brv.12647

Hering D, Haidekker A, Schmidt-Kloiber A, Barker T, Buisson L, Graf W, Stendera S et al (2010) Monitoring the responses of freshwater ecosystems to climate change. In: Climate change impacts on freshwater ecosystems, pp 84–118. https://doi.org/10.1002/9781444327397

Hessen DO, Andersen T, Armstrong McKay D, Kosten S, Meerhoff M, Pickard A, Spears B (2023) Lake ecosystem tipping points and climate feedbacks [Preprint]. Biosphere and ecosystems/Biosphere/atmosphere interactions/Idealized models. https://doi.org/10.5194/esd-2023-22

Ho JC, Michalak AM, Pahlevan N (2019) Widespread global increase in intense lake phytoplankton blooms since the 1980s. Nature 574(7780):667–670. https://doi.org/10.1038/s41586-019-1648-7

Inácio M, Barceló D, Zhao W, Pereira P (2022) Mapping lake ecosystem services: a systematic review. Sci Total Environ 847:157561. https://doi.org/10.1016/j.scitotenv.2022.157561

Inácio M, Das M, Burkhard B, Barceló D, Pereira P (2024) Mapping and assessment of lake ecosystem services in Lithuania. Ecosyst Serv 65:101589. https://doi.org/10.1016/j.ecoser.2023.101589

Janssen ABG, Hilt S, Kosten S, De Klein JJM, Paerl HW, Van De Waal DB (2021) Shifting states, shifting services: linking regime shifts to changes in ecosystem services of shallow lakes. Freshw Biol 66(1):1–12. https://doi.org/10.1111/fwb.13582

Jeppesen E, Kronvang B, Meerhoff M, Søndergaard M, Hansen KM, Andersen HE, Lauridsen TL, Liboriussen L, Beklioglu M, Özen A, Olesen JE (2009) Climate change effects on runoff, catchment phosphorus loading and lake ecological state, and potential adaptations. J Environ Qual 38(5):1930–1941. https://doi.org/10.2134/jeq2008.0113

Jeppesen E, Kronvang B, Olesen JE, Audet J, Søndergaard M, Hoffmann CC, Andersen HE, Lauridsen TL, Liboriussen L, Larsen SE, Beklioglu M, Meerhoff M, Özen A, Özkan K (2011) Climate change effects on nitrogen loading from cultivated catchments in Europe: implications for nitrogen retention, ecological state of lakes and adaptation. Hydrobiologia 663(1):1–21. https://doi.org/10.1007/s10750-010-0547-6

Jones L, Norton L, Austin Z, Browne AL, Donovan D, Emmett BA, Grabowski ZJ, Howard DC, Jones JPG, Kenter JO, Manley W, Morris C, Robinson DA, Short C, Siriwardena GM, Stevens CJ, Storkey J, Waters RD, Willis GF (2016) Stocks and flows of natural and human-derived capital in ecosystem services. Land Use Policy 52:151–162. https://doi.org/10.1016/j.landusepol.2015.12.014

Kadykalo AN, López-Rodriguez MD, Ainscough J, Droste N, Ryu H, Ávila-Flores G, Le Clec'h S, Muñoz MC, Nilsson L, Rana S, Sarkar P, Sevecke KJ, Harmáčková ZV (2019) Disentangling 'ecosystem services' and 'nature's contributions to people.' Ecosyst People 15(1):269–287. https://doi.org/10.1080/26395916.2019.1669713

Kagalou I, Latinopoulos D (2020) Filling the gap between ecosystem services concept and river basin management plans: the case of Greece in WFD 20+. Sustainability 12(18):7710. https://doi.org/10.3390/su12187710

Kath J, Boulton AJ, Harrison ET, Dyer FJ (2018) A conceptual framework for ecological responses to groundwater regime alteration (FERGRA). Ecohydrology 11(7):e2010. https://doi.org/10.1002/eco.2010

Keenan RJ, Pozza G, Fitzsimons JA (2019) Ecosystem services in environmental policy: barriers and opportunities for increased adoption. Ecosyst Serv 38:100943. https://doi.org/10.1016/j.ecoser.2019.100943

Klimaszyk P, Gołdyn R (2020) Water quality of freshwater ecosystems in a temperate climate. Water 12(9):2643. https://doi.org/10.3390/w12092643

Knoll LB, Sharma S, Denfeld BA, Flaim G, Hori Y, Magnuson JJ, Straile D, Weyhenmeyer GA (2019) Consequences of lake and river ice loss on cultural ecosystem services. Limnol Oceanogr Lett 4(5):119–131. https://doi.org/10.1002/lol2.10116

Kokkoris IP, Skuras D, Maniatis Y, Dimopoulos P (2023) Natura 2000 public awareness in EU: a prerequisite for successful conservation policy. Land Use Policy 125:106482. https://doi.org/10.1016/j.landusepol.2022.106482

Kortz AR, Magurran AE (2019) Increases in local richness (α-diversity) following invasion are offset by biotic homogenization in a biodiversity hotspot. Biol Let 15(5):20190133. https://doi.org/10.1098/rsbl.2019.0133

La Notte A, Liquete C, Grizzetti B, Maes J, Egoh BN, Paracchini ML (2015) An ecological-economic approach to the valuation of ecosystem services to support biodiversity policy. A case study for nitrogen retention by Mediterranean rivers and lakes. Ecol Ind 48:292–302

La Notte A, Marques A (2017) The role of enabling actors in ecosystem service accounting. One Ecosystem 2:e20834. https://doi.org/10.3897/oneeco.2.e20834
Laspidou C, Kofinas D, Mellios N, Latinopoulos D, Papadimitriou T (2017) Investigation of factors affecting the trophic state of a shallow Mediterranean reconstructed lake. Ecol Eng 103:154–163. https://doi.org/10.1016/j.ecoleng.2017.03.019
Latinopoulos D, Dimitriou A, Kagalou I (2020) Modelling water related ecosystem services in a heavily modified Mediterranean reservoir using the soil and water assessment tool (SWAT). In: The 4th EWaS international conference: valuing the water, carbon, ecological footprints of human activities, vol 2. https://doi.org/10.3390/environsciproc2020002002
Latinopoulos D, Koulouri M, Kagalou I (2021) How historical land use/land cover changes affected ecosystem services in lake Pamvotis, Greece. Hum Ecol Risk Assess Int J 27(6):1472–1491. https://doi.org/10.1080/10807039.2020.1855575
Latinopoulos D, Ntislidou C, Kagalou I (2020b) Relationships of environmental conditions and phytoplankton functional groups in a new re-constructed shallow lentic system in drought conditions. Ecohydrol Hydrobiol 20(3):369–381. https://doi.org/10.1016/j.ecohyd.2020.04.003
Leberger R, Geijzendorffer IR, Gaget E, Gwelmami A, Galewski T, Pereira HM, Guerra CA (2020) Mediterranean wetland conservation in the context of climate and land cover change. Reg Environ Change 20(2):67. https://doi.org/10.1007/s10113-020-01655-0
Lehmann J, Coumou D, Frieler K (2015) Increased record-breaking precipitation events under global warming. Clim Change 132(4):501–515
Liquete C, Cid N, Lanzanova D, Grizzetti B, Reynaud A (2016) Perspectives on the link between ecosystem services and biodiversity: the assessment of the nursery function. Ecol Ind 63:249–257. https://doi.org/10.1016/j.ecolind.2015.11.058
Liu X, Qian K, Chen Y, Gao J (2017) A comparison of factors influencing the summer phytoplankton biomass in China's three largest freshwater lakes: Poyang, Dongting, and Taihu. Hydrobiologia 792(1):283–302
Liu X, Liu W, Liu L, Tang Q, Liu J, Yang H (2021) Environmental flow requirements largely reshape global surface water scarcity assessment. Environ Res Lett 16(10):104029. https://doi.org/10.1088/1748-9326/ac27cb
Lynnyk PM (2023) Potential impact of climate change on mineralization and main ions ratio in surface fresh waters (a review). Hydrobiol J 59(4):83–97. https://doi.org/10.1615/HydrobJ.v59.i4.70
Maes J, Teller A, Erhard M, Liquete C, Braat L, Berry P, Bidoglio G et al (2013) Mapping and assessment of ecosystems and their services. In: An analytical framework for ecosystem assessments under action, vol 5. JRC publications, pp 1–58
May L, Taylor P, Gunn IDM, Thackeray SJ, Carvalho LR, Hunter P, Corr M, Dobel AJ, Grant A, Nash G, Robinson E, Spears BM (2022) Assessing climate change impacts on the water quality of Scottish standing waters. Scotland's Centre of Expertise for Waters (CREW)
Meerhoff M, Teixeira-de Mello F, Kruk C, Alonso C, González-Bergonzoni I, Pacheco JP, Lacerot G, Arim M, Beklioğlu M, Brucet S, Goyenola G, Iglesias C, Mazzeo N, Kosten S, Jeppesen E (2012) Environmental warming in shallow lakes. In: Advances in ecological research, vol 46. Elsevier, pp 259–349. https://doi.org/10.1016/B978-0-12-396992-7.00004-6
Mirli A, Kamidis N, Latinopoulos D, Kagalou I (2022) Assessing the ecosystem services of Nestos Delta Lagoons-Greece: coupling with sustainability indicators. Fresenius Environ Bull 31(8 A):8365–8371
Moss B (2011) Allied attack: climate change and eutrophication. Inland Waters 1(2):101–105. https://doi.org/10.5268/IW-1.2.359
Pascual U, Balvanera P, Díaz S, Pataki G, Roth E, Stenseke M, Watson RT, Başak Dessane E, Islar M, Kelemen E, Maris V, Quaas M, Subramanian SM, Wittmer H, Adlan A, Ahn S, Al-Hafedh YS, Amankwah E, Asah ST et al (2017) Valuing nature's contributions to people: The IPBES approach. Curr Opin Environ Sustain 26–27:7–16. https://doi.org/10.1016/j.cosust.2016.12.006
Raymond PA, Hartmann J, Lauerwald R, Sobek S, McDonald C, Hoover M, Butman D, Striegl R, Mayorga E, Humborg C, Kortelainen P, Dürr H, Meybeck M, Ciais P, Guth P (2013) Global

carbon dioxide emissions from inland waters. Nature 503(7476):355–359. https://doi.org/10.1038/nature12760

Reynaud A, Lanzanova D (2017) A global meta-analysis of the value of ecosystem services provided by lakes. Ecol Econ 137:184–194. https://doi.org/10.1016/j.ecolecon.2017.03.001

Reynolds CS (1984) The ecology of freshwater phytoplankton. Cambridge University Press

Rockström J, Gupta J, Qin D, Lade SJ, Abrams JF, Andersen LS, Armstrong McKay DI, Bai X, Bala G, Bunn SE, Ciobanu D, DeClerck F, Ebi K, Gifford L, Gordon C, Hasan S, Kanie N, Lenton TM, Loriani S et al (2023) Safe and just Earth system boundaries. Nature 619(7968):102–111. https://doi.org/10.1038/s41586-023-06083-8

Rosińska J, Kozak A, Dondajewska R, Kowalczewska-Madura K, Gołdyn R (2018) Water quality response to sustainable restoration measures—case study of urban Swarzędzkie Lake. Ecol Ind 84:437–449. https://doi.org/10.1016/j.ecolind.2017.09.009

Santos A, Godinho DP, Vizinho A, Alves F, Pinho P, Penha-Lopes G, Branquinho C (2018) Artificial lakes as a climate change adaptation strategy in drylands: evaluating the trade-off on non-target ecosystem services. Mitig Adaptation Strat Global Change 23(6):887–906. https://doi.org/10.1007/s11027-017-9764-x

Santos M, Cajaiba RL, Bastos R, Gonzalez D, Petrescu Bakış A-L, Ferreira D, Leote P, Barreto Da Silva W, Cabral JA, Gonçalves B, Mosquera-Losada MR (2022) Why do agroforestry systems enhance biodiversity? Evidence from habitat amount hypothesis predictions. Front Ecol Evol 9:630151. https://doi.org/10.3389/fevo.2021.630151

Savilaakso S, Storie J, Caitana B, Campagne CS, Depellegrin D, Geneletti D, Kagkalou I, Lacarac D, Leucci F, Luque S, Scott A, Zahvoyska L (2023) State of knowledge regarding how we can improve adherence to the Mitigation Hierarchy, with a particular focus on the avoid stage. Zenodo. https://doi.org/10.5281/ZENODO.7780894

Schirpke U, Ebner M, Pritsch H, Fontana V, Kurmayer R (2021) Quantifying ecosystem services of high mountain lakes across different socio-ecological contexts. Sustainability 13(11):6051. https://doi.org/10.3390/su13116051

Schirpke U, Ebner M (2022) Exposure to global change pressures and potential impacts on ecosystem services of mountain lakes in the European Alps. J Environ Manag 318:115606

Seebens H, Blackburn TM, Dyer EE, Genovesi P, Hulme PE, Jeschke JM, Pagad S, Pyšek P, Winter M, Arianoutsou M, Bacher S, Blasius B, Brundu G, Capinha C, Celesti-Grapow L, Dawson W, Dullinger S, Fuentes N, Jäger H et al (2017) No saturation in the accumulation of alien species worldwide. Nat Commun 8(1):14435. https://doi.org/10.1038/ncomms14435

Souissi I, Boisson JM, Mekki I, Therond O, Flichman G, Wery J, Belhouchette H (2018) Impact assessment of climate change on farming systems in the South Mediterranean area: a Tunisian case study. Reg Environ Change 18(3):637–650. https://doi.org/10.1007/s10113-017-1130-8

Souliotis I, Voulvoulis N (2021) Incorporating ecosystem services in the assessment of water framework directive programmes of measures. Environ Manag 68:38–52 (2021). https://doi.org/10.1007/s00267-021-01478-7

Spears BM, Chapman D, Carvalho L, Rankinen K, Stefanidis K, Ives S, Vuorio K, Birk S (2022) Assessing multiple stressor effects to inform climate change management responses in three European catchments. Inland Waters 12(1):94–106. https://doi.org/10.1080/20442041.2020.1827891

Stockwell JD, Doubek JP, Adrian R, Anneville O, Carey CC, Carvalho L, De Senerpont Domis LN, Dur G, Frassl MA, Grossart H, Ibelings BW, Lajeunesse MJ, Lewandowska AM, Llames ME, Matsuzaki SS, Nodine ER, Nõges P, Patil VP, Pomati F et al (2020) Storm impacts on phytoplankton community dynamics in lakes. Glob Change Biol 26(5):2756–2784. https://doi.org/10.1111/gcb.15033

Tammeorg O, Chorus I, Spears B, Nõges P, Nürnberg GK, Tammeorg P, Søndergaard M, Jeppesen E, Paerl H, Huser B, Horppila J, Jilbert T, Budzyńska A, Dondajewska-Pielka R, Gołdyn R, Haasler S, Hellsten S, Härkönen LH, Kiani M et al (2023) Sustainable lake restoration: from challenges to solutions. WIREs Water, e1689. https://doi.org/10.1002/wat2.1689

Taylor NG, Grillas P, Al Hreisha H, Balkız Ö, Borie M, Boutron O, Catita A, Champagnon J, Cherif S, Çiçek K, Costa LT, Dakki M, Fois M, Galewski T, Galli A, Georgiadis NM, Green AJ, Hermoso V, Kapedani R et al (2021) The future for Mediterranean wetlands: 50 key issues and 50 important conservation research questions. Reg Environ Change 21(2):33. https://doi.org/10.1007/s10113-020-01743-1

Vári Á, Adamescu CM, Balzan M, Gocheva K, Götzl M, Grunewald K, Inácio M, Linder M, Obiang-Ndong G, Pereira P, Santos-Martin F, Sieber I, Stępniewska M, Tanács E, Termansen M, Tromeur E, Vačkářová D, Czúcz B (2024) National mapping and assessment of ecosystem services projects in Europe—participants' experiences, state of the art and lessons learned. Ecosyst Serv 65:101592. https://doi.org/10.1016/j.ecoser.2023.101592

Vatitsi K, Ioannidou N, Mirli A, Siachalou S, Kagalou I, Latinopoulos D, Mallinis G (2023) LULC change effects on environmental quality and ecosystem services using EO data in two rural river basins in Thrace, Greece. Land 12(6):1140. https://doi.org/10.3390/land12061140

Verdonschot PFM, Spears BM, Feld CK, Brucet S, Keizer-Vlek H, Borja A, Elliott M, Kernan M, Johnson RK (2013) A comparative review of recovery processes in rivers, lakes, estuarine and coastal waters. Hydrobiologia 704(1):453–474. https://doi.org/10.1007/s10750-012-1294-7

Vigerstol KL, Aukema JE (2011) A comparison of tools for modeling freshwater ecosystem services. J Environ Manage 92(10):2403–2409. https://doi.org/10.1016/j.jenvman.2011.06.040

Vlachopoulou M, Coughlin D, Forrow D, Kirk S, Logan P, Voulvoulis N (2014) The potential of using the ecosystem approach in the implementation of the EU Water Framework Directive. Sci Total Environ 470–471:684–694. https://doi.org/10.1016/j.scitotenv.2013.09.072

Wangai PW, Burkhard B, Müller F (2019) Quantifying and mapping land use changes and regulating ecosystem service potentials in a data-scarce peri-urban region in Kenya. Ecosystems and People 15(1):11–32. https://doi.org/10.1080/21513732.2018.1529708

Wantzen KM, Rothhaupt K-O, Mörtl M, Cantonati M, Tóth LG, Fischer P (eds) (2008) Ecological effects of water-level fluctuations in lakes. Springer, Netherlands. https://doi.org/10.1007/978-1-4020-9192-6

Weyhenmeyer GA, Müller RA, Norman M, Tranvik LJ (2016) Sensitivity of freshwaters to browning in response to future climate change. Clim Change 134(1–2):225–239. https://doi.org/10.1007/s10584-015-1514-z

White M, Smith A, Humphryes K, Pahl S, Snelling D, Depledge M (2010) Blue space: the importance of water for preference, affect, and restorativeness ratings of natural and built scenes. J Environ Psychol 30(4):482–493. https://doi.org/10.1016/j.jenvp.2010.04.004

Woolway RI, Kraemer BM, Lenters JD, Merchant CJ, O'Reilly CM, Sharma S (2020) Global lake responses to climate change. Nat Rev Earth Environ 1(8):388–403. https://doi.org/10.1038/s43017-020-0067-5

Woolway RI, Sharma S, Smol JP (2022) Lakes in hot water: the impacts of a changing climate on aquatic ecosystems. Bioscience 72(11):1050–1061. https://doi.org/10.1093/biosci/biac052

Xu X, Jiang B, Tan Y, Costanza R, Yang G (2018) Lake-wetland ecosystem services modeling and valuation: progress, gaps and future directions. Ecosyst Serv 33:19–28. https://doi.org/10.1016/j.ecoser.2018.08.001

Zhou W, Wang L, Li D, Leung LR (2021) Spatial pattern of lake evaporation increases under global warming linked to regional hydroclimate change. Commun Earth Environ 2(1):255

Zhu X, Campanaro S, Treu L, Seshadri R, Ivanova N, Kougias PG, Kyrpides N, Angelidaki I (2020) Metabolic dependencies govern microbial syntrophies during methanogenesis in an anaerobic digestion ecosystem. Microbiome 8(1):22. https://doi.org/10.1186/s40168-019-0780-9

The Use of Methodological Participation Tools in the Investigation and Evaluation of the Development Perspectives of the Transboundary Prespa Park, Based on the Perceptions of the Area's Primary Sector Professionals

Anna Latsiou, Spyros Niavis, Theodora Kouvarda, Ifigenia Kagalou, and Olga Christopoulou

Abstract Shared by Greece, Albania and North Macedonia, the Transboundary Prespa Park is a protected area where the primary sector is the main source of income and lake water management in Prespa Lakes one of the most important issues. However, different levels of development between the three countries, exacerbated by the fact that Albania and North Macedonia are not EU member states, presents common water management with significant on-the-ground challenges. Cooperation between local communities, and their involvement in decision-making through participatory processes, emerges as a useful management tool, improving stakeholder awareness and information, promoting the assumption of personal responsibility and initiative, and leading towards the adoption of sustainable practices. For the purposes of the current study, 420 primary sector professionals from 19 park localities, spanning all 3 countries, were engaged in participatory processes, with the aim of assessing their behaviour and perceptions regarding the protection and management of area of residence. To process the results, the PCP method was developed, a tool for quantifying information drawn from questionnaires, based on selecting question segments and matching them to three axes of the participatory process: Participation,

Supplementary Information The online version contains supplementary material available at https://doi.org/10.1007/978-3-032-02168-7_10.

A. Latsiou (✉) · T. Kouvarda
Institute of Marine Biological Resources and Inland Waters, Hellenic Centre for Marine Research, Anavyssos, Attica, Greece
e-mail: alatsiou@hcmr.gr

A. Latsiou · S. Niavis · O. Christopoulou
Department of Planning and Regional Development, University of Thessaly, Volos, Greece

I. Kagalou
Department of Civil Engineering, Democritus University of Thrace, Kimmeria Campus, Xanthi, Greece

K. Stefanidis et al. (eds.), *Lakes of the Eastern Mediterranean and Balkans*,
https://doi.org/10.1007/978-3-032-02168-7_10

Cooperation and Preservation. Opinion analysis of the participants' responses, highlighted their willingness to be involved in decision-making on local issues concerning the management of their area and to cooperate in order to contribute to the sustainable development of the Park. Apart from being used for evaluating results in future similar research, the PCP method could also be a useful tool for public agencies and entities involved in making participatory-based management decisions.

Keywords Stakeholder's involvement · Transboundary areas · Prespa Park · PCP tool · Development perspectives · Primary sector professionals · Protected areas · Water resources management

1 Introduction

The term "Transboundary Protected Area" refers to a defined geographical area, which straddles one or more borders, stretching beyond the boundaries of national sovereignty or jurisdiction. These areas are extremely important, as the constituent national parts are ecologically connected to each other, with the aim of protecting and maintaining the biological diversity and the associated cultural resources. As a result, they should be managed cooperatively, through legal or other effective means (Sandwith et al. 2001).

A cross-border component increases the complexity of water bodies' management, as it creates hydrological, economic and cultural dependencies. In most cases, these lead to tensions among local communities, mainly due to political, cultural and religious differences, varying levels of development, as well as an absence of institutional and legal procedures to resolve disputes between the parties involved (Kolokytha and Mylopoulos 2005; Latsiou and Christopoulou 2014). Decisions made and actions taken by a single state, are likely to have a negative influence on the quality and availability of ecosystem services in other neighboring states, creating numerous problems with significant effects on both the areas themselves and the people who live and work there (Votruba et al. 2020).

Cross-border development, management and cooperation through the application of participatory methods and processes, seems to multiply the ecological and social benefits for local communities. For this reason, a large body of scientists and policy makers argue that the users of such areas should be included in cross-border decision-making, emphasizing on the advantages of utilizing local knowledge (Christopoulou and Tsachalidis 2002; Mylopoulos and Kolokytha 2008; Chen2013; Latsiou and Christopoulou 2018). Furthermore, this inclusion may increase the effectiveness and legality of transboundary protection and management procedures (Bruch and Feinber 2016). Finally, it improves the users' morale, reducing the sense of isolation, strengthening the spirit of togetherness, and promoting the assumption of personal responsibility towards the common good (Patera 2022).

Therefore, common plans, methodologies and management practices should be developed, which in turn would facilitate the users-residents to shift towards a common understanding in dealing with local issues, and mutually agreed upon ways to address local problems. The continuous awareness raising of the residents and the integration of local opinion and knowledge in the decision-making processes, are important tools in ensuring the sound management of cross-border resources, as well as the sustainable regional development of the protected areas.

The present study attempts to apply participatory processes to primary sector professionals in the transboundary Prespa Park, with the aim of examining their perceptions regarding the development perspectives of their area and highlighting the value of local community participation in the management of a transboundary protected area. The contribution of participatory processes to achieving a viable development pace in such areas, the investigation of the perceptions of professionals in the primary sector of the transboundary Prespa Park regarding the management of their protected area, the presentation of an innovative participatory tool for quantifying and evaluating the above attitudes and perceptions developed in the context of this research (PCP method), the awareness of the local community of the park and its familiarization with the use of new methodological tools, are some of the objectives of this study.

2 Study Area

The transboundary Prespa Park (hereinafter referred to as TPP) is an important area in the Balkans due to its geographical, historical, cultural and environmental value. Its designation as a protected area, on the one hand attests to its value, and on the other, serves as a tool for the conservation of its rare habitats and rich biodiversity. The area's human presence, mirrored by the settlements, agricultural lands and fishing grounds, is another dominant characteristic of the TPP. Due to human activity the area has been under significant pressure in recent years, and as a result, the most critical management issue is to align human activity with nature protection, in order to promote the area's sustainable development.

The TPP was founded in 2000, following a joint declaration by the Prime Ministers of Greece, Albania and North Macedonia (Declaration on the creation of the Prespa Park and the environmental protection and sustainable development of the Prespa lakes and their surroundings 2000). It occupies an area of approximately 2500 km^2 and includes the Prespa National Parks in Greece and Albania, the Ezerani Nature Park in N. Macedonia and parts of the Galičica and Pelister National Parks of the same country (Fig. 1).

According to the 2021 censuses, approximately 22,000 residents live within the TPP area (Prespa Forum Dialogue 2022; ELSTAT 2022). The North Macedonian part of the TPP is the most densely populated, with 16,500 inhabitants, followed by the Albanian part with about 5000 inhabitants, while the Greek part is inhabited by 1220 people (ELSTAT 2022). The primary sector is the largest in the area, supporting the majority of the local population throughout the TPP (Clarke et al. 2000), with

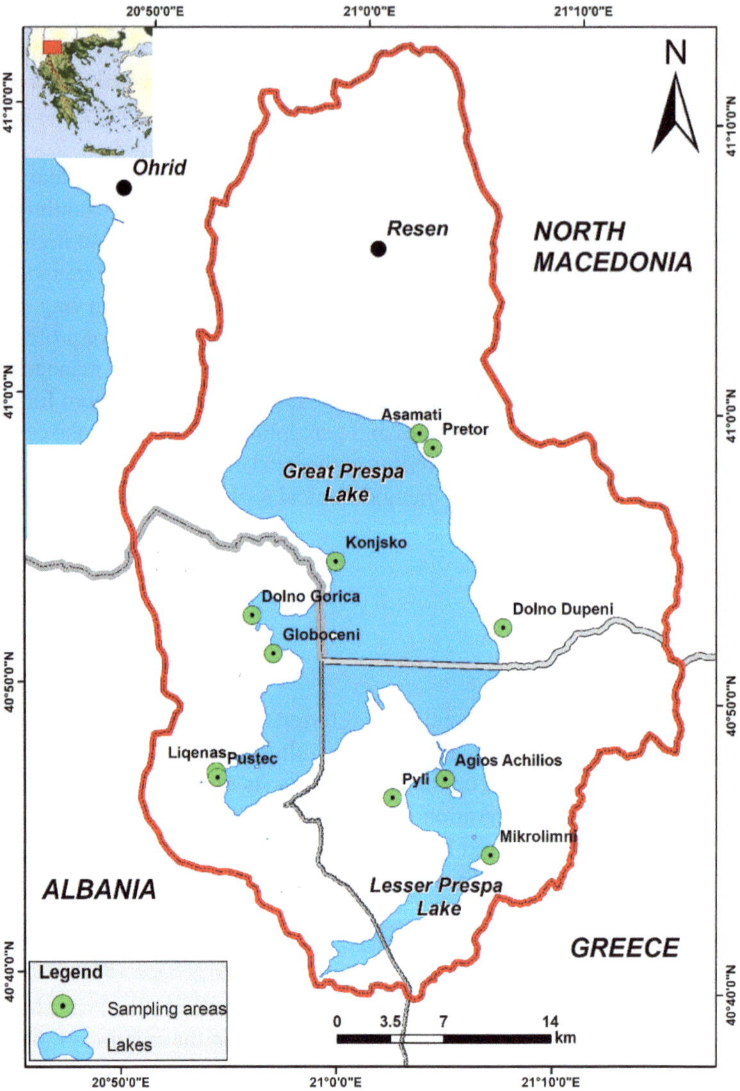

Fig. 1 The Transboundary Prespa Park and the sampling points

agriculture and fishing being the main professional activities among the residents (Catsadorakis et al. 2018, 2021a, b).

The founding of the TPP was the starting point of a coordinated approach towards the conservation and protection of the area's natural environment. The whole area has been recognized as an international biodiversity hotspot, not only due to the huge number of species and habitats that are concentrated in such a small area, but also due to their rarity and conservation importance (Petit et al. 2003). A second factor that led to its creation and the drawing up of common policies, was the need for common water management, one of the most important issues that concern its inhabitants.

The establishment of special committees played an important role in the management of the area, such as the Greek-Albanian permanent committee for transboundary freshwater issues (N3505/2005 FEK 264/A/25-10-2005), the Wetlands Management Committee, the Transboundary Committee of Wetlands, etc. Consisting of many partners and acting as management institutions, the above committees have contributed substantially to the definition of cooperation priorities for water management, to the protection of the rare biodiversity of the transboundary area, as well as to the exchange of information and scientific data between the three countries. The inclusion of the TPP residents in decision-making processes and the recognition of the value of local knowledge are important aspects of the functioning of these committees.

3 Materials and Methods

For the sampling needs of the present study, 13 localities were selected within the TPP (Fig. 1), based on their distribution in the lakeside area in order to fully represent all three states in the survey following (van den Broek et al. 2019), their proximity to the water body and the agricultural areas and their differing degree of social, economic, environmental and cultural development (Araviadis 2014; Michalatou 2023).

A structured questionnaire was created to collect the data, consisting of 4 parts and 43 questions, in three question types: close-ended (yes/no) questions, open-ended questions and multiple-choice questions (see Appendix).

The sampling was conducted during 8 weekly visits to the study area, with the participatory processes taking place either outdoors (during working hours), or in cafes and restaurants (in the afternoons and evenings).The examined sample consisted of 420 primary sector professionals (80 in Greece, 100 in Albania and 240 in North Macedonia). The applied sampling technique was the snowball sampling method (Bernard 2006), in which the participants themselves indicated the next candidates as soon as they completed filling in the questionnaire.

3.1 Quantification of the Results

According to De Koning and Avramoski (2021), the collaboration that comes from the participation of residents and users of the TPP, is vital to ensure the sustainable conservation of the protected area and can be based on previous experience and knowledge (Stratigea and Papadopoulou 2012). However, it remains "unofficial" with any efforts being limited to the local and national level, despite the fact that the institutional framework regarding the participation of citizens in issue-solving is constantly evolving. Furthermore, in the majority of transboundary areas in Europe, case studies usually develop their own tools, adopting specific approach methods (Rees et al. 2005), which can be a useful way of improving the participatory process. However, tools or methods that are designed for specific areas with specific characteristics are likely to be more effective.

For the current study, we developed a tool for data processing and result analysis. This is the PCP Method, a method of quantifying the responses of participants, which was based on three important axes of the participatory process: participation, cooperation and preservation (Brewer 2009; Chai 2016).

To begin with, the inclusion of residents and users of an aquatic transboundary area in decision-making processes is supported by the Water Framework Directive 2000/60/EC (Paletto et al. 2017). In the case of the TPP, the Directive's rationale is also adopted by Albania and North Macedonia, even if they are not EU members. In addition, participation is understood as an "open" process that includes the active involvement of individuals and groups in the decision-making process through the exchange of opinions, information and knowledge (Hatzimichailidou 2019). The concept of Cooperation highlights the important role of stakeholder composition in decision-making processes (IUCN 2013; Di Franco et al. 2020; Zikargae et al. 2022).

Finally, the definition of Preservation includes the participation of those involved in the management processes through cooperation, with a view to protecting their area. Thus, the participation process is essentially a collaboration platform, on which the public and different interest groups communicate and interact, with the aim of preserving and sustainably developing their area. The creation of the PCP method was based on matching the questionnaire questions, answered by primary sector professionals (farmers and fishermen) of the TPP, to the above axes. The specific questions examine the inclination for participation and shared use of park infrastructure, the adequacy of inter-country cooperation, and the environmental knowledge and awareness of the primary sector professionals of the area. Following this, and based on the content of these questions, 12 evaluation criteria were created, which were similarly assigned to the PCP axes. The questions per axis, as well as the evaluation criteria that were subsequently formed, are presented in Tables 1 and 2 respectively.

For the formation of the PCP tool, close-ended questions (YES/NO) were chosen. In these, the answer YES scored 1, while the answer NO scored 0. For questions that have a negative connotation regarding the attitude of the respondents towards the three axes, reverse coding was used so that the value one (1) would reveal a positive attitude and the total sum would be consistent with the direction: lower values—negative attitude → higher values—positive attitude. For each participant, the total score per axis (SUM) was calculated, with the lowest value indicating a lower willingness to engage in participation, collaboration and preservation processes, and hence a lower willingness to engage in the participatory process overall.

In order to be able to compare the scores per axis, and since each axis includes a different number of sub-axes, initially each participant's score is divided by the maximum possible score for each axis. Therefore, considering that the maximum value for each sub-axis is 1, for an axis of n sub-axes the maximum score (max score) is n (Eq. 1).

Table 1 Participatory process questions on which the PCP method was based

Participation
• Do you think you can contribute to the environmental management of the transboundary park by changing your everyday behaviour? • Do you think that your daily behaviour affects the ecosystem of the transboundary park? • Are you a member of a farmers' cooperative? • Would you like to get involved in the transboundary park management? • Are you involved in the transboundary park management? • Would you contribute financially to improve the operation and management of the protected transboundary ecosystem?
Cooperation
• Do you think it is right for each state to manage the lake independently, with its own legislation and rules? • Would you join a transnational partnership that would apply common water management rules for each use? • Would you be willing to share a common irrigation system with neighbouring countries? • Would you sign an agreement with the management bodies of the park regarding the conversion of your cultivation to organic farming? • Would you allocate part of your crop revenue to a transboundary body/cooperative for the distribution of agricultural products?
Preservation
• Did you know that the Lesser and Great Prespa lakes are integrated into the Natura 2000 Network? • Do you think that there is a benefit to the wider transboundary park area from this integration? • Do you know the main environmental issues in the transboundary park? • Have you identified any sources of pollution that are affecting the water resources in your area? • Is water in the transboundary park a natural resource?

Table 2 Evaluation criteria of the PCP method

Participation	• Change in everyday behaviour • Participation in the management of the transboundary park • Willingness to participate in the management of the transboundary park • Financial support for the sustainable development of the transboundary park
Cooperation	• Creation of a common water network • Creation of a common irrigation system • Drafting of common legislation for the use of water resources • Financial assistance to a transnational body for the distribution of primary sector products
Preservation	• Understanding of the term "Protected Area" • Treating water as a natural resource • Knowledge of the environmental conditions in the transboundary park • Knowledge of environmental issues in the transboundary park

$$Axis\ Score = \sum_{i=1}^{n} Score_i \Big/ n \quad (1)$$

Then, the final scores of each axis are normalized based on the following equation using the maximum (max) and minimum (min) values of the distribution (Eq. 2).

$$Final\ Axis\ Score = Score - Min\ Score \Big/ Max\ score - Min\ Score \quad (2)$$

The use of the PCP method can highlight the importance of local communities' participation in the management of protected areas, as the users who are engaged in them, possess important knowledge and experience that can be deemed useful for the decision-making process (Gulte et al. 2023).

The methodology and research planning stages are schematically presented in the flow chart shown in Fig. 2.

4 Results and Discussion

The use of the PCP axes allows for a comprehensive comparison of the views and perceptions of the 420 primary sector professionals concerning the TPP's sustainable management, based on their various socio-economic and demographic backgrounds.

However, in order for correlations with various qualitative parameters to be included (such as perceptions on the area's development, concerns about the future, daily practices at work, etc.), it was deemed necessary to use questions that were not introduced into the PCP evaluation framework. The evaluation was carried out by creating a Pivot table, which was a suitable tool through which the scores of each axis (per country) were correlated for each of the remaining research questions. The evaluation and scoring of these questions resulted in an average per country/ per axis, as well as in an average per country for the whole set of axes. The second average also constitutes the Total PCP Score, as shown in Table 3.

As shown in Table 3, Greek respondents scored a higher average in the Participation axis, which demonstrates a greater willingness to participate. For Albanian respondents, the highest score on the Preservation axis shows their willingness to be involved in the preservation of the region, while the Cooperation axis scores highest for the North Macedonian participants. Finally, respondents from North Macedonia scores the highest Total PCP.

The following results show the effect of primary sector professionals' demographic background in the shaping of their perceptions concerning the TPP's sustainability. These perceptions influence PCP components, and, in turn, these shape the participants' views and attitudes towards environmental preservation and the future development of the area of residence. Finally, a comparison is made between the professionals of the three countries, and the differences in their attitudes towards the TPP's sustainability are presented.

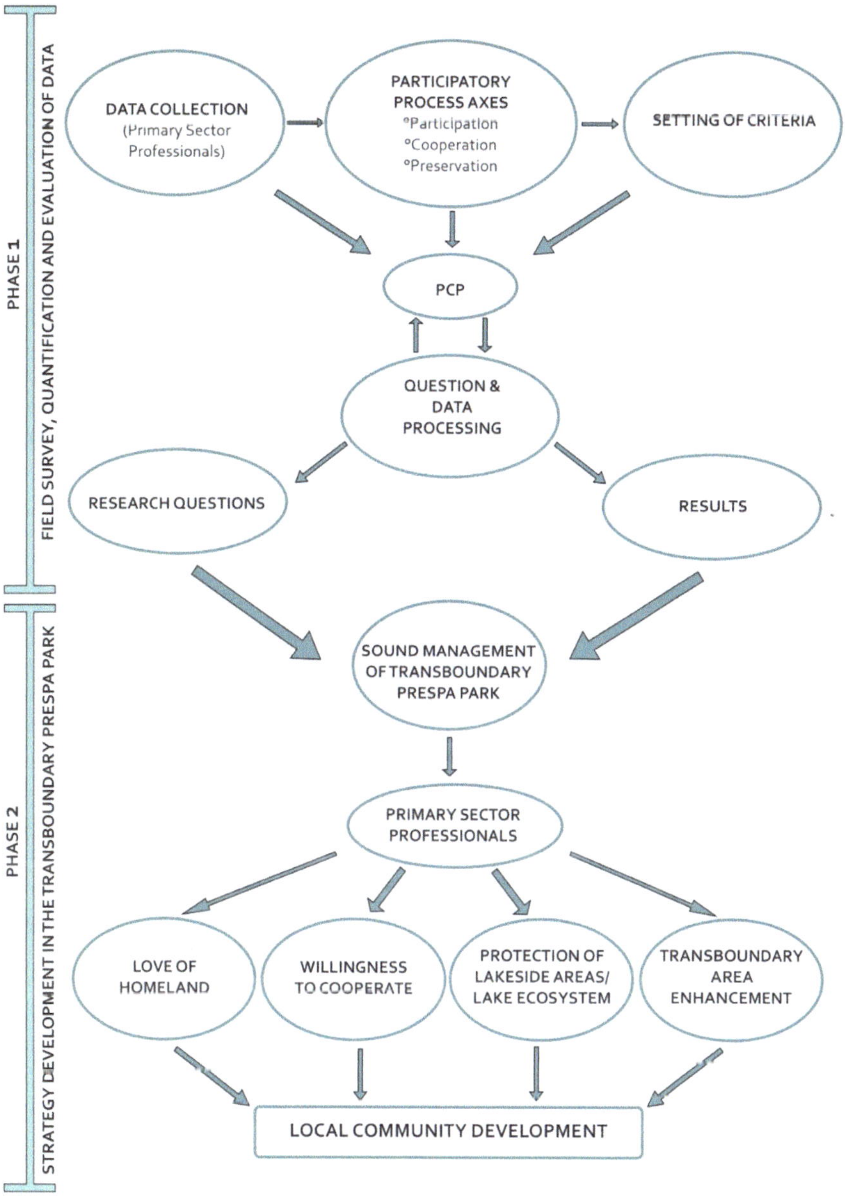

Fig. 2 Methodology and research planning stages

Table 3 PCP axes average and total PCP score per country

	Participation	Cooperation	Preservation	Total PCP
GR	0.660	0.460	0.631	0.584
AL	0.428	0.284	0.563	0.425
NM	0.660	0.678	0.623	0.653

4.1 The Role of Demographic Characteristics in Shaping Perceptions of Transboundary Park Sustainability

Demographic diversity in decision-making processes, has emerged as an important factor in investigating questions related to the challenges that face sustainable development in protected areas (Sampsonidou 2020). Observing changes in demographic characteristics enables the identification of trends and supports the development of management strategies that take those trends into account. The demographic characteristics that mainly determine the participants' profiles, are age, gender and education level, and those are the ones that the present study focuses on.

According to Fig. 3a, in all countries, the majority of participants were men, accounting for more than 78% of the total, while the percentage of women ranges from 9.5 to 22%, depending on the country. In related surveys that have been carried out in the Prespa area (Catsadorakis et al. 2018) as well as in other protected areas (Clusa et al. 2018), the presence of women engaged professionally in the primary sector is also low. In addition, according to the Mann–Whitney results (ESM_1), men are more receptive to the application of the PCP method (Mean Rank = 217.91) than women (Mean Rank = 164.24) in all cases. According to Louder (2020), this is because the majority of primary sector professionals are men who are better engaged with the TPP's ecosystem and understand or more often seek solutions to issues that arise and concern their daily lives. Population ageing is one of the biggest challenges facing rural areas. Also according to Fig. 3b, which shows the sample's age distribution, ageing among TPP's primary sector professionals was recorded at 36.25%, 47% and 48.75% for Greece, Albania and North Macedonia respectively. According to Kasimis and Papadopoulos (2013), incentives and support measures aiming to help young primary sector professionals are considered insufficient, and their impact on generational renewal remains limited, especially in areas that lack basic infrastructure and services. It is interesting that the Kruskal–Wallis test results, that correlate age data with the PCP axes (ESM_2), show that the participants aged 46–65 are more positively inclined to the Participation axis than those in the 26–45 age group, mainly due to a lack of regular financing and having to constantly deal with practical difficulties, such as the absence of infrastructure, the unclear ownership status and the lack of training for new farming practices (Gorner et al. 2012). On the contrary, participants who are younger than 25 years old show a more positive attitude towards the Cooperation and Preservation axes. It is important to mention that these findings are mostly assumptions since the results are not statistically significant. Finally, Fig. 3c illustrates that the majority of participants in the three countries have

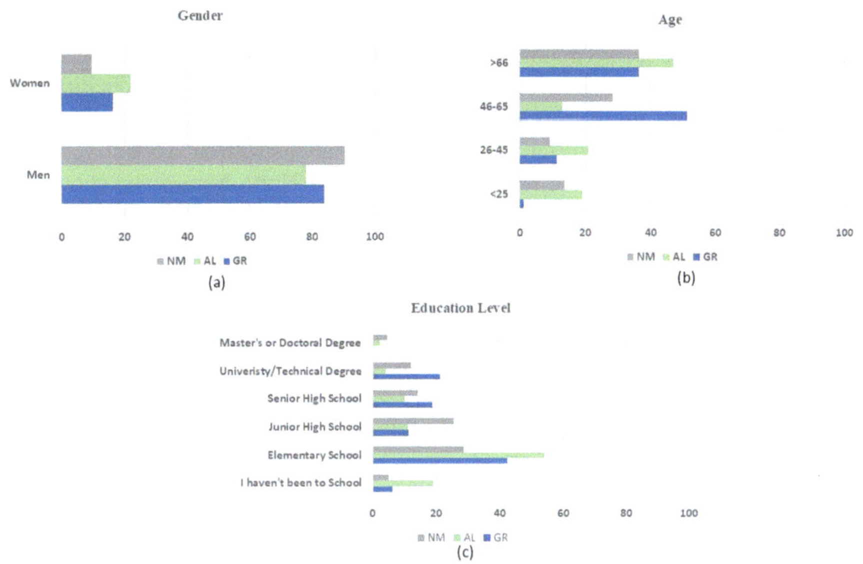

Fig. 3 Respondent gender, age and education level distribution

only completed elementary education, which explains their low inclination towards Participation, due to the lack of knowledge.

4.2 *Views and Perceptions of Primary Sector Professionals that Influence the PCP Index*

The following results examine whether the perceptions of primary sector professionals regarding water management issues, such as pollution, water availability and its appropriate use in agriculture, influence the PCP axes.

The majority of the participants consider that the issue of water pollution in the area is more important than that of water scarcity. This view affects their attitudes towards the Participation, but not with respect to the Cooperation and Preservation, nor to the Total PCP index. The Error Graphs (ESM_3,4,5) confirm the above, seeing that although the means of the Cooperation and Preservation axes differ, their confidence intervals overlap. Additionally, a statistically significant difference can be observed in the participation score of the North Macedonian participants, who seem negatively inclined regarding their Participation (ESM_3), as well as in the preservation score of the Greek participants, who seem positively inclined towards preservation (ESM_5). These results are in accordance with the value differences of the corresponding mean ranks of ESM_6. Concerning the participants who, contradictorily, consider that the issue of water scarcity is more important than that of pollution, it appears that their attitudes affect the Participation axis ($p = 0.030$), but not the Cooperation

($p = 0.488$) and Preservation ($p = 0.271$) axes, nor the Total PCP index ($p = 0.474$) (see ESM_7). In the Error Graphs (ESM_8,9) the above relationships are confirmed, despite the fact that the means of the Cooperation and Preservation axes differ, as the confidence intervals overlap. As a result, there cannot be a confident assessment of a statistically significant difference in their correlations with each country. Nevertheless, the Cooperation score for the participants of North Macedonia shows the highest value (ESM_9). From the overall results, it appears that the participants are positively inclined towards the Participation axis, i.e. they have the willingness to participate in decision-making processes regarding the issues of water pollution and scarcity. This is probably because the problems caused by the reduction of either the quality or the quantity of the water in agricultural activities, have a significant impact on their daily life.

According to the results of ESM_10, the participants' view that environmental degradation is primarily caused by agricultural cultivation (and not by increased water consumption, the reduction of water volume, the increase of soil erosion, etc.), significantly affects their predisposition towards the Participation axis ($p = 0.001$) but less towards the Cooperation and Preservation ($p = 0.000$ and $p = 0.010$ respectively) axes, nor towards the Total PCP index ($p = 0.019$). This shows that even though they believe that agricultural cultivation is responsible for environmental degradation, they are negatively inclined towards participating in an effort for sound management, as they know that, being professional farmers themselves, they are also responsible for this situation.

Regarding the issue of irrigation using lake water, the participants who know that irrigating in this way causes a problem in the water balance of the Prespa lakes, significantly influence the Participation axis ($p = 0.004$) and the Total PCP index ($p = 0.001$), but not the Cooperation ($p = 0.332$) and Preservation ($p = 0.236$) axes, as shown in ESM_11 Table. According to Geist and Hawkins (2016), participants who identify themselves or their activities as part (or even the cause) of a problem, show more openness in participating in various management bodies, looking to those third parties for assistance, in order to find the best possible solution for themselves and their region. Finally, the results of ESM_12 indicate that the participants who consider the use of water for agricultural purposes a more important ecosystem service than landscape aesthetics, significantly affect the Participation axis ($p = 0.005$) and the Total PCP index ($p = 0.004$), but not at all the Cooperation and Preservation axes ($p = 0.376$, $p = 0.379$).

4.3 How the PCP Index Affects Behavioural Change Among Primary Sector Professionals

The following chapter examines whether the PCP methodological tool can influence the daily behaviour and farming practices of primary sector professionals, and whether they are willing to commit to change, in order to contribute to a better management for the TPP.

The intention of the TPP users to legally commit (e.g. by signing a contract) to change their attitudes and behaviour in order to contribute to the area's conservation, may play an important role in management planning. Those participants who would be willing to sign such a contract are actually more than those who refuse to commit in this way, as they look forward to the possibility of financial benefits (ESM_13). This majority seems to be influenced by the axes of Participation ($p = 0.003$), Cooperation ($p = 0.015$) and Total PCP index ($p = 0.002$), but not by the Preservation axis ($p = 0.734$).

At the same time, those participants who are positively inclined towards converting their cultivation to organic farming, are more than those who refuse to do so. As it appears, even marginally ($p = 0.057$), the Preservation axis is the only one that seems to be affected by this majority of participants (ESM_14), which is also confirmed by high scores for all three countries in the Error Graph of Fig. 4. As organic farming is utilizing natural resources and producing through natural processes, it is fully harmonized with the natural environment (Dimitrovski et al. 2012; Europark Federation 2023). The participants seem to be influenced by the axis of Preservation, as they are convinced that by converting their crops to organic, they will manage to maintain the exceptional attributes of the transboundary area. They also express the opinion that through organic cultivation, they will be able to take part in more European programmes and receive larger subsidies. In addition, they aspire for higher quality products giving them access to a wider market and allowing for better selling prices, thereby stimulating the economy. The fact that both the production and consumption of organic products is constantly increasing in all western countries, reinforces this view (Haider et al. 2016; Pawlewicz et al. 2022; Pawlewicz and Pawlewicz 2023).

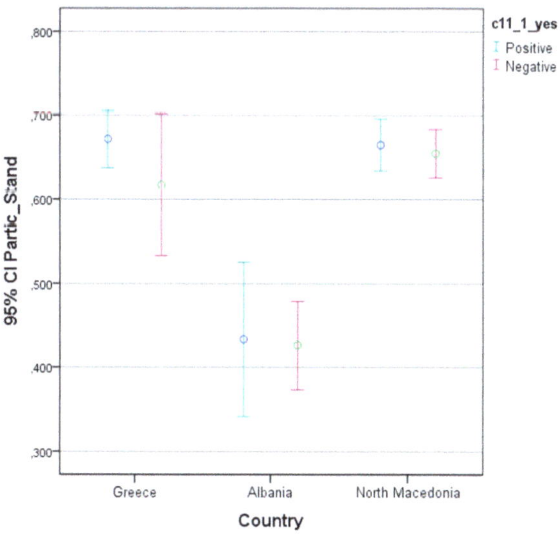

Fig. 4 Comparison of mean values of the "Preservation" axis by country and confidence intervals (95%)

4.4 Shaping the Views of the Primary Sector Professionals Regarding the Future Development of the Area, According to the PCP Method

The following chapter examines whether and how the PCP method can shape participants' views on the future development of the primary sector, as well as how the respondents view the perspectives of a rise in trade and in exports of local agricultural products.

Rural lakeside zones in protected areas can get on the fast track for economic growth, if their users are willing to participate in the management of these areas (Kazoglou and Vrahnakis 2008). According to the results presented in ESM_15, the majority of participants believe that the area's rural development will follow a stable or upward course. This majority are positively inclined towards the Participation axis ($p = 0.000$), whereas no particular correlation was found with regards to the Cooperation and Preservation axes, nor with the Total PCP index ($p = 0.483$, 0.465 and 0.062 respectively). At the same time, as shown in ESM_16, most participants hold a positive view regarding the development of trade (and especially export trade) of local agricultural products Participation ($p = 0.000$), Cooperation ($p = 0.000$) and Total PCP index ($p = 0.000$)]. It seems that the majority of primary sector professionals believe that participation and cooperation between them can achieve positive results towards the growth of export trade, something that is expected to further economic development in general.

4.5 Inter-country Differences in the Views of Primary Sector Professionals Towards the Area's Sustainability, According to the PCP Method

ESM_17, shows the inter-country differences in the views of primary sector professionals regarding the implementation of the PCP index, which reflects their willingness to engage in participatory processes. Generally, the differences observed in all PCP axes are statistically significant. Regarding the Participation axis, mean ranks for Albania are lower (118.97) than those for Greece and North Macedonia (238.85 and 239.55 respectively). In the Cooperation axis, it is noted that the North Macedonian participants are particularly positive towards using the tool in order to achieve cooperation at a transboundary level. In relation to the Preservation axis, the Greek participants seem the most positive; having realized the importance of the TPP's sensitive ecosystem, they understand the necessity of its protection. The presence and actions of the Society for the Protection of Prespa seem to have been a catalyst for this. Finally, regarding all PCP axes, North Macedonians are the most receptive towards their use, with Greeks and Albanians following up respectively. The overall high scores of North Macedonian participants are probably due to their great desire to join the European Union, as they seem to believe that their cooperation with neighbouring countries, as well as with entities operating at a transboundary level, will strengthen their country's integration processes (Damjanovski 2022).

Finally, Table 4 illustrates the formation of the participants' views regarding the usefulness of the Participation, Cooperation and Preservation axes, as well as the Total PCP index, in relation to issues of management, protection and sustainability within the TPP. All the answers provided by the participants in all three countries, mostly affect (or are affected by) the Participation axis and the Total PCP index, rather than the Cooperation and Preservation axes. This is probably due to the fact that public participation is the first stage of the participatory process, where the necessity of user involvement in environmental management and protection is still a theoretical process. However, things become more challenging when users are called upon to implement in practice specific management strategies, through cooperation, as they strive towards the sustainable development of the protected area.

5 Advantages and Disadvantages of the PCP Assessment Tool

The TPP area is characterized as one of the most ecologically valuable areas in Europe. Balancing between the human element and the protection of nature, or otherwise, maintaining a harmonious coexistence with the aim of achieving sustainable development, is the most critical issue in the area's management. The present study attempted to quantify and assess the beliefs and behaviors of primary sector professionals in the TPP, using the PCP method, an innovative tool designed and created for this purpose.

The application of such a methodology aims to create a model in which the participants may evaluate the area and its functions, but also conduct self-criticism regarding their perceptions and behavior towards the protected ecosystem. It is a simple tool that provides a basis upon which the application of participatory processes can be built and therefore, it can be used by a multitude of agencies and entities. Furthermore, it provides the ability to prioritize management efforts, highlights the interaction opportunities between countries at multiple levels (economic, environmental, developmental, educational, etc.), explores interactions between different occupational groups (both within a country and at a cross-border level) and finally, facilitates management decision-making.

The participants' views and perceptions affect (and are affected by) the use of the PCP assessment tool. The application of the participatory processes concerns, firstly, user participation (Participation)—based on the necessity of adopting measures to protect and preserve the natural environment (Preservation) in combination with the area's economic development—and secondly, the cooperation between users (Cooperation). The reservations or receptiveness expressed by the users towards ecosystem protection can negatively or positively affect their participation and cooperation. In

Table 4 Summary table of participants' views in relation to the PCP axes

Research questions	Participation	Cooperation	Preservation	Total
Views and perceptions affecting the PCP index				
Does the respondents' view that farmers are the group that causes the most important problems in the TPP ecosystem, affect the PCP axes?	●	▲	●	▲
Does the respondents' view that water pollution in the TPP is primarily caused by agriculture, affect the PCP axes?	▲	▲	▲	▲
Does the fact that respondents consider the issue of water pollution more important than that of water scarcity in the TPP, affect the PCP axes?	▲	●	●	●
Does the respondents' perception that farmers are primarily responsible for the reduction in available water in the TPP, affect the PCP axes?	●	●	●	●
Does the respondents' knowledge that irrigating using lake water disturbs the water balance of the Prespa lakes, affect the PCP axes?	▲	●	●	▲
Do respondents who consider the use of water for agriculture a more important ecosystem service than landscape aesthetics, affect the PCP axes?	▲	●	●	▲
How PCP index influences behaviour change				
Does the PCP index influence participants towards contractually committing to a better management for the TPP?	▲	▲	●	▲
Can the PCP index influence farmers to convert to organic farming?	●	●	▲	●
Does the PCP index influence one country's participants to sign a contract involving changes in land use, if they know that their colleagues from the other two countries have already done so?	▲	▲	●	▲
Formation of opinions of primary sector professionals towards the future development of the area according to the PCP method				
According to the PCP index, what do the participants think about the rural development of their area?	●	▲	●	●

(continued)

Table 4 (continued)

Research questions	Participation	Cooperation	Preservation	Total
According to the PCP index, what do the participants think about rising trade and local product exports?	▲	▲	●	▲
According to the PCP index, what opinion do the participants form about the development of ecotourism in their region?	▲	▲	●	▲

▲ It affects the axes and the overall PCP frame
● It does not affect the axes and the overall PCP frame

addition, the PCP method can help the decision-making processes cover the users' various and constantly changing needs, enhancing the participatory process.

In summary, the following are highlighted:

- The PCP method is a useful, simple-to-use, cost-effective and innovative participatory tool. Its application allows for local communities' views on complex issues to be classified, evaluated and supported.
- The PCP method can be used to assess the sustainability of specific management measures and policies, taking into account three important axes of participatory processes: Participation, Cooperation and Preservation.
- The implementation of the PCP method can encourage scientific practices among TPP residents and professionals.
- Using the PCP method, cross-border communication and agreement between residents and professional groups can be strengthened, for the purpose of conservation.
- Primary sector professionals are arguably able to perceive environmental changes within the TPP ahead of other organized user groups, as they are directly affected by the ecosystem's functions. Through the use of the PCP method, they could perform the important role of bringing these changes to everyone's attention.
- Close cooperation between users and the TPP management bodies is considered essential.
- Scientific studies can strengthen local community participation in the area's management, for the purpose of conservation.
- Participatory processes can promote the implementation of new tools for the effective use of local knowledge, contributing to the achievement of sustainable development.
- More environmental education and awareness raising projects should be implemented in all three TPP countries, emphasizing the importance of the protected area and the cooperation between citizen groups. These should primarily focus in the areas closer to the lakes, where primary sector professionals are mostly active.
- Continuous monitoring of Prespa's ecosystem functions, of human activities and interventions, as well as of the local community's participation in the area's management, is essential.

- Adopting and implementing sound management measures based on local knowledge can contribute towards the effort of achieving a common growth rate within the TPP.
- The implementation of more studies in the TPP area focusing on the relationships between different social groups, is considered crucial, as continuous information feedback is necessary at every management level. The development and promotion of methodologies that strengthen local community participation and contribute to the implementation of participatory processes, is a key element in this direction.

The creation of local communities' opinions assessing tools is necessary as public participation in the decision-making process is proved to be extremely useful.

The application of such tools, especially when they are used for the first time, depends mostly on the sample of participants. It is important for the reliability of the method to include representatives from all social groups of interest to ensure that their views reflect the real needs and attitudes of the population. Another major issue is the availability of time given to the participants on order to answer the questions posed.

The sample used in this study in which the PCP method was created, was representative and there was no time limitation for the participants. However, in the case of a transboundary area like the TPP, other issues such as the different cultural perceptions, the different rates of development between countries and the conditions of political instability that may appear from time to time in the region, are likely to make PCP inadequate in detecting such types of inequalities as it is based on criteria that respond better to more homogeneous samples.

Furthermore, the functionality of these novel tools such PCP, can be assessed more objectively after being tested in different surveys and over a larger period of time.

References

Araviadis RV (2014) Investigation of pressures on the environment and especially on water resources in the Prespa Basin and proposals for sustainable economic development in the area with an emphasis on the protection and security of water resources. Master's Thesis, Faculty of Engineering, School of Civil Engineering, Aristotle University of Thessaloniki (in Greek)

Bernard R (2006) Research methods in anthropology: qualitative and quantitative approaches. Rowman Altamira

Brewer J (2009) Public participation in environmental assessment and decision making. National Academies Press. ISB: 0-309-12399-2

Bruch E, Feinberg F (2016) Decision-making processes in social contexts. Ann Rev Sociol 43:1–21. https://doi.org/10.1146/annurev-soc-060116-053622

Catsadorakis G, Papadopoulou E, Petrakos M, Koutseri I (2018) Status of fisheries at Megali Prespa Lake and Mikri Prespa Lake, Greece, based on a census of fishermen's opinions. Environ Ecol Res 6:583–592. https://doi.org/10.13189/eer.2018.060608

Catsadorakis G, Roumeliotou V, Koutseri I, Malakou M (2021a) Multifaceted local action for the conservation of the transboundary Prespa lakes Ramsar sites in the Balkans. Mar Freshw Res. https://doi.org/10.1071/MF21123

Catsadorakis G, Mougiakou E, Kizos Th (2021) Ridge-and-furrow agriculture around Lake Mikri Prespa, Greece, in a European perspective. J Eur Landsc 2:7–20
Chai T (2016) The importance of public participation. Environmental Justice and Sustainability Clinic. Retrieved from https://ejsclinic.info.yorku.ca/2016/04/the-importance-of-public-participation/
Hatzimichailidou A (2019) Collective administration and the decision-making process in school units of Primary Education. Master's Thesis, School of Human Movement and Quality of Life Science, Department of Sport Management, University of the Peloponnese (in Greek)
Chen J (2013) Voter partisanship and the effect of distributive spending on political participation. Am J Polit Sci 57:200–217. https://doi.org/10.1111/j.1540-5907.2012.00613.x
Christopoulou O, Tsachalidis E (2002) A contribution to the management of protected areas: Comparative study of wetlands residents' opinions in Northern Greece. In: Proceedings of the international conference: protection and restoration of the environment, Skiathos Island, Greece
Clarke R, Kolemisevska-Gugulovska T, Dimirovski G, Gjiknuri L, Grupce L, Andreesk C (2000) Conservation and conflict resolution in transboundary protected areas: Prespa as a European case study. IFAC Proc Vol 33(8):127–134. https://doi.org/10.1016/S1474-6670(17)35432-0
Clusa L, Miralles L, Fernández S, Garcia-Vazquez E, Dopico E (2018) Public knowledge of alien species: a case study on aquatic biodiversity in North Iberian rivers. J Nat Conserv 42. https://doi.org/10.1016/j.jnc.2018.01.001
Damjanovski I (2022) Analysis of public opinion on North Macedonia's accession to the European Union (2014–2021). Public Opinion Analysis Paper No. 04/2022. Konrad Adenauer Foundation in the Republic of North Macedonia, Institute for Democracy "Societas Civils"—Skopje
Declaration on the creation of the Prespa Park and the environmental protection and sustainable development of the Prespa lakes and their surroundings (2000). Retrieved from https://spp.gr/wp-content/uploads/2023/04/2000-Declaration-on-the-creation-of-the-Prespa-Park-and-the-environmental-protection-and-sustainable-development-of-the-Prespa.pdf
De Koning M, Avramoski O (2021) The importance of partnerships for effective protected area management. In: Protected area management—recent advances. https://doi.org/10.5772/intechopen.99595
Di Franco A, Hogg K, Calò A, Bennett NJ, Sévin-Allouet M, Alaminos O, Lang M, Koutsoubas D, Prvan M, Santarossa L, Niccolini F, Milazzo M, Guidetti P (2020) Improving marine protected area governance through collaboration and co-production. J Environ Manage 269:110757. https://doi.org/10.1016/j.jenvman.2020.110757
Dimitrovski D, Todorović A, Valjarević A (2012) Rural tourism and regional development: case study of development of rural tourism in the region of Gruţa, Serbia. Procedia Environ Sci 14:288–297. https://doi.org/10.1016/j.proenv.2012.03.028
ELSTAT (2022) Census 2022. Retrieved from https://elstatsources.statics.gr/Census2022GR.pdf
Europarc Federation (2023) by Espen Quinto-Asman. Conservation management practices in three different European protected areas—with focus on hydrology and grazing. Alfred Toepfer Natural Heritage Scholarship, Ottenby Bird Observatory, Sweden. Retrieved from https://www.europarc.org/wp-content/uploads/2023/10/ATS-final-report_Espen-Quinto-Ashman_smaller_compressed.pdf
Geist J, Hawkins SJ (2016) Habitat recovery and restoration in aquatic ecosystems: current progress and future challenges. Aquat Conserv Mar Freshwat Ecosyst 26:942–962. https://doi.org/10.1002/aqc.2702
Gorner T, Najmanova K, Cihar M (2012) Changes in local people's perceptions of the Šumava National Park in the Czech Republic over a ten-year period (1998–2008). Sustainability 4:1354–1370. https://doi.org/10.3390/su4061354
Gulte E, Tadele H, Haileslassie A, Mekuria W (2023) Perception of local communities on protected areas: lessons drawn from the Bale Mountains National Park, Ethiopia. Ecosyst People 19(1). https://doi.org/10.1080/26395916.2023.2227282
Haider U, Mostegl NM, Kelemen-Finan J, Haider W (2016) Farmers' preferences for future agricultural land use under the consideration of climate change. Environ Manage 58:446–464

IUCN (2013) Stakeholder participation toolkit for identification, designation and management of marine protected areas. Ed. RAC/SPA, Tunis, 30 pp
Kasimis C, Papadopoulos A (2013) Rural transformations and family farming in contemporary Greece. https://doi.org/10.1108/S1057-1922(2013)0000019013
Kazoglou Y, Vrahnakis M (2008) Networking of five LIFE-Nature projects in Greek Ramsar Wetlands: lessons learnt from vegetation management and monitoring activities. In: Proceedings of the 6th European conference on ecological restoration, Society for Ecological Restoration, Europe, 8–12 Sept 2008, Ghent, Belgium
Kolokytha E, Mylopoulos Y (2005) Hydrodiplomacy: a ground for action, transboundary cooperation and economic development in Nestos River Basin, Greece. In: Proceedings of the 31st IAHR world congress
Latsiou A, Christopoulou O (2014) Rural development perspectives, based on the rural population's perceptions of the cross-border Lake Prespa (North Greece). In: Proceedings of the 15th world lake conference, Perugia
Latsiou A, Christopoulou O. (2018) The reduction of water resources as a distinctive pressure in the transboundary Prespa: views of the rural population. In: Proceedings of the 9th congress of the Hellenic Ecological Society, Heraklion, Greece
Louder E (2020) Leadership styles and barriers to leadership for women in agriculture: a mixed methods study, Master's thesis, Utah State University, Logan, Utah
Michalatou P (2023) Transboundary cooperation in the field of water resources management in South East Europe: the case of transboundary rivers and lakes of Greece: legal and political parameters, Ph.D. Thesis, Department of Balkan, Slavic and Oriental Studies, University of Macedonia (in Greek)
Mylopoulos Y, Kolokytha E (2008) Integrated water management in shared water resources: the EU Water Framework Directive implementation in Greece. Phys Chem Earth. https://doi.org/10.1016/j.pce.2008.02.031
Paletto A, Guerrini S, De Meo I (2017) Exploring visitors' perceptions of silvicultural treatments to increase the destination attractiveness of peri-urban forests: a case study in Tuscany Region (Italy). Urb for Urban Green. https://doi.org/10.1016/j.ufug.2017.06.020
Patera A (2022) Development of tools to support decision making in coastal zone. School of Environment. Ph.D. Thesis, Department of Marine Sciences, University of the Aegean
Pawlewicz A, Gotkiewicz W, Brodzińska K, Pawlewicz K, Mickiewicz B, Kluczek P (2022) Organic farming as an alternative maintenance strategy in the opinion of farmers from Natura 2000 areas. Int J Environ Res Public Health. https://doi.org/10.3390/ijerph19073793
Pawlewicz A, Pawlewicz K (2023) The risk of agricultural land abandonment as a socioeconomic challenge for the development of agriculture in the European Union. Sustainability. https://doi.org/10.3390/su15043233
Petit R, Aguinagalde I, de Beaulieu JL, Bittkau C, Brewer S, Cheddadi R, Ennos R, Fineschi S, Grivet D, Lascoux M, Mohanty A, Müller-Starck G, Demesure-Musch B, Palmé A, Martín JP, Rendell S, Vendramin GG (2003) Glacial refugia: hotspots but not melting pots of genetic diversity. Science 300(5625):1563–1565. https://doi.org/10.1126/science.1083264
Prespa Forum Dialogue (2022). Retrieved from https://pfd.mk/, Ohrid
Rees JT, Searle B, Pahl-Wostl C (2005) Social learning in public participation in river basin management—early findings from HarmoniCOP European case studies. Environ Sci Policy 8(3):287–299. ISSN 1462-9011. https://doi.org/10.1016/j.envsci.2005.03.003
Sandwith T, Shine C, Hamilton L, Sheppard D (2001) Transboundary protected areas for peace and co-operation. IUCN, Gland, Switzerland and Cambridge, UK. Xi +111 p
Sampsonidou A (2020) Participatory planning and inclusion of groups at risk of social exclusion. Master's Thesis, Aristotle University of Thessaloniki. https://doi.org/10.26262/heal.auth.ir.320888 (in Greek)
Society for the Protection of Prespa (SPP), WWF-Greece, Protection and Preservation of Natural Environment in Albania (PPNEA), Macedonian Alliance for Prespa (MAP) (2005) Strategic

action plan for the sustainable development of the Prespa Park. Executive summary. Aghios Germanos, p 76

Stratigea A, Papadopoulou CA (2012) The role of higher education in regional economic development. Regional Science Texts III(1):59–73. https://econpapers.repec.org/RePEc:hrs:jrnkpe:v:iii:y:2012:i:1:p:59-73

van den Broek J, Benneworth P, Rutten R (2019) Institutionalization of cross-border regional innovation systems: the role of university institutional entrepreneurs. Reg Stud Reg Sci 6(1):55–69. https://doi.org/10.1080/21681376.2018.1562367

Votruba N, Koschorke M, Thornicroft G (2020) Public attitudes and the challenge of stigma. In Geddes J, Andreasen NC, Goodwin GM (eds) New Oxford textbook of psychiatry (2 edn). Oxford University Press, Oxford, USA

Zikargae MH, Woldearegay AG, Skjerdal T (2022) Assessing the roles of stakeholders in community projects on environmental security and livelihood of impoverished rural society: a nongovernmental organization implementation strategy in focus. Heliyon 8(10):e10987. https://doi.org/10.1016/j.heliyon.2022.e10987

Cultural-Historic Heritage of the Lake Ohrid (Albanian Part)

Laura Shumka

Abstract The Lake Ohrid basin is widely well recognized and it harbours a rich cultural and historical heritage of different historical periods and forms. The communities that surround Lake Ohrid are some of the oldest in Europe. At least six underwater locations that depict lakeshore villages spanning a very long time span (from the Neo-Eneolithic period to the present) have been located on the littoral areas of the lake. The most significant example of a tri-conch building in Albania is the early Christian church of Lin, which, with certain simplifications, replicates the model of the comparable construction in Ohrid (Lychnidos). It was most likely constructed at the start of the sixth century CE and remained in use until the Early Middle Ages, when it was dramatically destroyed. White, black, and red are the primary colors utilized in the mosaics that cover the church's seven floors. Nevertheless, orange, yellow, green, and brown hues are also used. Lin mosaics, some of which are in excellent condition, are similar in style to the early medieval mosaics in Macedonia (basilicas in Ohrid, Radolisht, and Studencisht) located just across the lake. They feature biblical themes (Eucharistic), flowers, animals, svastica, and numerous other religious symbols. So, the traces of the first human communities appeared way back in the Palaeolithic period, while the recent archaeological founds imply on the first traces of spiritual life from the Neolithic period. The monumental heritage of the Middle Ages are also preserved, architectonically the most important and stylistically the most authentic. Under the influence of different political and social circumstances, the area of the Lake Ohrid becomes the place of interference of stylistic and architectonical headings. So, the wider area of the Albanian part of Lake Ohrid presents one of the most significant regions in developments of Albania. The historical background of those processes is explained in this chapter, which also looks back at the cultural artifacts and sacred and profane structures of past civilizations that contributed to the Lake Ohrid's richness and diversity of cultural components.

Keywords Lychnidos · Illyrians · Lake Ohrid · Albania · Pogradec · Tombs

L. Shumka (✉)
Faculty of Applied Sciences and Creative Industries, Barleti University, Selite, Tirana, Albania
e-mail: shumkalaura@gmail.com

K. Stefanidis et al. (eds.), *Lakes of the Eastern Mediterranean and Balkans*,
https://doi.org/10.1007/978-3-032-02168-7_11

1 Introduction

By its natural and cultural values, the region of Ohrid and Prespa is one of the most representative regions in the transboundary area in between Albania, North Macedonia and Greece. In 2010, International Astronomical Union scientists gave the name *Ohrid* to a lake on *Titan*, a moon that revolves around Saturn! For those who have visited the Lake Ohrid the home of numerous endemic species and phenomenon of speciation, two characteristics quickly become evident (Fig. 1). First, lake itself with spectacular panoramic view—a natural dreamscape of water surface surrounded by high mountains and second, the long history co-existence among lake and settlements (Kostoski et al. 2010; GIZ 2015). This has been recently re-confirmed by scientists that have uncovered what may be one of Europe's earliest sedentary communities, so a stretch of the Albanian shore of the lake once hosted a settlement of stilt houses some 8000 years ago. The archaeologists believe, making it the oldest lakeside village in Europe discovered to date.

Because of the ideal climate, advantageous location, and plenty of plant and animal life, the larger Lake Ohrid region was populated in prehistoric times. According to the results of the archaeological tests conducted so far, there are established archaeological findings in the basin around the lake that suggest the presence of human settlement throughout the Palaeolithic era. Since hunting and fishing are the primary production activities of Palaeolithic groups, it is undeniable that the settlements in the surrounding area were directed toward the shores and hinterland of Lake Ohrid.

The high richness of animal and plant species on the Ohrid and Prespa region represents a unique ecological and bio-geographical phenomenon in Europe. The variety of climate, complex geological history, and interactions between populations, species, and ecosystems have all resulted in enormous diversity and abundance of plants, animals, and ecosystems in this area of South Eastern Europe (Stankovic 1960; Savić 2008). Along with a large number of endemic, rare and relict species highly

Fig. 1 View of Ohrid Lake

significant in themselves, their different ecological characteristics, distribution and origin has contributed to the region's uniqueness.

Albania's side of Lake Ohrid was also designated UNESCO world heritage status in 2019 as an extension mixed property following North Macedonia's side of Lake Ohrid that was declared a World heritage in 1979, with the site being extended to also include the cultural and historic area of Ohrid in 1980. So, for 45 years, UNESCO's venerable list has recognized this place of "outstanding universal value" for protection. But in the recent two decade it comes with challenges from development to over tourism, where the endemic Ohrid trout is standing as a symbol of threats and survival. So, the romance of global scientific significance value has slowly becoming incubus as a recent development plans (*sometimes no plans, or quickly modified, no unified among littoral countries*!) and inadequate use of natural resources (fish primarily) leads freshwater diversity towards impoverishment, that on the other hand is rapidly enhanced by climate changes with scenarios that considers Mediterranean freshwater bodies (including Lake Ohrid) among most fragile ecosystem. The nominated property encompasses basically the Albanian part of Lake Ohrid region and an extension including the Lin Peninsula and coastal strip north to the Macedonian border and the Drilon springs and the watercourses linking them to the World Heritage Property "Natural and Cultural Heritage of the Ohrid Region" in North Macedonia inscribed in the World Heritage List as a natural site in 1979, and extended to create a mixed (natural and cultural) in 1980.

The rapid changes and developments after '90 and increasing human population in the Lake catchment areas (almost tripled), eutrophication of their waters, and ongoing use of water for agriculture are currently adversely affecting the ecosystems (Matzinger et al. 2006, 2007; Shumka et al. 2024). In addition to the water bodies, the terrestrial part of the lakes system is equally important and the existence of a wide spectrum of protected areas (including natural monuments) represents a clear, formal recognition of these assets (Fig. 2). In November 1999 the Albanian side of Lake Ohrid (Decision of Council of Ministers No. 80, 18.2.1999) was proclaimed as a Protected Area—Pogradec Terrestrial/Aquatic Territory Protected Landscape Area of Terrestrial/Aquatic territory with a surface area of 27,323 ha. This area covers both water and terrestrial ecosystems.

Lake Ohrid is a 290 m deep oligotrophic lake, shared between Albania and North Macedonia, oldest in Europe (age of several million years) and with tectonic origin. It has catchment area of over 2.600 km^2 and surface of 358 km^2 (Albrecht and Wilke 2008; Wagner et al. 2009; Wagner and Wilke 2011). Lake Ohrid has four permanent tributaries: Cerava, Velgoska, Koselska, Sateska and more than 30 temporary creeks and numerous underwater (sublacustrine) springs. It has a surface outlet, River Drim/Drin which belongs to the Adriatic drainage system. Three cities are situated around the lake, two in North Macedonia, Ohrid and Struga, and Pogradec in Albania with in total 150.000 inhabitants which are including also more than 10 villages. Tourism, small and medium enterprises are main source of income for the majority of the citizens and there is no intensive industry or agriculture in the region in the last three decades.

Fig. 2 Satellite image of the World Heritage property

Situated on the shores of Lake Ohrid, the town of Ohrid is one of the oldest human settlements in Europe. Built mostly between the seventh and nineteenth centuries, Ohrid is home to the oldest Slav monastery (dedicated to St. Pantelejmon) and more than 800 Byzantine-style icons of worldwide fame dating from the eleventh century to the end of the fourteenth century. Ohrid's architecture represents the best preserved and most complete ensemble of ancient urban architecture of this part of Europe (Meksi 1985). Slav culture spread from Ohrid to other parts of Europe. Seven basilicas have thus far been discovered in archaeological excavations in the old part of Ohrid. These basilicas were built during the fourth, fifth and beginning of the sixth centuries and contain architectural and decorative characteristics that indisputably point to a strong ascent and glory of Lychnidos, the former name of the town. The structure of the city nucleus is also enriched by a large number of archaeological sites, with an emphasis on early Christian basilicas, which are also known for their mosaic floors. Special emphasis regarding Ohrid's old urban architecture must be given to the town's masonry heritage. In particular, Ohrid's traditional local influence can be seen among its well-preserved late-Ottoman urban residential architecture dating from the eighteenth and nineteenth centuries (Shumka 2019, 2023; Shumka et al. 2020). The limited space for construction activities has led to the formation of a very narrow network of streets.

2 Cultural and Historical Richness

The sites identified through archaeological investigation have witnessed human presence in the region since the early Neolithic and some of them, e.g. Maliq, attest to the continuity of settlement along the Late Neolithic, Copper and Bronze Ages in the region (Figs. 3 and 4). A number of hill fortifications (Zvedza, Gradishta of Symiza) dated to the Early Iron Age have been studied with major density of traces of fortifications dating back to the Late Antiquity period (kalaja Dishnice, Mborja, Kloca, Gopesh, Podgorie).

In the Pogradec area, research still awaits systematic reactivation: in most cases, archaeological sites have been investigated around the 1960s–1970s and here the vestiges of prehistoric human occupation appear less dense than around Korça but promising in terms of future investigations and certainly interesting in relation to their potential for attracting visitors and to their ability to shed light on the history of human settlement pattern within the Lake Ohrid Region. However, it should also be noted that the intense building activity in Pogradec has limited the possibility for archaeological investigations (Anamali 1969, 1971).

In the immediate vicinity of the Lake Ohrid, some archaeological sites have been detected and, in some cases, investigated. Some pile dwelling settlements were discovered along the shore of the lake Ohrid in Zagradie, Lin, Udenisht dhe Pogradec. Further archaeological surveys need to be conducted in order to fully understand the extent of evidences in the overall landscape. Important archaeological traces of human settlement in the Lake Ohrid area are located in the inland west of the lake.

Fig. 3 **a** Early Christian church on the Lin promontory. **b** Pile dwelling on lake shore Buqeza. **c** Prehistoric elements, mainly Bronze and Iron Age. **d** Location of pile dwelling on lake shore near Hudenisht

These include important remains of the Illyrian Monumental Tombs of Selca (fourth to third centuries BC), with the town of Lower Selca, and the fortified settlement of Gradishta e Sllabinje (Ceka and Baçe 1972). The monumental Tombs of Selca are located in the hills west of the lake, south of Qafe Thane, not far from the route followed by the ancient Via Egnatia, in the hill known as "Gradishtë" where also lie the remains of the ancient Illyrian town of Lower Selca. They have been in use during the third century BC, a period which corresponds to the flourishing period of the development of the city, and stand as the only known architectural feature of this urban ensemble. The tombs of Lower Selca were proclaimed Monument of 1st Category in 1948 and then were subject to extensive archaeological excavations in 1970–1972. The tombs of lower Selca, are since 1994 in the WH Tentative List of Albania.

The Castle of Pogradec located in the top of the hill overlooking the city of Pogradec, belongs to the group of the early Illyrian-Albanian settlements. The site itself has been used by the local population for more than 1000 years. The settlement has been first inhabited in the sixth century BC. During the mid fourth century BC this Illyrian settlement was fortified with walls. From the archaeological findings, it was concluded that the local community used to practice agriculture, fishing and various crafts and had relations with the neighboring settlements (Anamali 1973, 1979).

Fig. 4 **a** Tumulus burial of the Iron Age-chance finds of Roman and medieval periods in Verdova village. **b** Belongings to Roman and medieval-Geshtenjas. **c** Finds reported from village Tushemisht. **d** Chance finds—bronze spearheads of the late Bronze Age in village of Peshkepi

The castle of Pogradec continued to be inhabited also during the roman period, but without the protective walls and the intensity of life experienced before. By the end of the fourth century AD new defensive walls were erected and the fortress was used by the locals during the barbarian invasions. In the sixth century AD the walls were once again reconstructed. The fortress, with a strong strategic position was transformed again in an important residential centre and remained in this position until the invasions of the ninth century.

Important traces of the Christianisation of the region, initiated in the first century AD, are found also along the today Albanian side of the Lake Ohrid: on the top of the peninsula of Lin, the remains of the Early Christian church of Lin dated back to fifth to sixth centuries AD, one of the most important archaeological sites in the area, have been discovered in 1968 by a researching expedition of the local museum of Pogradec. It was dated at the sixth century AD, but it seems to have been in use until the medieval period (Çikopano 2013). On the stratigraphy of his period it is well evidenced the presence of fire indicating that the church was burned and its walls destroyed. The church has seven premises paved with mosaics where white, black and red are the main used colours, but also brown, green yellow and orange. Similar in style to the early medieval mosaics in Macedonia (basilicas in Oktis, Radolisht and Studencisht) just across the lake, biblical scenes (Eucharistic), flowers, animals, svastica and many other objects are depicted on the church floor, some in good condition.

3 Description of Sites Shown in the Cultural Heritage Within Albanian Littoral Side

The most prominent sites are those considered as underwater site. Pile dwelling, prehistoric one on the lake shore and discovered in the area of Lin village followed by other underwater sites represented by remains of probably modern period (Fig. 5). Additional underwater site has been discovered in the lake shore of Boqeza village. Pile dwelling on lake shore. Very good preservation and material culture of Neo Eneolithic periods are easily identifiable on the lake floor. It reflects a good potential for future research of this type of heritage. The underwater site in Hudenisht and Pogradec through pile dwelling on lake shore are of prehistoric, mainly Bronze and Iron Age materials identified.

Lin Church belongs represents an Early Christian church on the Lin promontory. It is one of the most important monuments in the region. Tri-conch ground plan and architecturally a very complex building. High quality mosaic floors are also incorporated with church assemble (Fig. 5).

The castle of Pogradec (besides limited truck transmitted to nowadays) represents highly important monument located on hill fort site to the north of the city of Pogradec. Following Anamali (1979), it has been inhabited from the late prehistory to the medieval period.

The Tumulus burial of the Iron Age are reported for in village of Verdova, along with a number of chance finds of Roman and medieval periods. Byzantine church, probably of the seventh century, reconstructed in the fifteenth century, while the most significant discoveries are those as Illyrian helmet shown at the local Pogradec museum that is well preserved.

Fig. 5 Plan of the Lin church (left) and partial view of its mosaic floors (right)

Fig. 6 Rock-cut chamber tombs of Selca e Poshtme

Several finds are reported for Tushemisht village. Mosaic remains of an early Byzantine church. Surrounding walls reported on the low hill known as *kalaja* (the castle), together with many finds of pottery and storage pithoi. Other finds include late Antique grave offerings (Fig. 6).

A very important site in southeastern Albania is found close to the village of Selca e Poshtme. Rock-cut chamber tombs unparalleled elsewhere in the country are found there (Ceka 1972, 1985). The settlement has been occupied in five stages, according to excavations. Different pottery forms are used to represent the late Neolithic, early Bronze Age, and late Bronze Age periods of Selcë I through III. From the Bronze Age to the Iron Age, the settlement was constantly populated (Fig. 6).

Traces of burned homes, pottery, including imports from Corinth in the lower horizon, and some Ionian wares show that the Selcë IV period started around 570 BC/ 550 BC. Local reddish-brown painted ceramics, two-handled wheel-made pottery, and Ionic and Attic items were discovered in the upper horizon. Greek models were imitated and their style was also influenced by the local potters. An encircling wall of well-cut stone defended the acropolis in the fourth century. As a military and administrative hub, Selca was re-fortified with stone walls secured with mortar in the fourth century AD. Repurposed Roman and Illyrian stonework was utilized to build the houses (Ceka 1972). Two phases of building can be identified based on coin findings. The first spans the years 364–375 under Valentinian I, while the second spans the years 547–548 under Justinian I (518–565). Before the Slavs overran the city and destroyed the last remnants of the Illyrian city, the city was significant politically and economically.

Together with other manifestations of intangible heritage, the region's traditional vernacular architecture and town planning represent another facet of its cultural traditions. Resources that need methodical investigation and study are represented by the village of Lin and, to a lesser extent, the historic center of Pogradec. A key component of Lin town design is the effective use of the natural terrain for a functional community life, resulting in a harmonious combination of the lake scenery and built architecture.

4 The Area of Cultural Continuity and Interaction with Natural Components

A rich and natural backdrop for the lake is created by the gradation of ecosystems and cultural land use from the lake coast to the agricultural lowlands, the forested hills, and the rocky upland pastures. Most of the time, the villages' settlement structures are still mostly intact, and the residential architecture still uses traditional materials, layouts, and features, which adds to the landscape's overall atmosphere and attests to the long-standing and peaceful relationship between the local people and the land.

However, this unique location is susceptible to high-rise structures, improper restoration, excessive lakefront development, landscape fragmentation, and construction on open hillsides. On a smaller scale, improper shoreline development, solid waste, and air and water pollution all degrade the quality of the land and waterscape. The biological connections between the lake and its surroundings are being severed by coastal development; the remaining undeveloped regions along the coast, including between the Lin Peninsula and the North Macedonian border, are especially crucial for conservation and responsible management.

4.1 Local Customs and Traditions

Albanian side the Lake Ohrid in distinguished by a rich intangible cultural heritage, which needs further investigation and considerations. The most significant values include:

- The city of Pogradec is known as the "city of poets", being the native city of the poet Lasgush Poradeci, numerous painters and photographers.
- Performing arts. Several local festivals maintain local traditions of music, performing arts, handicrafts, foods and produce.
- Notable traditional crafts of the area include boat building and decorative wood carving. Traditional stonework and building skills still persist, but few craftspeople remain.
- The area is known for cultivation of various distinctive local crops and for culinary produce such as mountain cheeses, honey and other bee products, wine and liquor, herbs, herbal teas and traditional medicinal plants. Traditional knowledge still persists about the healing properties of local plants.

5 State of Conservation of Cultural Heritage

In general, the Albanian side of the Lake Ohrid Region's cultural heritage and values could be regarded as vulnerable, and its level of conservation is variable. This is caused by a variety of circumstances, chief among them being the absence of or low level of common awareness, comprehension, and appreciation of the uniqueness of this region, which has been shaped over the years by geographic, historic, cultural, and political elements.

Particularly in the Korça district and more recently in thc Pogradcc municipal territory, this process of inquiry, documenting, and recognition—as well as legal designations and conservation intervention—has lately been resurrected in the region. In general, built or archaeological heritage in public ownership suffers from poor maintenance and neglect due to a lack of funds and enough professional staff, despite recent efforts to restore and rehabilitate historic neighborhoods, which could be made possible by financial resources made available through EU or other donors' funding. When land plots where heritage properties exist have development potential, built heritage in private hands may be subject to demolition as well as neglect and over-care, which could lead to unsuitable or ostentatious interventions. This includes archaeological resources that are unknown or unmapped, as well as structures with historic or landscape significance that are not officially identified (Fig. 7).

In the vicinity of towns or more developed villages, the rural landscape is vulnerable to urban encroachment; in more remote areas, however, poverty and a lack of opportunities encourage young people to emigrate, which lay the foundation for the abandonment of villages and farming activities that maintain the land.

Fig. 7 View of Lake Ohrid landscape

To determine the priority of intervention and the most pressing needs to be met, a methodical, updated evaluation of the conservation status of cultural heritage resources is required, particularly of those that are already officially protected and inventoried.

6 Factors Affecting the Integrity of the Cultural Components

Unplanned urban growth, population growth, aging infrastructure, improper solid waste and wastewater treatment, unlawful water spring interventions, and tourism pressure are the main risks to the property's integrity.

The overall coherence of the UNESCO site property is impacted by the inadequate control of new development, especially the interaction between urban buildings and the surrounding landscape.

Many residential structures have had their external and interior out-fittings updated to improve living circumstances over time, despite the fact that the buildings' initial residential use has remained constant. To a considerable degree, the original characteristics and volumes have been preserved.

The authenticity of the property is at risk since reconstructions frequently use materials that are exactly the same as those used during construction, as well as new ones. Due to the conversion of the area's agricultural land, there are an increasing number of new residential and commercial structures, which is raising the level of influence on the lakeshore, especially in the summer.

Numerous historic structures in Pogradec's center badly need to be restored or repaired.

Covering the mosaics with sand and other methods has greatly decreased active degradation since the Early Christian Church on the Lin Peninsula abandoned a permanent shelter as a tactic. Better management of the immediate area, including the removal of certain unsuitable walls and structures, the use of alternative fence options, and other enhancements, is necessary, as are additional conservation efforts for the archaeological remains. Additionally, there may be more undiscovered ruins in the vicinity and larger context of the early Christian church on the Lin Peninsula.

There is currently little information available regarding the precise location and size of the prehistoric pile dwellings along the lakeside. The locations have not yet been thoroughly explored for more sites, which may exist both on land and underwater, and not all of the recently found sites are officially protected.

The Lake Ohrid Region's unique identity and its rich local customs, crafts, and traditions could be undermined by the region's rapid growth and the expected rise in tourists. There is a risk that the region would turn into a generic, typical beach resort, losing its unique local products, customs, and vernacular architecture. Even though many locals still cherish their traditions, the outflow of young people from the area poses a threat to the transmission of local knowledge and customs between generations, and as property values rise, locals may be forced to leave their traditional dwellings.

7 Conclusions

The Lake Ohrid is one of the most important locations in the development of culture and civilization in both Albania and North Macedonia. A favorable natural environment has created the preconditions for forming the first human community on the shores of the lake in prehistoric times. An essential condition for the preservation and integrity of this property is the preservation of the biological and natural processes of the Lake Ohrid area. Thus, it is crucial to manage protected areas effectively in order to conserve aquatic, littoral, and other lake ecosystems. An ecosystem strategy that combines the management of the current protected areas with other important activities taking place on the property, such as fishing, agriculture, tourism, etc., must also be ensured by the land's efficient management. Maintaining and improving the level of financial and human resources necessary for the property's efficient management is one of the specific objectives for the long-term conservation and management of the property, as outlined in the management plan. Adequate resources must also be made available for research, monitoring, and evaluation of the protected regions that comprise the property. The involvement of many interested groups in the planning and management of the property through consultation and participatory approaches is a specific necessity.

It is worth to mention the fact that until the twentieth century, this will be the place of dynamic political processes and civilization conflicts that will decisively influence the formation of cultural and historical heritage. The Prehistoric, Illyrians, Romans, Ottomans, and other are among the cultural remnants of material civilization that have been maintained on the lake's shores and in its near hinterland. Regretfully, only shards of material culture and rich heritage have survived. Additionally, a large portion of this heritage has not yet been fully revealed due to incomplete and inadequate archaeological research. As a result, the biggest challenges in the coming period will be devoted preservation of the current heritage and efforts to conduct additional research on the cultural and historical heritage of Lake Ohrid. The preservation of these distinctive areas of civilization can only be ensured by a thorough and committed approach to the preservation and enhancement of this cultural and historical legacy.

References

Albrecht C, Wilke T (2008) Ancient Lake Ohrid: biodiversity and evolution. Hydrobiologia 615:103–140

Anamali S (1969) Rezultatet e gërmimeve në kalanë e Pogradecit (Les resultats des fouilles dans la citadelle de Pogradec). Bul Arkeol 1969:89–101

Anamali S (1971) Rezultatet e gërmimeve në kalanë e Pogradecit (1970–1971) [Les resultats des fouilles dans la citadelle de Pogradec (1970–1971)]. Bul Arkeol 120–132

Anamali S (1973) Rezultatet e gërmimeve të vitit 1973 në kalanë e Pogradecit (Fouilles archeologiques 1973-Pogradec). Iliria (1974) 469–474; Bul Arkeol (1974) 4:111–118; Iliria (1975) III:481–486

Anamali S (1979) Kështjella e Pogradecit (La forteresse de Pogradec). Iliria (1970–1980) 9–10:211–258

Ceka N (1972) Qyteti ilir ne Selces se Poshtme (La cité illyrienne près de Selca e Poshtme). Tirana, 121 pp

Ceka N (1985) Qyteti ilir prane Selces se Poshtme (La cité illyrienne près de Selca e Poshtme). Tirana, 121 pp

Ceka N, Baçe (1972) Vrojtimet arkeologjike në rrethin e Pogradecit, luginën e Devollit dhe Shkumbinit (Observations archeologiques faites dans la region de Pogradec et dans la vallees du Devoll et du Shkumbin). Monumentet 3:205–214

Çikopano U (2013) Mozaikët paleokristianë dhe Bazilika e Linit (Paleochristian Mosaics and Lin's Basilica), Tirana, Albania, 120 pp

GIZ (2015) Indicial lakes characterization report. Conservation and sustainable use of biodiversity at Lakes Prespa, Ohrid and Shkodra/Skadar (CSBL), Tirana, 230 pp

Kostoski G, Albrecht C, Trajanovski S, Wilke T (2010) A freshwater biodiversity hotspot under pressure—assessing threats and identifying conservation needs for Lake Ohrid. Biogeosciences 7:3999–4015

Matzinger A, Spirkovski Z, Patceva S, Wuest A (2006) Sensitivity of ancient Lake Ohrid to local anthropogenic impacts and global warming. J Great Lakes Res 32(158–179):10

Matzinger A, Schmid M, Veljanovska-Sarafilovska E, Patcheva S, Gusheska D (2007) Eutrophication of ancient Lake Ohrid: global warming amplifies detrimental effects of increased nutrient inputs. Limnol Oceanogr 52(1):338–353

Meksi A (1985) Arkitektura paleokristiane në Shqipëri. Monumentet 2:13–44

Savić I (2008) Diversification of the Balkan fauna: its origin, historical developments and present status. Adv Arachnol Dev Biol Monogr 12:57–78

Shumka L (2019) Comparison of indoor climate features following different climate guidelines in conservation examples of selected churches in Albania. Int J Conserv Sci (IJCS). 10(4):623–630

Shumka L (2023) Considering landscape and water in the Dumrea Region: challenges for integrated planning and sustainability. J Thalassia Salentina 40(supplement 1):147–154. e-ISSN: 1591-0725

Shumka L, Peri L, Lato E (2020) The needs for determining degradation risks from temperature and relative humidity of post-Byzantine church indoor environment. J Environ Manage Tour (JEMT) 11(3):601–606

Shumka S, Shumka L, Špoljar M, Shuka L (2024) Evidence of climate change and the conservation needed to halt the further deterioration of small glacial lakes. Climate 12:124. https://doi.org/10.3390/cli12080124

Stankovic S (1960) The Balkan Lake Ohrid and its living world. W Junk Den Haag Monogr Boil 9:357

Wagner B, Wilke T (2011) Evolutionary and geological history of the Balkan lakes Ohrid and Prespa. Biogeosciences 8:995–998

Wagner B, Lotter AF, Nowaczyk N, Reed JM, Schwalb A, Sulpizio R, Valsecchi V, Wessels M, Zanchetta G (2009) A 40,000-year record of environmental change from ancient Lake Ohrid (Albania and Macedonia). J Paleolimnol